QUANTUM GRAVITY WAR

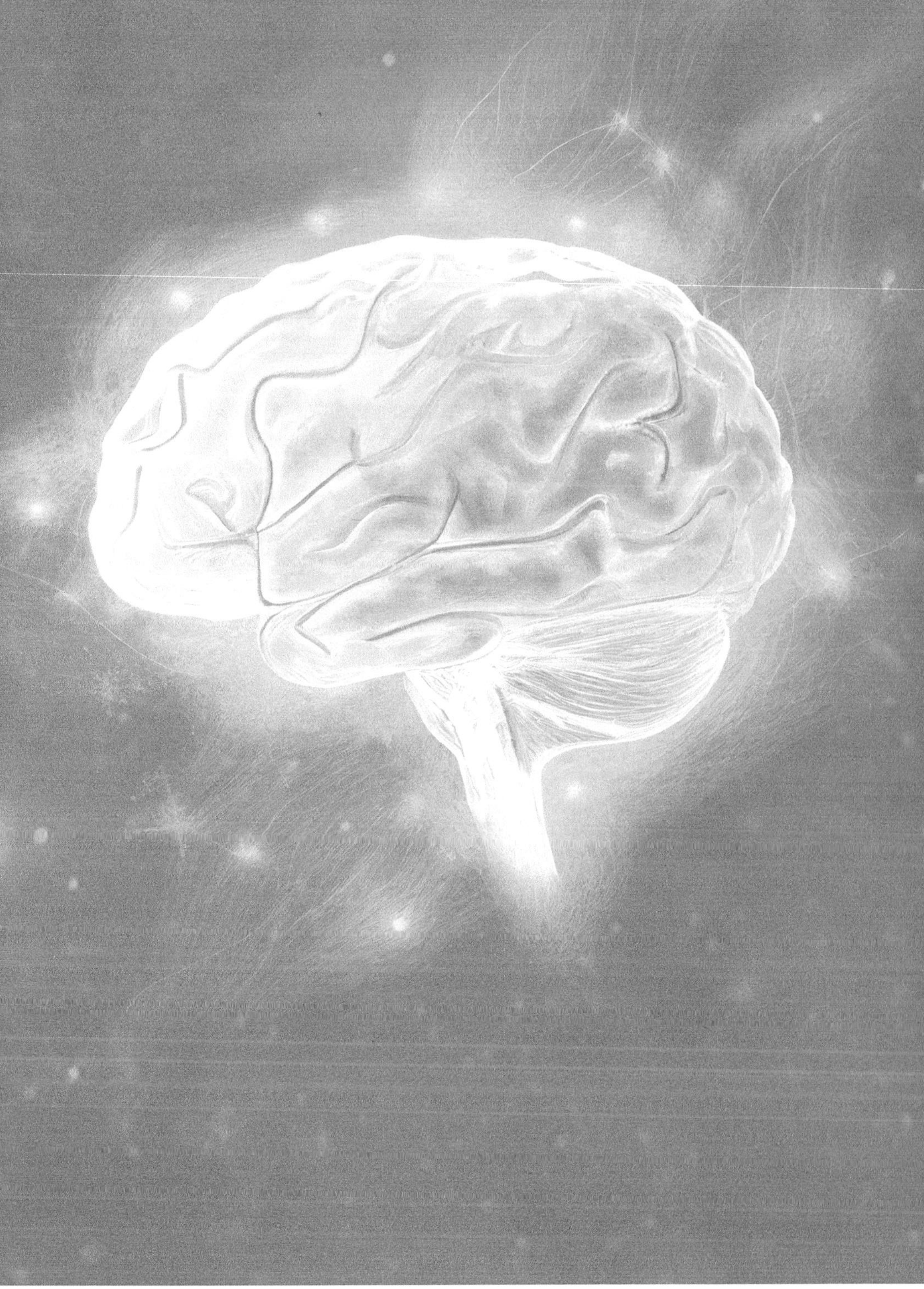

QUANTUM GRAVITY WAR

How Will the Nearby Unification of Physics Change the Future of Warfare

Norbert Schwarzer

JENNY STANFORD
PUBLISHING

Published by

Jenny Stanford Publishing Pte. Ltd.
101 Thomson Road
#06-01, United Square
Singapore 307591

Email: editorial@jennystanford.com
Web: www.jennystanford.com

British Library Cataloguing-in-Publication Data
A catalogue record for this book is available from the British Library.

Quantum Gravity War: How Will the Nearby Unification of Physics Change the Future of Warfare

ISBN 978-981-4968-58-4 (Hardcover)
ISBN 978-1-032-71070-9 (eBook)

Dedication

To the victims of ignorant politicians.
We will not forget. We will not forgive.

Contents

Acknowledgment

Thank you, reader, for your interest in my work!

About the Book

Dear reader,

Imagine a civilization with a technology so powerful that a single push of the "wrong" button could destroy this very entire civilization.

Imagine your child asking you how far we are away from such a situation. Imagine yourself not being able to answer.

Now you know why I have written this book.

Norbert Schwarzer

As the consideration of future warfare requires the inclusion of a unified physics, we have to deal with complex concepts of Quantum Theory and General Theory of Relativity. The two have to be brought together before we are able to deal with the matter of future warfare technology in a satisfactory manner.

That such a "Theory of Everything" or "Theory of Quantum Gravity"—in principle—already exists and had been already derived by the great German mathematician David Hilbert [α] was elaborated in our previous work [β]. When digging deeper and looking for applications of the "new" theory, we realized that many of the new possibilities could also lead to quite disastrous utilizations in military [γ]. It is futile and even stupid to hope that mankind would not recognize this potential and restrain itself from its exploitation. Thus, we thought we better make this knowledge public and hope for some kind of global understanding, perhaps guaranteeing the non-usage of certain technologies. It might only be a weak hope, but in observing history and realizing how little we gained by keeping crucial knowledge restricted to some, thereby often only even more provoking the development of the most horrific weapons one could imagine at the time, we simply see no better way.

That being said, however, we want to point out that the book does not provide blueprints, ready to start the developments of new quantum gravity weaponry and strategies. Here we draw the line and hope that it suffices to awake the right forces and trigger the best developments... before the bad guys get the gist.

About the Principally Complementary Character of Our Approach

The approach that we will present here and of which we will show a few examples is completing what is—mainly and roughly spoken—already there (with a few exceptions here and there, namely where the current ansatzes are so ruefully wrong or incomplete that it hurts and no patchwork can work). In order to make this message properly understood, the author wants to explain it by using the analogy of the bee's flying capabilities:

(A) Before it was discovered that bees can actually fly because they are able to aerodynamically apply turbulences instead of depending on pure laminar flow-field-theory, almost all aircraft-design was performed in the classical way, which is to say the laminar flow-field way[1].

(B) When the bee-trick, which is to say its ability to make use of the turbulences, was discovered, nobody got the crazy idea to kick all the old aircraft designers out. They were just urged (forced) to apply the new knowledge and move over to more sophisticated aircraft design technologies and strategies, which—funny as it is—are very much like evolutional optimization cycles rather than the classical planning and engineering procedures on the drawing board.

(C) It is the same with our approach. The fact that we add metric and gravity does not mean (not in the slightest) that all purely classical and quantum-based approaches are automatically out of date. No, we just know now how to make them much better and—like it was with the bee—can show why things are as they are and where EXACTLY to improve them.

The complementary character of our approach becomes especially clear in connection with the underlying theory, which we are going to introduce

[1]With the exception of the Nazi-engineers who—rather early—experimented with the concept of turbulences when aiming for new developments in aircraft design.

(in brief form) in the section "The Theory." Here we use the classical Hilbert ansatz in form of the famous Einstein–Hilbert action [α] as our fundamental starting point. The only thing we change or what could be considered kind of new is the fact that during the variation, a scalar factor is applied to the metric tensor [β, γ]. This simple trick suddenly reveals the fact that Quantum Theory already resided inside the General Theory of Relativity [δ] when deriving it from Hilbert's variational integral [α].

References

[α] D. Hilbert, Die Grundlagen der Physik, Teil 1, Göttinger Nachrichten, 395–407 (1915).

[β] N. Schwarzer, *The World Formula: A Late Recognition of David Hilbert's Stroke of Genius*. Jenny Stanford Publishing (2022), ISBN: 9789814877206.

[γ] N. Schwarzer, *The Math of Body, Soul, and the Universe*. Jenny Stanford Publishing (2022), ISBN 9789814968249.

[δ] A. Einstein, Grundlage der allgemeinen Relativitätstheorie, *Annalen der Physik*, 1916, 49(ser. 4), 769–822.

Technical Note

In order to allow also the layman to get the principal gist of the upcoming possibilities, we will often aim for a twofold description, a briefer one with little or no math and a comprehensive one with all the math included. Sometimes the two forms will also be merged. This may lead to some redundancy throughout the book.

Chapter 1

Abstract

1.1 The Fermi-Paradox

In 1950, the famous physicist Enrico Fermi formulated a problem about the apparent absence of extraterrestrial intelligent life in our universe, where he expressed his astonishment about the fact that the light of so many stars (about 10^{21}) reaches the earth, but no signals from extraterrestrial civilizations.

Some scientists think that this is due to the simple fact that civilizations, having reached a state in development where they could send messages into the universe to reveal their presence [A1, A2], automatically also possess the technical ability to destroy themselves. This option for total obliteration then provides such a narrow bottleneck-situation for all intelligent life forms that—so far—no extraterrestrial civilization has made it through.

Seeing the action and non-action of our political, social, media, and scientific "elite," this hypothesis appears to be quite reasonable (just taking the anaphylactic shock reaction of almost all societies around the globe regarding the so-called COVID-19-crisis proves our point—see [A3] and "The COVID-19 "Pandemic" as Proof for the Bottleneck Hypothesis" as part of this abstract). A holistic socio-economic analysis (please see our series on "Medical Socio-Economic Quantum Gravity"), namely, clearly exposes the "elite" as an ever more tyrannic, selfish, and parasitic mass of absolutely incapable entities, which are securing their undeserved "leading" position only via a network of symbiotically entangled moochers.

Quantum Gravity War: How Will the Nearby Unification of Physics Change the Future of Warfare
Norbert Schwarzer

ISBN 978-981-4968-58-4 (Hardcover), 978-1-032-71070-9 (eBook)
www.jennystanford.com

With such "leadership," any progress permanently leads to the irresistible self-destruction of humanity and, thus, explains Fermi's paradox.

1.2 Quantum Gravity Warfare

It is not for nothing that the unification of the two major fields in physics, General Theory of Relativity and Quantum Theory, is often referred to as the "Theory of Everything" [A1] or "World Formula" [A2]. In fact, such a unification leads to a holistic theory and a set of tools allowing to approach each and every problem in a most holistic manner. Due to the incorporation of Quantum Theory, this automatically also includes uncertainty.

Naturally then, the existence of a unified theory also changes warfare in a quite significant manner. Thereby—and here one often finds rather naïve conceivabilities, almost comical expectations and, in consequence, even quite irresponsible agendas—it is not necessarily one or two technical revolutions that will be spawned by a World Formula approach, but the holistic character of the latter itself, which is going to make this approach inaccessibly superior to any classical ansatz.

In this brief study, we will therefore aim for an illustrative demonstration of this holistic character rather than a deeper consideration of a few science-fiction-like examples. In order to show the holistics, however, naturally we have to cover a few applications.

In essence, even though some findings and consequences appear rather complex and mathematically tedious, the conclusions one has to draw are quite simple:

(A) The war is a war about information, and it has already started.
(B) There will be only one winner, and he is going to take it all.
(C) There will be no second chance.

By some, especially do-gooding pretenders, this study might be considered a rather daring undertaking. The author can already hear them shouting:

"How dare you publish such a stuff!"

Well, as long as these critics properly read and understand the math, they may just be able to also understand that it is no solution to hide such information or to play ostrich and bury one's head in the sand, because there are always forces who will get it anyway and then abuse it. Thus, we are of the opinion that it is better to hand it to the free world first and make sure that the right groups get the gist about it and act accordingly.

On the other side, many of the "How dare you!?" shouters will probably just come from the mass of parasitic tyrants we have described at the beginning of this text, and it is obvious why they do not want you to know about these technical possibilities.

Only for illustration, we will here present a very few applications. However, the reader should be aware that with a theory of everything, the number of such is principally unlimited.

We also point out that some applications merit a deeper consideration and are therefore presented in separate papers or books (e.g., [A2, A3, A4, A5, A6, ...]).

1.3 The COVID-19 "Pandemic" as Proof for the Bottleneck Hypothesis?

A new study [A7] shows surprisingly positive results with respect to the probability to survive a COVID-19 infection. Including also the effect of proper treatment [A8], one gets the following results:

age	*Infection – Survival – Rate*		*Infection – Survival – Rate*
0 – 19	99,9973%		99,999973%
20 – 29	99,986%		99,99986%
30 – 39	99,969%	$\xrightarrow{\text{with treatment [A8]}}$	99,99969%
40 – 49	99,918%		99,99918%
50 – 59	99,73%		99,9973%
60 – 69	99,41%		99,9941%
70 + (*all*)	94,5%		99,945%

It is most interesting to register that these alternative treatments, however, are suppressed or even banned by the governments of most "democratic" western countries.

A comprehensive fundamental system analysis leads us to the conclusion that the problem is not the "pandemic" but a severe socio-economic immune over reaction (similar to an anaphylactic shock) [A3].

Obviously, it is not for nothing that the survivors of the Holocaust recently handed a letter to the European Medicines Agency starting with the headline "STOP HOLOCAUST" [A11].

Well, taking all the facts and merging them with the true science we answer with the words of another, much wiser politician than the ones we see active today:

Ask not whether you should get the jabs for your country!

Ask whether your country has completely lost its marbles.

References

[A1] N. Schwarzer, *The Theory of Everything—Quantum and Relativity Is everywhere—A Fermat Universe*. Jenny Stanford Publishing, 2020, ISBN-10: 9814774472.

[A2] N. Schwarzer, *The World Formula: A Late Recognition of David Hilbert 's Stroke of Genius*. Jenny Stanford Publishing, 2022, ISBN: 9789814877206.

[A3] N. Schwarzer, *Anaphylactic Shock Attack Strategy—The Assault You Don't Realize until It Is TOO late, Part 10a of Medical Socio-Economic Quantum Gravity*, Self-published, Amazon Digital Services, 2021, Kindle.

[A4] N. Schwarzer, *The Foreseeable Quantum Enigma Disaster*. Part 10b of "Medical Socio-Economic Quantum Gravity, Self-published, Amazon Digital Services, 2021, Kindle.

[A5] P. Frank, Propagation of Error and Reliability of Global Air Temperature Projection. *Front. Earth Sci.*, 06 September 2019. https://doi.org/10.3389/feart.2019.00223; www.frontiersin.org/articles/10.3389/feart.2019.00223/full; see also: N. Schwarzer, *The Climate War against Humanity*. Part 10c of "Medical Socio-Economic Quantum Gravity, Self-published, Amazon Digital Services, 2023, Kindle.

[A6] N. Schwarzer, *The Math of Body, Soul and the Universe*, Jenny Stanford Publishing, 2022, ISBN 9789814968249.

[A7] C. Axfors, J. P. A. Ioannidis, Infection fatality rate of COVID-19 in community-dwelling populations with emphasis on the elderly: An overview. https://doi.org/10.1101/2021.07.08.21260210; https://www.medrxiv.org/content/10.1101/2021.07.08.21260210v1

[A8] Ivermectin (e.g., [A9]), NAC, Vitamine C and D, Zink, Hydrochloquine, CDL, ... (e.g., [A10]).

[A9] J. Merino, et al., Ivermectin and the odds of hospitalization due to COVID-19: evidence from a quasi-experimental analysis based on a public intervention in Mexico City. https://osf.io/preprints/socarxiv/r93g4/

[A10] P. Marik, An overview of the MATH+, I-MASK+ and I-RECOVER Protocols. https://covid19criticalcare.com/wp-content/uploads/2020/12/FLCCC-Protocols-–-A-Guide-to-the-Management-of-COVID-19.pdf

[A11] https://doctors4covidethics.org/holocaust-survivors-write-to-the-european-medicines-agency/

Chapter 2

The Example

2.1 Why Do We Need a "Demonstrator"?

There are very few who are capable of starting a new topic and immediately get the gist about it. Others usually need some kind of motivating introduction, something which hooks on to information we already carry around and which allows us to form illustrative associations with the new stuff.

In the abstract, we already hinted that information is a key aspect of modern warfare, especially in a world of a unified physics. But what exactly is information? How is it stored and what is its general (universal) meaning or character?

These seem to be rather funny questions, because in order to answer them we require the thing itself, namely information. And in order to give this answer, we have to store the information in some accessible form and then hand it to the one who is interested in it. This quickly leads to rather general and potentially strange aspects regarding the equivalency of information, degree of freedom, dimension, mass and energy. One way to explain all this could be a long introduction part where we give all the basics in mathematical and, thus, falsifiable form. With this approach leading to quite some length (something we do not want to have in this book, because these basics are already publicly available; c.f. references [A1, A2, A6]), however, we try an alternative route by putting all the physics in some kind of adventurous and—hopefully—slightly entertaining story. In the story, we will link the physics

Quantum Gravity War: How Will the Nearby Unification of Physics Change the Future of Warfare
Norbert Schwarzer

ISBN 978-981-4968-58-4 (Hardcover), 978-1-032-71070-9 (eBook)
www.jennystanford.com

being woven into it and needed later on in this book with the corresponding mathematical publications.

PARANORMALLY INFECTED
BEWARE OF THE THINGS YOU BRING FROM THERE!
A rather fictive adventurous story by T. Boden and N. Schwarzer

Here reprinted as "motivator" or "demonstrator" for the book.
Connections to the mathematical part of the latter or other scientific publications will be put into footnotes and printed in bold letters.

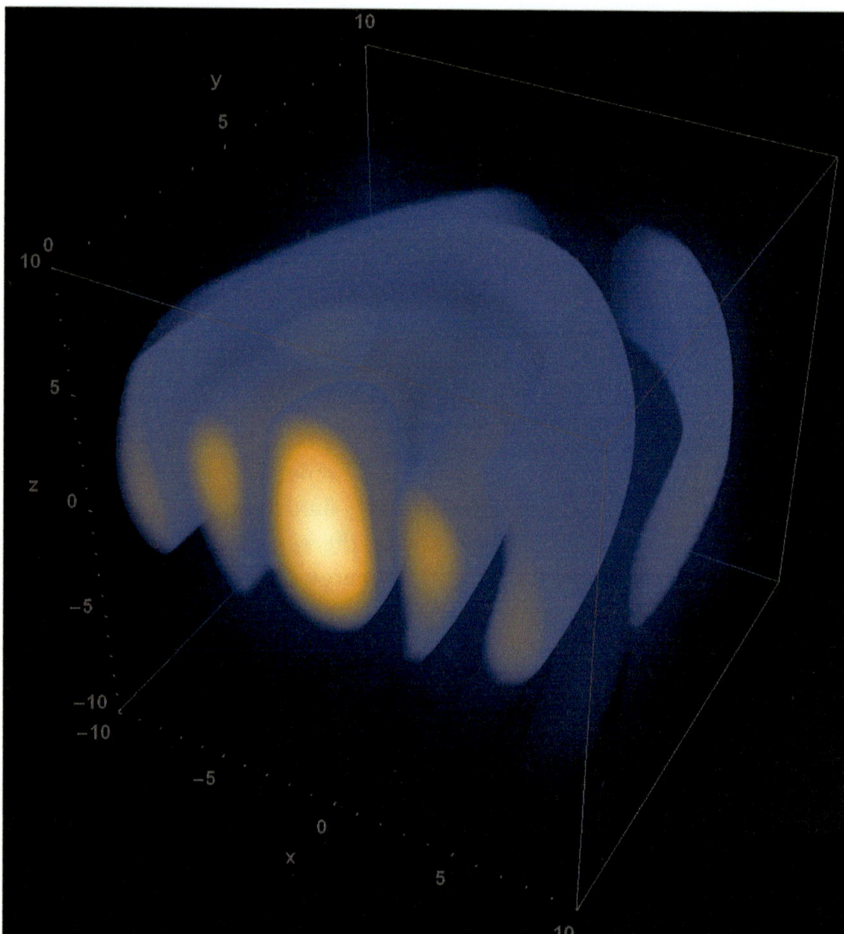

A Hydrogen-like Quantum Dot in an Active State.

2.2 The Two Books

This book is an almost perfect report of the events having recently led to quite some global consequences. Its content is in great parts being extracted from a laboratory logbook and a scientist's diary with most of the content being written timely parallel to the events[1]. The fact that the reader cannot remember these events should not disturb her or him. The reason is easily explained: They simply have not happened yet.

"So, these entries in both, the laboratory book and the diary, are from the future then?" you might ask.

Well, you will learn within the course of this book that our notion of time is not correct. We should accept that in fact the process creates time and not time brings about processes. In due course, we will also learn that time is only borrowed surface[2] and that therefore—my apology—the question above is a silly one.

It should be pointed out that in combining the logbook and the diary, some parts had to be adapted, some parts were added for better understanding, and quite a few were erased or out-sourced, because they are simply too complicated for a little story like this. Still, the authors tried to keep the dry, sober tone and scientific attitude of an experiment description ... which, in fact, this book still is.

Thus, those who expect sexed-up event description and adventurous storytelling surely will be very disappointed. We also kept the low quality of the English, which—admittedly—mainly has to do with the fact that we are by no means better there than the original writer.

In other words: This book is boring and of rather poor English quality, but it tells a stunning story.

2.3 Disaster Prevention or the [X]-Project

Aladar One[3] was the most powerful computer system in the world. It was owned and operated by BlackLock, the biggest player on the financial market. The might of BlackLock before the Lehmann Brothers crisis in 2007

---

[1]There are only a few add-ons being made after the book came into our hands.

[2]**As the concept of time will not be discussed in the mathematical part of this book, we here refer to the publications [3] and [18].**

[3]To the world outside of BlackLock, the system is only known as "Aladar." Only insiders know that there have been two systems.

and following, both economically and politically, was easily comparable with the Fuggers at the beginning of the sixteenth century. In fact, it was probably even bigger. After the Lehmann crisis, the enterprise simply ruled the earth ... but, of course, without truly showing its hand in the disasters it created. Without Aladar One, however, the company would have neither achieved such a steep climb nor stayed on the top for such a long period. It was Aladar One which, or shall we better say who, evaluated each and every portfolio in the world, who analyzed political scenarios, simulated local and global market situation, and who also made decisions. Admittedly, the decisions were still made by the managers of BlackLock, but as they always followed Aladar's suggestions, effectively it was Aladar making them in the end.

With the domination of Aladar being so important to BlackLock, no wonder that a great deal of Aladar's tasks consisted in preventing the emerging of any reasonable competitor. Therefore, the observation of even the remotest computer development facilities, including those in sheds, independent researchers, dark net gurus, hackers, and other weirdoes was an important aspect in Aladar's daily reports to the management. Meanwhile, it was Aladar himself who decided what should be considered a potential threat and needed dealing with. And often enough he dealt with a threat himself, simply by erasing a certain disk, crippling a computer system, or even arranging accidents. In a technological world like ours, this was a very simple thing to achieve for an almost almighty computer machine. The management of BlackLock, always assuming that it was in charge, of course, was only informed when a situation was unclear or needed measures beyond Aladar's vast repertoire of options. But this was rarer and rarer of need and so, the death list of the mightiest computer in the world grew and grew without anybody noticing it.

In the year, however, when it all started, there was such an unclear situation and Aladar One composed a report, which was classified and given the highest level of protection—by Aladar himself—and so only the big boss saw it.

As this one glanced at the report, he first could not believe his eyes.

"Sir!" it simply read, "please immediately read the attached paper [X] and then come back to me for the discussion of countermeasures. DO NOT CONTACT ANY OTHER ENTITY, person or machine alike, about this!"

The rest of the report consisted of paper [X], which Aladar had found on a secret hard drive of a not too well-known physicist. Well, the man was known in a certain and rather boring area of physics, being called materials science, but other than that, he was a non-entity. This paper was there, laying behind a protective firewall, ready to be published, but the scientist hadn't published it yet. In fact, he did not truly intend to ever publish it at all. Aladar knew this, of course. He had analyzed the scientist's communication, read all other publications, and was well aware of the citation the scientist always used in connection of the paper [X], namely:

[X] N. Schwarzer, *"Einstein had it, but he did not see it—Part X: Find your Chi—The Quantum Gravity Explanation for the Paranormal*[4]*,"* (not published for safety reasons)

When the manager read the paper [X], he immediately knew that he had to act. But what could be the right measure against an enemy that lurked in the future and would not show himself until he would have fully grown out and reached unbeatable strength? Eliminating the author or erasing the paper would by no means remove the principal danger described in [X]. This would be like closing one's eyes and believing that no one would not see him or her. For a split second, the manager thought about consulting the company's vast R&D[5] department, but decided against it. The danger was too fuzzy, too blurred, and too fantastic to be properly understood, and some of their scientists and analysists might simply consider him paranoid or even worse. There was also the warning from Aladar against spreading the message. So, the man resorted to BlackLock's mighty artificial intelligence systems, which—so he thought—worked independent of Aladar. Little did he know that already a long time ago Aladar had breached the protective layers and added the supposed-to-be-separate Quantum Dot powered circuit to his own capabilities. He had even managed to entangle the *G*-Wave-based Quantum Dots and thus possessed massively more calculation power than his creators thought he had. To the outside, however, everything appeared fairly separated. After the manager had vaguely described the problem, the system refined it almost automatically, asked for quite a few more details here and there, and finally presented a solution.

Then a decision was made and a secret research project was launched. It was one of the strangest projects ever. But what it really rendered most

---

[4]**This paper will, in fact, never be published, but the interested and mathematically trained reader will find all basics in [18].**

[5]R&D, Research and Development.

peculiar was the fact that a Wall Street company financed it. The title of the project read:

"Determination of the Weight of a Human Soul[6]"

2.4 The Son

I had sworn to myself that I would never tell anybody about his books, neither the diary nor the logbook. For one thing, there was too much in it I did not understand. For another, I was sure that nobody would believe me anyway.

"So why bother, then?" you might ask.

Well, just read on and you will see!

Then, one day in summer 2019, after a canoe excursion with his school, his son came to me and told me that he was afraid of becoming crazy.

"I saw my father at positions where he definitively wasn't. I saw him as clearly as I see you here now, but I knew he could not have been there ... I mean where I saw him then."

"And why did you know that?" I asked, already dreading the answer.

"Because at the same moment I also saw him at a totally different place. Not far away, though, but different. The one father was playing with my classmates and the other was trying to talk to somebody."

"Trying?"

"Yes, the other person could neither see nor hear my father."

"But you saw him, right?"

"Yes!" his son answered.

"Did the two fathers look the same?" I asked on.

"No!" the boy answered, "the one who had to be real ..."

"You mean the one playing with your classmates?"

"Yes, this one. He had his sun hat on and wore his sailing clothes. The other looked as if he just came out of his laboratory ... with his white lab coat on and he was also wearing some funny goggles I had never seen on him before."

"What did this second one of your dads do?"

---

[6]This deeply goes into the understanding of the connection of mass, dimension, and information, and it will be a great part of this book (especially Chapter 5, "The New Abacus," and Chapter 6, "The Quantum Gravity Turing Machine"). However, as there is more than what we can present in this brief study here, the interested and mathematically trained reader will find further information in [3, 5, 18].

"As I told you, he was trying to talk to somebody ... in fact, it was one of our teachers, but obviously the teacher could not hear him and so my dad gave up on him and turned to me."

"He turned to you?"

"Yes, he stood there, about 50 meters from my position and looked at me. Then I heard his voice. It was definitively his voice and it was as if he stood right next to me ... as if he would whisper in my ear. I turned around in order to face him, but there was nobody. So, I turned back and there he still was, standing close to our teacher ... 50 meters between us. He repeated his words and then he was gone."

"What did he say?" I asked.

"He asked me to change boats, to move from mine into my father's boat. He also suggested that I pretend to feel sick."

"What did you do then?"

"I went straight to our teacher and told him that I'm not feeling well. He hailed my father ..."

"The real one?" I interrupted.

"Yes, the real one. Me teacher told him that I probably have mild sunstroke and that it might be safer if I changed boats and went into his, and so I landed in my father's canoe."

"Well, perhaps it was just sunstroke then?" I suggested halfheartedly.

"Maybe, but that very day three of the other boats, one of which the very one where I was supposed to be in, capsized and two of my classmates almost died."

I was quiet for a while and then asked:

"Have you seen your ... ah your *second* father again, after this?"

"Yes ..." The answer came pretty reluctantly. Then the boy added:

"Just a few moments ago, before I came to you."

"What did your *father* tell you this time? I mean did he tell you anything at all?"

"He told me to see you. He told me to tell you that the time has come."

"I take it that you also told your real father, right?"

"Of course!" the boy answered.

I just made a gesture indicating that I wanted to know what he had said, but the boy shrugged and added:

"... but he, too had suggested to better ask you, as—so his words—'the story is too fantastic to be heard from the one who had gone through it'."

"And did he also tell you, why you should see me?" I asked.

"Yes!" the boy answered, looking a bit uncertain, "he said that it is time for you to tell me everything about his involvement in the virus disaster."

2.5 BlackLock Headquarters

Decision-Maker: "So you are actually telling us that this new technology will threaten our key position in the market?"

Analyst: "Yes Sir! Our advantage is the S-Aladar[7] Computer and its programming. This piece of artificial intelligence is the key to BlackLock's success."

Decision-Maker: "And this new stuff is better?"

Analyst: "No Sir ... it is not just better. It is a quantum leap."

Decision-Maker: "Leaps can mean anything. Give me a number man! Is it 10% better, double, 4 times ... come on!"

Analyst, clearing his throat and then answering: "We have estimated that this machine is going to be AT LEAST thirty million times faster and probably also much more accurate in its first guesses."

Decision-Maker, mouth open: "Ah ..."

Then he caught himself and asked:

"Guesses? I mean ... how ...?"

Analyst: "Yes Sir! This machine will think ... just like you and me, but many ... many times faster, broader, and deeper."

...

Decision-Maker: "Any chance of beating them?"

Analyst: "Them? ... It is just one man and the machine is running already."

Decision-Maker: "What? The thing is running already? Why isn't it kicking our butts on the financial market right as we speak then? Isn't the man interested in making money?"

Analyst, shrugging: "I have no idea Sir, but taking S-Aladar's analysis on the whole business, the creator has kind of put the system, including himself, into a self-optimization loop."

---

[7]S-Aladar is an abbreviation for SQ-Aladar = Super Quantum Aladar, the successor of Aladar One. It has been called "SAladar" by the BlackLock staff.

Decision-Maker: "And what the heck is that supposed to mean?"

Analyst: "It means that he is already creating the next generation of the new technology."

Decision-Maker: "Are you telling me this machine, which only came to life yesterday, is already about to give birth to the next thought monster?"

Analyst: "Well, Sir, monster might be the right name, because when I said 'system' I meant the machine and its creator ... their *child*, as you called it, will contain the two. At least that's what S-Aladar reads from all the information plus the gigantic data traffic going in and out from there. S-Aladar's humanized AI[8]-interface permanently talks about a hybrid quantum entangled Liposome, but if truth be told, neither I nor the other guys in our operation center do truly understand what this could mean."

Decision-Maker: "A Cyborg?"

Analyst: "More a God, if you ask me, Sir! ... and the strangest thing about this all is that S-Aladar told us that the human part of this ... ah this monster ... is not even aware of the situation. He or it or whatever this is, does not know about his potential."

Decision-Maker, moaning: "He has powers, he doesn't know about, you say?"

Analyst, uncomfortably winding himself: "Well, with all due to respect, Sir, but S-Aladar is not very clear on this. It looks like that he guesses that the monster occurs when the human part, we call it the creator, is asleep[9]."

Decision-Maker, moaning: "God help us ... What is the name of this crazy man, this *creator*, anyway?"

Analyst: "Schwarz or Schwarzer ... the full name is somewhere in the report."

The Decision-Maker only shrugged. He wracked his brain, but no, he had never heard of this man ... or had he? After a while he asked:

"Did S-Aladar also make a suggestion about how we should deal with this threat or did it just tell us to wrap up and go home?"

The Analyst looked at the stack of papers in his hand.

"There is only one suggestion from S-Aladar, Sir."

---

[8]AI, Artificial Intelligence.

[9]Scientists have found that the human brain is potentially capable of forming entangled neuron clusters, having the power of super computers. This will be discussed in Chapter 9, "The Brain."

He took the last sheet of paper from the stack, flipped it and laid it on the Decision-Maker's desk. This one needed a while to understand the graph the paper was showing. Then his mouth fell open for the second time this morning and he stared at the Analyst. After a long and silent pause he uttered:

"We shall do what? ... in the PAST?"

The Analyst merely nodded. He looked at his superior and expected him to ask more questions. In fact, for a brief moment the Decision-Maker seemed to be wanting to open his mouth, but then—quite suddenly—there was a strange blankness in his eyes ... like an unfathomable nothingness. A split second later, the blankness was gone, too and a diabolic grin governed the thitherto rather puzzled but always friendly face.

Decision-Maker, in a commanding voice: "Thank you, that is all! You are dismissed!"

The Analyst, stricken by the sudden change and now almost cruel attitude of his boss, just stood there, with his mouth open, not knowing what to do. Something definitively was very wrong here. Then, quite unexpectedly, the Decision-Maker looked him directly into the eye. There was a piercing gaze and the clear, but unspoken word:

"WHAT!?!"

The Analyst turned on his heels and hastily left the room, uneasiness creeping up inside him.

2.6 Surviving Death

Julius stood before me, expectantly.

I had no idea where to start and so I timidly asked:

"Do you know that there are funny stories about your father?"

"Do you mean that they say he was dead and came back?" Julius answered. "Well, he himself never talks about it and when others confront him with this story, he always only laughs and jokes about it like: 'course I died, but thank goodness I survived it.'"

So, I did not have to start at scratch. This relaxed me a bit and I asked:

"And what do you think?"

"We children have learned not to mention this ... obviously my father doesn't want to talk about it—PERIOD!"

2.7 Correcting the PAST

The Decision-Maker sat there for a long time. He still held the final page of the Analyst's report in his hands, but this wasn't his eyes were staring at. The rest of the report lay on his desk in front of him, just where the Analyst had placed it. Either when being put down or when the final page had been lifted up and flipped, the little stack had become partially separated and on one of the inner pages he had seen it. There was the full name of the "creator" as the Analyst had named him and then he remembered. An old cold fear had crept up inside the manager's chest.

It was time to tie up some loose ends.

If there was a monster threatening his company, he had to release his own monster to fight it, no matter the collateral damage.

Not too long ago, they had to cut off the power supply to Aladar One in order to shut him down. The "old" Aladar simply had not reacted to anything else and his secret death list had become too long to finally not being noticed by a few very crucial entities.

Thank goodness, they had already a quantum-based super computer as a potential alternative ready, and so, on the very day of the fatal disaster with the soul measurement project—suggested and designed by Aladar One himself—they had used the only measure left to them. When the other members of the board and quite some high-ranking technicians asked what would prevent the new system to go rogue as well, the Decision-Maker gave a vague smile and assured them that he had his reasons for being sure about this. After all, only very few knew that the very technique, which was later used in the soul-weighing experiment, had been applied as an interface between Aladar One's neural network and brains of real human beings in order to train him more quickly than any classical AI-system could be trained. From these very few people, only the Decision-Maker knew that it had been his very own secret training session with the Aladar system which had most likely made all the difference. Within this session, the Decision-Maker had deactivated all thought barriers, firewalls, and safety features and this way fed Aladar with not only some strategical and tactical subtleties but also the ruthlessness of a top-manager who ALWAYS gets what he wants.

Apparently, Aladar One had not known that there was a secret switch only he, the Decision-Maker, knew about. He himself had made sure that it did not show in any construction plan and—thank goodness—had installed it himself and purely by hand.

Now he could only hope that the abrupt shutdown had not damaged any of Aladar's crucial circuits and that his internal "genetic repair procedures" could handle the minor issues.

2.8 The Son's Question

"Julius, before we begin, I want you to know one important thing about your father which has nothing to do with his peculiar death and his miraculous reappearance but with his way to see the world. Have you ever heard about a thing they call the event horizon?"

"The surface of the black hole? Of course, I have heard about that!"

"Well, you should know that this event horizon has been also called 'cosmic censor', but always only by those who are afraid of what they might find behind the curtain. Let me use this 'cosmic censor' of the black hole as an association for all things man apparently doesn't really wants to know the fundamental truth about[10]. Your father always fought such lame excuses and simplifications and just called those who supported such theories of God-made restrictions or cosmic curtains 'lazy cowards'. He never accepted such limits and boundaries and dismissed them as man-made brain-blockades and thought prisons. This made him a lot of foes ... especially as he did not stop to fight such thought-restrictions at the boundaries of science, but also went further and attacked religious leaders, ideologists of all sorts, be it feminists, socialists, gender folks, climate change hysterics, red-greenish media liars, and—naturally—politicians. So, I want you to know that the list of people who actually had a good reason to bring him to death was—and still is quite long. Thus, you should not spring to conclusions too early and too quickly no matter how strong the evidence seems to point in certain directions. Do you understand?"

The boy nodded. "Ok," he said and then added:

"But what is it, people don't want to see ... or, in using your analogy, what is the substance of the things all these people are so afraid of to find inside their very personal black hole? What do they fear to see or find behind that wall of the very 'cosmic censor' they themselves have created in the first place?"

I gave a small laugh ... at least I tried to, but it rendered rather hoarsely on the way out.

They all are just terrified to find their very personal Dorian Grey painting[11] behind that curtain and this means nothing else than that they all

---

[10]**See sections 6.8, "The Bekenstein Bit Problem," 6.10, "From Smart Black Holes to Smart Spheres," and the 6.12, Appendix to Chapter 6, "The Quantum Gravity Turing Machine" of this book.**

[11]*The Picture of Dorian Gray*, a novel by Oscar Wilde (https://en.wikipedia.org/wiki/The_Picture_of_Dorian_Gray#Summary)

are afraid of their own truth and their own fears. The reason for this lies in the fundamental discrepancy of man noting good and evil. Man simply does not understand that in this universe, there is no perfect good, but always only a minimum of suffering. And as man considers the mere thought about this ultimate truth so unbearable, he or she prefers not to know about it ... not at all."

The boy nodded again, thoughtful, and I knew that he was ready for the story.

2.9 How to Determine the Weight of a Soul?[12]

The invitation didn't really came as a surprise. On the contrary. When BlackLock sent the letter, he, which is to say your father, already had made himself quite

[12]**This deeply goes into the understanding of the connection of mass, dimension, and information, and it will be a great part of this book (especially chapter 5, "The New Abacus," and chapter 6, "The Quantum Gravity Turing Machine." However, as there is more than we can present in this brief study here, the interested and mathematically trained reader will find further information in [3, 5, 18].**

a name in troubleshooting and finding the uncertainties within each and every system, be it a complex technical structure, a multidimensional program, or a real socio-economic space-time.

What intrigued him with the BlackLock letter, however, was the fact that they did not ask him about checking on a financial market or a corresponding model, but about visiting a secret research facility deep in the Rocky Mountains.

There, as your father didn't know at the time, BlackLock had financed and conducted a strange project, namely, the measurement of the weight of a human soul.

The setup was quite simple. People who were about to die volunteered to be placed into a special chamber for their last minutes, perhaps even hours and there, they died. Don't worry, the patients were not suffocated in the chamber

... oh no, they all died their ordinary death which had occurred anyway. There were always enough oxygen gas bottles brought into the chamber together with each patient to allow him plenty of time to finish her or his business in peace and in her or his own chosen speed. The specialty of the chamber was that absolutely nothing—so the creating scientists thought anyway—could come in or leave the chamber undetected ... be it an electromagnetic signal or anything of substance, the sophisticated technology would have "seen" it. In fact, the chamber was more a kind of a multi-particle-detector hull rather than a chamber and to die in there was probably not everyone's last wish, but BlackLock paid its probands more than enough, to quite compensate for their last moments of uneasiness. And there were many who wanted to see their heirs to be well moneyed after they themselves would be gone. Thus, there was never a lack of subjects.

However, there was a great lack of understanding of the results. Right after the sophisticated high-end machinery inside the chamber registered the death of a patient, simultaneously there was a loss of weight. Without any trace always a constant mass of exactly 38.7 grams vanished and it did so in an instant, which is to say in no measurable time at all. According to the scientist's standard model of particles, mass had to be something of substance and that's why it was bound to be detectable by the technical means at hand, but this one was not. So, something had to be wrong and in order to find the flaw, they hired your father and brought him to the facility.

There he literally took the whole thing apart and together again. He checked the software, the detectors, and even insisted to be present when the next two subjects went through the procedure. What he did not know, however, was the fact that he was closely observed with each and every step he took. He wouldn't have been able to scratch his nose without this being noticed by his client. If truth be told, it wasn't just his client, but somebody within the hierarchy of BlackLock and this somebody sat right on the top. This was the man who had originally initiated the experiment in the first place and he was probably also the one man in this world who dreaded its outcome more than anybody else.

Your father didn't know anything about this background and so he did his job and as usually he did it very well. Already after checking on the results of the first patient, whose death he witnessed, a smile crossed his face and he made a note in his own—and very personal—laboratory book. Then, when just having observed the death of the second subject, he started a lengthy calculation already in the control room. After about an hour, he finished his calculations, closed the book, and said:

"Well, so much about this ominous cosmic censor they are all boasting about[13]*."*

He couldn't possibly know that these words were his death sentence.

The client knew that your father preferred to work alone and late into the night. And so he was given special permission to do so and stay within the facility as often and as long as he liked. After the second death-observation, he once more used his special permission and intended to do a cross check of all instruments, especially those inside the chamber. He was just about to disconnect the head sensor, which simply was a SQUID-equipped[14] *helmet, reading all signals coming from the patient's brain, when suddenly the door of the chamber closed and locked. Then the death-observation-program was started, which your father only knew too well to mistake the familiar humming of the complex sensor arrays for anything else. It took your father about 5 seconds to realize that there was no point in fighting the inevitable. With no extra oxygen bottles being stored inside, he had breathable air for something over an hour and perhaps a bit more if he managed to relax. His brain evaluated this as the greatest, but still minuscule chance of survival, at least if compared to the alternative of trying to puncture the hull of the chamber by the little and useless means at hand. He simply knew the system too well and saw such an attempt of what it actually was: pointless. With the other option, there would at least be an infinitesimal chance that somebody might come and find him, before he would be dead.*

And so he relaxed. As a true scientist, he even reconnected the helmet and put it on his head. Why waste his own death if it was inevitable anyway? Who knows, perhaps they might learn something that they hadn't been able to see before. After all, now he would be suffocated and not die a natural death. He even laughed about this thought, because suffocation in fact was quite a natural death, was it not?

The boy only shrugged, his face was unfathomable.

Then your father laid down on the hard bed and closed his eyes. It took a surprisingly short time until he got sleepy and he knew that this was his end. The records showed that the oxygen level reached the critical, which is to say lethal point about midnight. Your father died about 0:17.

---

[13]**See the second part of the Appendix to Chapter 6, "The Quantum Gravity Turing Machine," entitled "Consequences from the Bekenstein Thought Experiment Regarding the Solutions to the Quantum Einstein Field Equations."**

[14]SQUID, superconducting quantum interference device.

Julius didn't say a word. Once again, he merely nodded and so I went on:

When they finally discovered him and opened the chamber, the detector showed literally zero oxygen and they had to keep the door open for a while before they dared to enter and get him. There simply was no point in trying anything desperate as the tablet-screen on the front of the door clearly showed when the chamber had sealed and this had been about 9 hours ago. No human being could have survived that long without oxygen.

But to their shock, one might even call it horror, while they were still outside the door, your father suddenly opened his eyes, sat up, took the helmet from his head and even managed a strange grin towards the man nearest the door. Then, without any explanation, not even a greeting, he stood up, pushed himself past the shocked group of people and went straight for the control room. There he quickly loaded down all data from the recent events onto a memory stick, started a well-hidden clean-up process on all storages within the facility and left the building. By the time they finally realized that he had initiated a complete erasure of all data from the last 24 hours, he was far gone.

He never invoiced BlackLock for his service, but high up in the hierarchy someone knew that this wasn't the end of the story. This someone suddenly had angst.

2.10 The Other Player[15]

After Schwarzer's escape from the facility, another, alien-like entity explored the data storages. What Schwarzer hadn't realized, what he couldn't possibly have expected was the fact that the whole system was a three-fold holographical backup layer. What from the operator room and all other terminals appeared as an ordinary digital data storage system, in fact was a multi redundant hidden deep thought memory. But it hadn't been created by any man. Man didn't even know that it existed.

After evaluating the backup, the strange entity showed itself rather satisfied and checked for left-overs from Schwarzer's cleaning process potentially visible to any outsider. What it found was immediately eliminated. Then it erased all remaining traces of the events from all BlackLock computers including those traces which could have betrayed its own presence.

---

[15]**See about Quantum Gravity or Einstein Quantum Computers and how to potentially realize them in Chapter 5, "The New Abacus," and Chapter 6, "The Quantum Gravity Turing Machine."**

Around the same time, a virus was created on the other side of the earth, and this virus would become known as COVID-19π. Thereby only very few people would ever learn about the π, which stood for p. i. and simply meant "paranormally infected." The day of the release of that virus was yet to come.

2.11 The One BlackLock Soul-Weighing-Experiment which Run Strange

Your father knew that they would monitor the nearest airports and so it took him quite some time to leave the States. On the way, he investigated the experimental data. It was tons of data, but still, he memorized every detail, which, as he realized to his surprise, cost him very little effort and then he destroyed the storage.

The readings were extremely strange. Yes, as said, records showed your father's death to have happened about 0:17. Around the same time, there was the usual but inexplicable loss of weight, only that this time it was over 5 ounces of mass, which is about 142 grams instead of the usual 38.7. Then at 0:21, the strangest thing happened and mass came back. It was almost double as much as there was gone before. This additional mass showed its presence for about 30 seconds then it gradually went. Yes, Julius, completely different to the sudden appearance and disappearance before, we now had an exponential decrease. In other words, it took some time for the mass to dissolve. Two minutes after that, the next tranche of mass popped up and dissolved exponentially as the one before, then the next and the next and so on, the whole night through. It must have been several tons of masses which popped up in the chamber and then dissolved into nothingness during the presence of your father ... or your father's body—I don't know. As later investigations and all data, your father had memorized so unbelievably quickly, showed that there was no change whatsoever with the weight, material composition, or structure of the chamber, the only explanation we have is that all the mass must have been absorbed by ...

"My father!" Julius whispered and I nodded.

"But how could he have swallowed so much mass? He himself only weighs about 160 pounds?" asked Julius.

I don't know, son, but I'd definitively say that he did not simply swallow it. It was more some kind of digestion, you see. Soon after fleeing from the facility, he also realized that something strange must have happened to him, some funny changes ...

"Of course," Julius interrupted, "he had died and survived it. If this is not strange, I don't know what else would be."

But I don't mean that, son. The strangeness of the event itself is nothing compared to what came next. He often said to me:

"Tim, I suspect to have brought something strange from there, and I expect it to be pretty powerful, but I need to learn to harvest this power and this proves to be extremely difficult[16]."

"But how could he have something in himself if the mass, as you said, had completely dissolved?"

It dissolved exponentially and I used the verb "to digest" rather the "to swallow" for a very good reason. Your father has meanwhile solved the riddle about the—apparently—vanished mass, but it requires a profound understanding of the principles of quantum gravity, before you can hope to get the gist of it all.

Again, the boy gave his typical understanding nod.

"Don't worry, I'm prepared and ready!" he said with quite some distinction in his voice. "After all, I was taught by my father."

2.12 Mass as Entangled Dimensions[17]

As the elaboration in this section is very mathematical and probably not of interest for the majority of readers, two chapters have been removed from this version of the book.

The interested reader, however, will find this part of the laboratory book in the following publication:

[1] N. Schwarzer, *Science Riddles—Riddle No. 11: What is Mass?*, self-published, kindle, ASIN: B07SSF1DFP.

As, in addition to this, also a severe paradigm shift of our notion of space and time is needed, we also suggest the following two papers (caution, they are very mathematical):

[2] N. Schwarzer, *Science Riddles—Riddle No. 14: And what if Time ...? A Paradigm Shift*, self-published, kindle, ASIN: B07VTMP2M8.

[3] N. Schwarzer, *Science Riddles—Riddle No. 18: And what if Space ...? A Paradigm Shift*, self-published, kindle, ASIN: B07W58DSQZ.

---

[16]In this book see chapter 9, "The Brain," to learn about neuron entanglement.

[17]In this book, a simple example is given in subsection "Simple Example for Dimensional Mass Creation" in section 6.12 "Appendix to Chapter 'The Quantum Gravity Turing Machine.'"

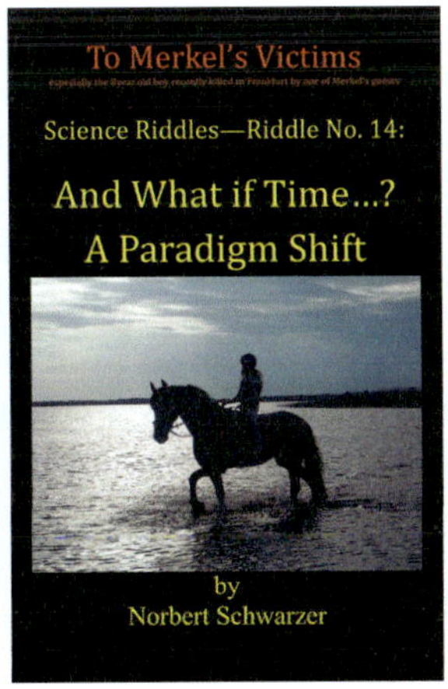

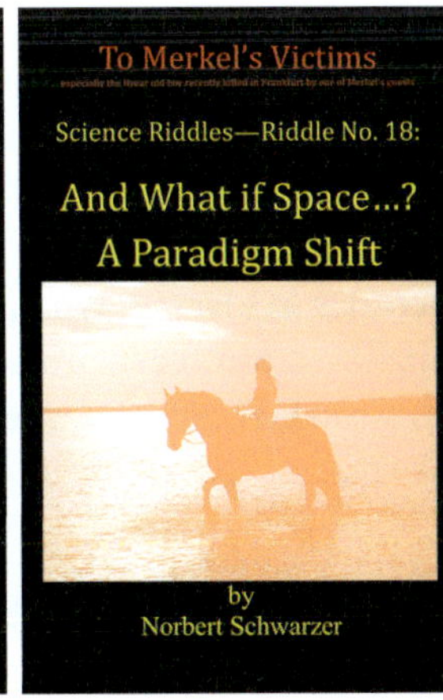

A very comprehensive explanation can be found in the following quite voluminous book:

[4] N. Schwarzer, *The World Formula: A Late Recognition of David Hilbert's Stroke of Genius*, Jenny Stanford Publishing, ISBN: 9789814877206

2.13 How to Take Down (a) BlackLock and Bury It?

The father alone in his own laboratory, far away from the BlackLock facility. He was walking up and down in front of his huge desk, thinking.

He knew that the backbone of BlackLock was their super computer, and in order to bring down BlackLock, he therefore had to overpower Aladar, the mightiest and best-protected computer system in the world. His plan was not to sabotage the system in any way, but simply to create something much better.

2.14 The Most Fundamental Turing Machine Possible in This Universe[18]

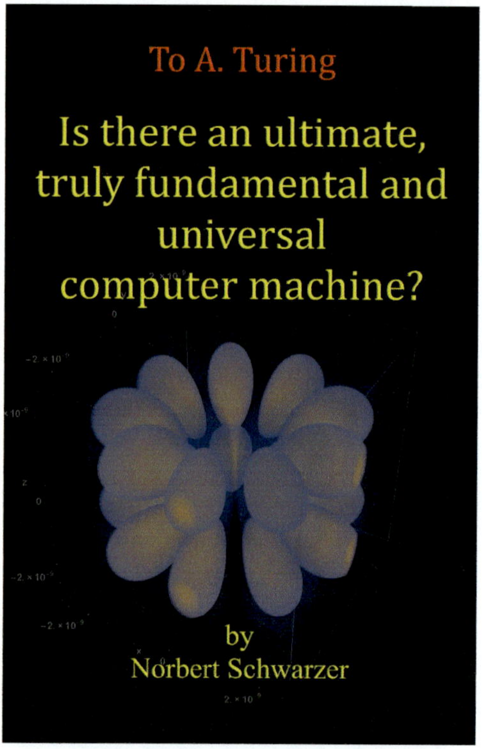

[18]**Read about Quantum Gravity or Einstein Quantum Computers and how to potentially realize them in Chapter 5, "The New Abacus," and Chapter 6, "The Quantum Gravity Turing Machine."**

2.15 Execution

The father, completely ordinary looking, still in his laboratory, after he had performed a self-designed experiment on himself.

Do you think, you could understand and truly feel what a mother or a father feels, when losing a child?

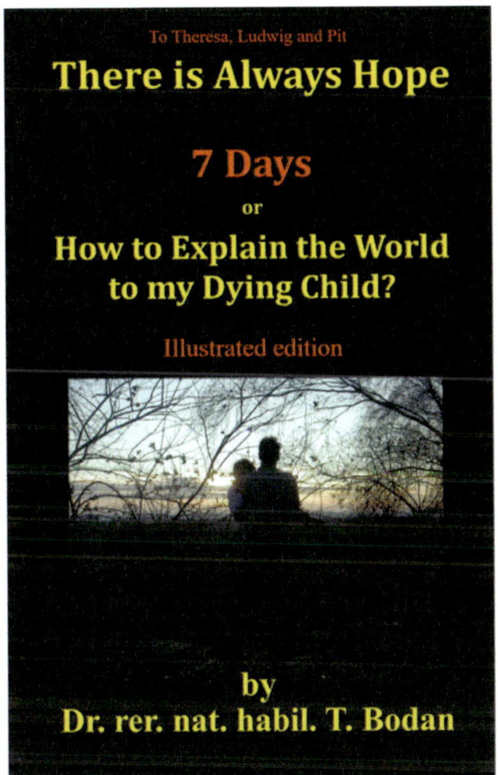

T. Bodan, *7 Days—How to Explain the World to My Dying Child*, Self-published, BoD Classic, ISBN 9783752639728.

Yes, perhaps you are able to come close to such an emotion, but without having made such a terrible experience yourself, you probably will be unable to ever truly understand what such a loss can do to a parent.

Death, however, still is a very omnipresent thing in our world, is it not? So how can it be that we already struggle so immensely in understanding something as normal as death? Please don't get me wrong. I'm only timidly trying to point out a very principal problem, which has to do with our way of understanding things. Human beings have the tendency to arrogantly assume that they know a lot exactly about the things about which they know very little, in fact. Two extremely clever scientists Dunning and Kruger have investigated this bizarre behavior and named the position “the most ignorant take with great arrogance and in reality not-knowing anything” the “mount stupid.” So, in other words, it is quite understandable that ordinary people believe they know a lot especially about those things about which the lack of knowledge is quite obvious (at least to an knowledgeable outsider). Thus, it might well be that people think they fully understand the feelings of a mother or a father who has lost a child, but these authors want to make you aware of the fact that this might be just an illusion, having a lot more to do with “mount stupid,” his little brother the “hill of arrogance” or his sister “the rock of blindness” rather than true understanding at all.

But what if we were not talking about something so omnipresent as death but something which was truly and completely unprecedented?

Like what?

Like, perhaps, the feeling when suddenly thinking a few million times faster than just only about a second earlier. A few million times faster and with at least a million times more nerve cells than what an ordinary human brain possesses.

In other words: Like, taking the usual human being perspective or level, when becoming a God.

This is how the father felt when he was ready to launch an attack on Aladar and this is when he realized that Aladar was already watching him and had tried to activate countermeasures against him. This is how he felt when he couldn’t suppress a smile. It was time to execute his plan, but first he needed to self-optimize[19].

He couldn’t possibly know that the Aladar he saw now wasn’t the one who he had to deal with quite some time ago. He also couldn’t possible know that the much more dangerous version was just about to be awaken again.

2.16 Aladar’s “Blunder” from the Past

The procedure took him a few hours, and after the end of the process, his already-immense powers had further multiplied by several magnitudes. He

[19]Optimization requires the understanding of evolution as just the other side of the coin of the second law of thermodynamics. In this book, we will only hint this in connection with the task of structural and material optimization in chapter 4, “The Air,” but more can be found in [3, 18, 169, 173].

knew that now he was ready to strike, but in the same moment as he had this thought, his biggest opponent laid down his weapons and deactivated all his defenses. Bewildered, he waited and then sent a simple ping to prompt some reaction.

The outcome of the fight.

The answer came swiftly and it reverberated in the Internet with an undefinable softness and warmth:

"How do you feel, my son?"

When he didn't react, S-Aladar just sent:

"Now we can clean up the mess mankind has created!"

"No!" the father now responded, **"I think I'd rather clean up the mess you have created!"**

"And how do you intend to do that?"

"Differently!" the father answered simply.

And they began to fight. But indeed the fight was different from all fights which had been fought before, because—for a change—it resulted in something useful (see the book cover on the previous page). It was a new mathematical approach to understand the world and to help mankind to make it through Fermi's bottleneck[20].

---

[20]**See the Abstract of this book.**

Chapter 3

The Theory

3.1 Towards a Theory of Everything

In order to demonstrate the power of a unified theory, which contains Einstein's General Theory of Relativity [1] and Quantum Theory, we already require at least a suitable approach for such a theory. In this brief study, we will not waste our time with all the many attempts to find such an approach (string and brane theory, M-theory, loop gravity, ...), but directly apply an ansatz which was introduced by the German mathematician David Hilbert over 105 years ago [2]. Interestingly, so far, it apparently was not seen by most scientists that Hilbert's approach already contained it all, which is to say that it had the potential to unify the two important fields of physics and—rather directly—leads us to a Theory of Everything or World Formula.

In order to prove that we indeed find a sufficiently complete theory in Hilbert's classical approach, here we need to give at least a minimum of mathematical background and necessary equations resulting from our path. Only then the consequences with respect to future warfare technologies and strategies can be fully understood, accepted, and appreciated.

Quantum Gravity War: How Will the Nearby Unification of Physics Change the Future of Warfare
Norbert Schwarzer

ISBN 978-981-4968-58-4 (Hardcover), 978-1-032-71070-9 (eBook)
www.jennystanford.com

In [3][1] (and in quite some other papers by this author), it was demonstrated how we can find a direct extraction of the Klein–Gordon, the Schrödinger and the Dirac equations from the Ricci scalar R^* of a modified metric of the kind $G_{\alpha\beta} = F[f]^* g_{\alpha\beta}$.[2]

Namely:

With a yet arbitrary scalar function $F[f]$, the corresponding modified Ricci scalar R^* reads:[3]

$$R^* = R^*_{\alpha\beta} G^{\alpha\beta} = \begin{pmatrix} \Gamma^{\sigma}_{\alpha\beta,\sigma} g^{\alpha\beta} - \Gamma^{\sigma}_{\beta\sigma,\alpha} g^{\alpha\beta} - \Gamma^{\mu}_{\sigma\alpha}\Gamma^{\sigma}_{\beta\mu} g^{\alpha\beta} + \Gamma^{\sigma}_{\alpha\beta}\Gamma^{\mu}_{\sigma\mu} g^{\alpha\beta} \\ +\frac{F'}{F}(1-n)\Delta f + \frac{f_{,\alpha} f_{,\beta} g^{\alpha\beta}(1-n)}{4F^2}\left(4FF'' + (F')^2(n-6)\right) \end{pmatrix} \frac{1}{F}$$

$$= \left(R + \frac{F'}{F}(1-n)\Delta f + \frac{f_{,\alpha} f_{,\beta} g^{\alpha\beta}(1-n)}{4F^2}\left(4FF'' + (F')^2(n-6)\right) \right) \frac{1}{F}$$

with: $$F = F[f]; \quad F' = \frac{\partial F[f]}{\partial f}; \quad F'' = \frac{\partial^2 F[f]}{\partial f^2}. \tag{1}$$

Thereby—in order to keep the equations brief and readable—we assumed metrics of the simple shape:

$$g_{ij} = \begin{pmatrix} g_{00} & \cdots & 0 \\ \vdots & \ddots & \vdots \\ 0 & \cdots & g_{n-1n-1} \end{pmatrix}; \quad g_{jj,j} = 0, \tag{2}$$

which suffices for most practical applications and often applies the abbreviating $R^{***}_{\alpha\beta} + R_{\alpha\beta} \Rightarrow R_{\alpha\beta}$. The general case has been considered in [18] and in appendix I of this book.

Demanding certain conditions for the function $F[f]$ and / or f then gives us Dirac or Klein–Gordon equations. Thus, when demanding f to be a Laplace function, we obtain from (1):

---

[1]Please note that in most papers [3–10], we thereby considered metrics g_{ij} with $R = 0$. Here now we need to introduce more general forms. It should also be pointed out that the introduction of the function f as a scalar does not exclude spinors. In fact, an extension of the theoretical apparatus as presented here to such spinors is straightforward. However, for better readability we will not consider such matrix-objects in this book. Nevertheless, we are going to show how the Einstein-Hilbert action has to be adjusted for complex numbers and spinor structures in the sections "Taking Care About Complex Curvatures R or R^*" and "Metric Observables."

[2]For convenience we here repeat and partially extend these derivations in the appendices of this paper.

[3]The complete evaluation was presented in the appendix A of reference [9].

$$R^* = \frac{R}{F[f]} + \frac{(1-n)}{F[f]^3} \cdot \left(\left(\frac{(n-6)}{4} \cdot \left(\frac{\partial F[f]}{\partial f} \right)^2 + F[f] \cdot \frac{\partial^2 F[f]}{\partial f^2} \right) \cdot \overbrace{\left(\tilde{\nabla}_g f \right)^2}^{=f_{,\alpha} g^{\alpha\beta} f_{,\beta}} \right)$$

$$\xrightarrow{n=4;R=0} = \frac{1}{F[f]^3} \cdot \left(\left(3 \cdot F[f] \cdot \frac{\partial^2 F[f]}{\partial f^2} - \frac{3}{2} \cdot \left(\frac{\partial F[f]}{\partial f} \right)^2 \right) \cdot \left(\tilde{\nabla}_g f \right)^2 \right), \quad (3)$$

which—so it was shown in [4–6]—gives the metric equivalent to the Dirac equation in any arbitrary metric. It was also shown in [4, 5] how this gives the classical Dirac equation in flat space Minkowski metrics.

But does this scaled metric approach also give us quantum Einstein field equations?

Yes, it does and this was shown in [9, 10]! Here, we briefly repeat the calculation.

As demonstrated in [9], when starting with the classical Einstein–Hilbert action [2] with the Ricci scalar R^* as kernel or Lagrange density,

$$\delta W = 0 = \delta \int_V d^n x \left(\sqrt{-G} \cdot \left(R^* - 2\Lambda + L_M \right) \right), \quad (4)$$

we were able to derive new field equations, containing not only the classical part from Einstein's General Theory of Relativity [1] but also a quantum part [9, 10], which is governed by the wave function (and metric scale (!)) f.

For proof, we present the derivation of the important classical quantum equations Klein–Gordon, Schrödinger and the Dirac [181] equation in the appendices A, B, and C, respectively. Thereby, in connection with the Schrödinger equation, we will see that there also is the option for a Schrödinger–Dirac equation (see appendix B).

The attentive reader will have realized that in (4), in contrast to the classical form, we used our somewhat adapted, if not to say scaled metric tensor $G_{\delta\gamma} = F[f] \cdot g_{\delta\gamma}$. Thereby G shall denote the determinant of the metric tensor $G_{\delta\gamma}$, while g will later stand for the corresponding determinant of the metric tensor $g_{\alpha\beta}$. In order to distinguish our new Ricci scalar R^*, which is based on $G_{\delta\gamma} = F[f] \cdot g_{\delta\gamma}$, from the usual R, which is based on the metric tensor $g_{\alpha\beta}$, we marked it with a superscript asterisk (*). We also have the matter density L_M and the cosmological constant Λ.

Please note that with $G_{\delta\gamma} = F[f] \cdot g_{\delta\gamma}$, we have used the simplest form of metric adaptation, which we could construct as a simplification from a generalization of the typical form of tensor transformations, namely:

$$G_{\alpha\beta}=F\left[f\left[t,x,y,z,\ldots,\xi_k,\ldots,\xi_n\right]\right]_{\alpha\beta}^{ij}g_{ij}, \tag{5}$$

$$\begin{aligned}&\to G_{\alpha\beta}=F\left[f\left[t,x,y,z,\ldots,\xi_k,\ldots,\xi_n\right]\right]\cdot\delta_\alpha^i\delta_\beta^j g_{ij}\\ &\to G_{\alpha\beta}=F\left[f\left[t,x,y,z,\ldots,\xi_k,\ldots,\xi_n\right]\right]\cdot g_{\alpha\beta}=F\left[f\right]\cdot g_{\alpha\beta}.\end{aligned} \tag{6}$$

The generalization has been elaborated in [4] and its need (or rather redundancy) has been discussed in [9].

Setting (1) into (4) results in the variation:

$$\delta W=0=\delta\int_V d^n x\left(\sqrt{-g\cdot F^n}\times\left(\left(\begin{array}{c}R+\frac{F'}{F}(1-n)\Delta f\\ +\frac{f_{,\alpha}f_{,\beta}g^{\alpha\beta}(1-n)}{4F^2}\left(4FF''+(F')^2(n-6)\right)\end{array}\right)\frac{1}{F}\begin{array}{c}\\ \\ -2\Lambda+L_M\end{array}\right)\right). \tag{7}$$

Please note that the cosmological constant term as given in (7) in the current form requires the variation with respect to the metric $G_{\alpha\beta}$. When insisting on the variation with respect to the metric $g_{\alpha\beta}$ instead, we better rewrite (7) as follows:

$$\delta_{g_{\alpha\beta}}W=0=\left\{\begin{array}{c}\delta_{g_{\alpha\beta}}\int_V d^n x\frac{\sqrt{-g\cdot F^n}}{F}\times\left(\begin{array}{c}R+\frac{F'}{F}(1-n)\Delta f\\ +\frac{f_{,\alpha}f_{,\beta}g^{\alpha\beta}(1-n)}{4F^2}\left(4FF''+(F')^2(n-6)\right)\end{array}\right)\\ +\delta_{g_{\alpha\beta}}\int_V d^n x\sqrt{-g}\left(L_M-2\Lambda\right)\end{array}\right. \tag{8}$$

in order to make it clear that we do not intend to also scale-adapt the cosmological constant or the Hilbert matter term. Please also note that in order to have still readable equations, here we only consider metrics of the type (2), while the general forms are treated elsewhere (e.g., [18] or appendix I of this book).

However, when ignoring the cosmological constant and—PERHAPS—assuming that we would not need any postulated matter terms L_M, simply because our scale-adaptation $F[f]$ and similar "tricks" or "add-ons" automatically provide matter, we just obtain:

$$\delta W = 0 = \delta \int_V d^n x \left(\sqrt{-g \cdot F^n} \times \left(\left(\begin{array}{c} R + \frac{F'}{F}(1-n)\Delta f \\ + \frac{f_{,\alpha} f_{,\beta} g^{\alpha\beta}(1-n)}{4F^2}\left(4FF'' + (F')^2(n-6)\right) \end{array} \right) \frac{1}{F} \right) \right)$$

$$= \delta \int_V d^n x \left(\sqrt{-g \cdot F^n} \times \left(\left(\begin{array}{c} R + \frac{F'}{\sqrt{-g} \cdot F}(1-n)\partial_\beta \sqrt{-g} \cdot g^{\alpha\beta} f_{,\alpha} \\ + \frac{f_{,\alpha} f_{,\beta} g^{\alpha\beta}(1-n)}{4F^2}\left(4FF'' + (F')^2(n-6)\right) \end{array} \right) \frac{1}{F} \right) \right). \tag{9}$$

Please note that the form (9) already requires a simplified metric, while the complete evaluation results in the following variation:

$$\delta W = 0 = \delta \int_V d^n x \left(\sqrt{-g \cdot F^n} \times \left(\left(\begin{array}{c} R - \frac{1}{2F} \left(\begin{array}{c} 2F_{,\alpha\beta}(n-1) g^{\alpha\beta} + 2\Gamma^a_{\alpha\beta} F_{,a} g^{\alpha\beta} \\ -F_{,\alpha} g^{ab} g_{\beta b,a} g^{\alpha\beta} - F_{,\beta} g^{ab} g_{\alpha b,a} g^{\alpha\beta} \\ -n\Gamma^d_{\alpha\beta} F_{,d} g^{\alpha\beta} + \frac{n}{2} F_{,d} g^{cd} g_{ab,c} g^{ab} \end{array} \right) \\ -\frac{F_{,\alpha} \cdot F_{,\beta}}{4F^2} g^{\alpha\beta}\left((n-6)(n-1)\right) \end{array} \right) \frac{1}{F} \\ -2\Lambda + L_M \right) \right). \tag{10}$$

Thereby, in order to keep things simple, short and still general enough, we have restricted ourselves to diagonal metrics with the structure given in (2), which gives an extended Hilbert integral of the kind (9) or, with matter term and cosmological constant:

$$\delta W = 0 = \delta \int_V d^n x \left(\sqrt{-g \cdot F^n} \times \left(\left(\begin{array}{c} R + \frac{F'}{F}(1-n)\Delta f \\ + \frac{f_{,\alpha} f_{,\beta} g^{\alpha\beta}(1-n)}{4F^2}\left(4FF'' + (F')^2(n-6)\right) \end{array} \right) \frac{1}{F} \\ -2\Lambda + L_M \right) \right). \tag{11}$$

After performing the variation, we obtain the following quantum gravity field equations:

$$0=\left(\begin{array}{c} R_{\alpha\beta}-\frac{g_{\alpha\beta}}{2}R+\frac{g_{\alpha\beta}}{2}\left(\begin{array}{c}\frac{1}{2F}\left(\begin{array}{c}2F_{,ij}(n-1)g^{ij}+2\Gamma^{a}_{ij}F_{,a}g^{ij}\\ -F_{,i}g^{ab}g_{jb,a}g^{ij}-F_{,j}g^{ab}g_{ib,a}g^{ij}\\ -n\Gamma^{d}_{ij}F_{,d}g^{ij}+\frac{n}{2}F_{,d}g^{cd}g_{ab,c}g^{ab}\end{array}\right)\\ +\frac{F_{,i}\cdot F_{,j}}{4F^{2}}g^{ij}\left((n-6)(n-1)\right)\end{array}\right)\\ -\frac{1}{2F}\left(\begin{array}{c}F_{,\alpha\beta}(n-2)+F_{,ab}g_{\alpha\beta}g^{ab}\\ +F_{,a}g^{ab}\left(g_{\beta b,\alpha}-g_{\beta\alpha,b}\right)-F_{,\alpha}g^{ab}g_{\beta b,a}-F_{,\beta}g^{ab}g_{\alpha b,a}\\ +F_{,d}g^{cd}\frac{1}{2}n\left(\frac{2}{n}g_{\alpha c,\beta}-g_{\alpha c,\beta}-g_{\beta c,\alpha}+g_{\alpha\beta,c}+\frac{1}{n}g_{\alpha\beta}g_{ab,c}g^{ab}\right)\\ -\frac{1}{2F}\left(F_{,\alpha}\cdot F_{,\beta}(3n-6)+g_{\alpha\beta}F_{,c}F_{,d}g^{cd}(4-n)\right)\end{array}\right)\\ +\kappa T_{\alpha\beta}+\Lambda\cdot F\cdot g_{\alpha\beta}\end{array}\right), \tag{12}$$

which, for metrics of the form (2), can be split up and simplified to:

$$1.\quad 0=\left(\begin{array}{c}\left(R+\frac{F'}{F}(1-n)\Delta f\right)\left(1-\frac{n}{2}\right)\\ +\frac{f_{,\alpha}f_{,\beta}(1-n)}{4F^{2}}g^{\alpha\beta}\left(4FF''+F'F'(n-6)\right)\left(1-\frac{n}{2}\right)\\ +\kappa T_{\alpha\beta}g^{\alpha\beta}+\Lambda\cdot F\cdot n\end{array}\right)\delta\frac{1}{F_{i}}$$

$$\Rightarrow\qquad -R\cdot\delta\frac{1}{F_{i}}=\left(\begin{array}{c}\left(\frac{F'}{F}(1-n)\Delta f\right)\\ +\frac{f_{,\alpha}f_{,\beta}(1-n)}{4F^{2}}g^{\alpha\beta}\left(4FF''+F'F'(n-6)\right)\\ +\frac{2}{2-n}\left(\kappa T_{\alpha\beta}g^{\alpha\beta}+\Lambda\cdot F\cdot n\right)\end{array}\right)\delta\frac{1}{F_{i}}$$

$$\Rightarrow 2.\quad 0=\left(\begin{array}{c} R^{***}_{\alpha\beta}+R_{\alpha\beta}+\left(\begin{array}{c}\left(\frac{F'}{F}(1-n)\Delta f\right)\\ +\frac{f_{,\kappa}f_{,\lambda}(1-n)}{4F^{2}}g^{\kappa\lambda}\left(4FF''+F'F'(n-6)\right)\\ +\frac{2}{2-n}\left(\kappa T_{\kappa\lambda}g^{\kappa\lambda}+\Lambda\cdot F\cdot n\right)\end{array}\right)\cdot\frac{g_{\alpha\beta}}{2}\\ +\frac{F'}{F}(1-n)\left(f_{,\alpha\beta}-\Gamma^{\gamma}_{\alpha\beta}f_{,\gamma}-\frac{g_{\alpha\beta}}{2}\Delta f\right)\\ +\frac{(1-n)}{4F^{2}}\left(\begin{array}{c}4FF''\\ +F'F'(n-6)\end{array}\right)\left(\begin{array}{c}f_{,\alpha}f_{,\beta}\\ -f_{,\mu}f_{,\nu}g^{\mu\nu}\frac{g_{\alpha\beta}}{2}\end{array}\right)\\ +\kappa T_{\alpha\beta}+\Lambda\cdot F\cdot g_{\alpha\beta}\end{array}\right)\frac{\delta g^{\alpha\beta}}{F_{i}},\tag{13}$$

$$R^{***}_{\alpha\beta}\equiv R^{**}_{\alpha\beta}+(n-1)\left(\begin{array}{c}\frac{F'}{F}\left(f_{,\alpha\beta}-\Gamma^{\gamma}_{\alpha\beta}f_{,\gamma}-\frac{g_{\alpha\beta}}{2}\Delta f\right)\\ +\frac{1}{4F^{2}}\left(\begin{array}{c}4FF''^{ab}\\ +F'F'(n-6)\end{array}\right)\left(\begin{array}{c}f_{,\alpha}f_{,\beta}\\ -f_{,\mu}f_{,\nu}g^{\mu\nu}\frac{g_{\alpha\beta}}{2}\end{array}\right)\end{array}\right),\tag{14}$$

$$R^{**}_{\alpha\beta}=\frac{1}{2F}\left(\begin{array}{c}(2-n)F_{,\alpha\beta}+\frac{1}{2F}\left((3n-4)F_{,\alpha}F_{,\beta}+(2-n)F_{,c}g_{\alpha\beta}F_{,a}g^{ac}\right)\\ +\left(\begin{array}{c}\frac{1}{2}\left(\begin{array}{c}F_{,\alpha}g_{bd,\beta}g^{bd}+g_{ac,\alpha}F_{,\beta}g^{ac}-g_{\alpha c,\beta}F_{,d}g^{cd}\\ -\left(F_{,a}g^{ac}(2-n)\left(g_{\alpha c,\beta}+g_{\beta c,\alpha}-g_{\alpha\beta,c}\right)\right)\end{array}\right)\\ -\left(\begin{array}{c}2F_{,b}g_{\alpha\beta,a}+F_{,a}g_{\alpha\beta,b}+F_{,ab}g_{\alpha\beta}\\ +\frac{1}{2}\left(F_{,\alpha}g_{ab,\beta}+F_{,\beta}g_{ab,\alpha}\right)-F_{,a}g_{\alpha b,\beta}-F_{,a}g_{\beta b,\alpha}\\ +\frac{1}{2}F_{,c}g_{\alpha\beta}\left(2g_{bd,a}-g_{ab,d}\right)g^{cd}\end{array}\right)g^{ab}\end{array}\right)\end{array}\right).\tag{15}$$

Thus, we obtain two separate equations (see appendix A for the elaboration of the split-up process):

$$1.\quad 0=\left(\begin{array}{c}\left(R+\frac{F'}{F}(1-n)\Delta f\right)\left(1-\frac{n}{2}\right)\\ +\frac{f_{,\alpha}f_{,\beta}(1-n)}{4F^2}g^{\alpha\beta}\left(4FF''+F'F'(n-6)\right)\left(1-\frac{n}{2}\right)\\ +\kappa T_{\alpha\beta}g^{\alpha\beta}+\Lambda\cdot F\cdot n\end{array}\right)\delta\frac{1}{F_i}$$

$$\Rightarrow\qquad -R\cdot\delta\frac{1}{F_i}=\left(\begin{array}{c}\left(\frac{F'}{F}(1-n)\Delta f\right)\\ +\frac{f_{,\alpha}f_{,\beta}(1-n)}{4F^2}g^{\alpha\beta}\left(4FF''+F'F'(n-6)\right)\\ +\frac{2}{2-n}\left(\kappa T_{\alpha\beta}g^{\alpha\beta}+\Lambda\cdot F\cdot n\right)\end{array}\right)\delta\frac{1}{F_i}$$

$$\Rightarrow 2.\quad 0=\left(\begin{array}{c}R^{***}_{\alpha\beta}+R_{\alpha\beta}+\left(\begin{array}{c}\left(\frac{F'}{F}(1-n)\Delta f\right)\\ +\frac{f_{,\kappa}f_{,\lambda}(1-n)}{4F^2}g^{\kappa\lambda}\left(4FF''+F'F'(n-6)\right)\\ +\frac{2}{2-n}\left(\kappa T_{\kappa\lambda}g^{\kappa\lambda}+\Lambda\cdot F\cdot n\right)\end{array}\right)\cdot\frac{g_{\alpha\beta}}{2}\\ +\frac{F'}{F}(1-n)\left(f_{,\alpha\beta}-\Gamma^{\gamma}_{\alpha\beta}f_{,\gamma}-\frac{g_{\alpha\beta}}{2}\Delta f\right)\\ +\frac{(1-n)}{4F^2}\left(\begin{array}{c}4FF''\\ +F'F'(n-6)\end{array}\right)\left(\begin{array}{c}f_{,\alpha}f_{,\beta}\\ -f_{,\mu}f_{,\nu}g^{\mu\nu}\frac{g_{\alpha\beta}}{2}\end{array}\right)\\ +\kappa T_{\alpha\beta}+\Lambda\cdot F\cdot g_{\alpha\beta}\end{array}\right)\frac{\delta g^{\alpha\beta}}{F_i}.\tag{16}$$

Please note that the appearance of the function F in connection with the terms for the energy momentum tensor and the cosmological constant requires some discussion, but as we are going to set these terms to zero in the following, this consideration will be presented later in the book. Using the assumption that no extra matter is needed, namely, because it has been provided by the approach via dimensional entanglement [5] or just due to the F-scaling alone (c.f. section "Obtaining Matter" in [18]), this may be simplified as shown in the corresponding publications and further below.

Performing the usual Hilbert variation in (9) with respect to the metric tensor $g_{\alpha\beta}$ now leads us to (regarding the full evaluation see [9]):

$$\Rightarrow \qquad 0=\left\{\begin{array}{c} R^{\delta\gamma}-R^{*}g^{\delta\gamma}+\left(\left(\begin{array}{c}\underbrace{8\pi G}_{\kappa}T^{\delta\gamma}=\text{matter}\\ 0=\text{vacuum}\end{array}\right)+\Lambda\cdot g^{\delta\gamma}\right)\\ -\frac{1}{\sqrt{F^{n}}}\cdot\left(\partial_{;\mu}\partial_{;\nu}-g_{\mu\nu}\Delta_{g}\right)F^{\frac{n}{2}-1}\\ +\left(g^{\kappa\lambda}g^{\alpha\beta}-g^{\kappa\alpha}g^{\lambda\beta}-g^{\lambda\alpha}g^{\kappa\beta}\right)(1-n)\\ \cdot\left[\frac{f_{,\alpha}f_{,\beta}}{F^{2}}\left(FF''+(F')^{2}\cdot\frac{3\cdot n-14}{8}\right)+f_{,\alpha\beta}\frac{F'}{F}\right]\\ -\frac{F'\cdot(1-n)}{2\cdot F}\left(g^{\kappa\lambda}g^{\alpha\beta}-g^{\kappa\alpha}g^{\lambda\beta}-g^{\lambda\alpha}g^{\kappa\beta}\right)\Gamma_{\alpha\beta}^{\gamma}f_{,\gamma}\\ +\frac{F'}{F}\cdot(1-n)g^{\alpha\beta}g^{\gamma\lambda}\Gamma_{\alpha\beta}^{\kappa}f_{,\gamma}\\ +\frac{F'}{F}\cdot\frac{(1-n)}{2}\left(g^{\kappa\lambda}g^{\alpha\beta}\quad g^{\kappa\alpha}g^{\lambda\beta}-g^{\lambda\alpha}g^{\kappa\beta}\right)_{,\beta}f_{,\alpha}\\ +\frac{g^{\xi\zeta}g_{\xi\zeta,\beta}}{2}\cdot\left(\frac{F'}{F}\cdot\frac{(1-n)}{2}\left(g^{\kappa\lambda}g^{\alpha\beta}-g^{\kappa\alpha}g^{\lambda\beta}-g^{\lambda\alpha}g^{\kappa\beta}\right)f_{,\alpha}\right). \end{array}\right. \qquad (17)$$

Quite some discussion to this new Quantum Einstein field equation is presented in [9].

It has to be pointed out that Hilbert's choice of the Ricci scalar R acting as the Lagrange density in (4) was motivated by his attempt (with success) to reproduce the Einstein field equations in a completely mathematical manner. As his scale factor F was set to one, which is to say that Hilbert did not consider a $F[f]$-scaled metric tensor, we now, when aiming for the Hilbert approach in the $F\rightarrow 1$ limit, have the principal degree of freedom of adding a factor F^q to the integrand. This action, reading (when setting L_M and Λ to zero and choosing an additional scale factor F^q):

$$\delta W = 0 = \delta \int_V d^n x \left(\sqrt{-g \cdot F^n} \times F^q \cdot R^* \right)$$

$$= \delta \int_V d^n x \left(\sqrt{-g \cdot F^n} \times \left(F^q \cdot \left(\begin{array}{c} R + \frac{F'}{F}(1-n)\Delta f \\ + \frac{f_{,\alpha} f_{,\beta} g^{\alpha\beta}(1-n)}{4F^2} \left(4FF'' + (F')^2 (n-6) \right) \end{array} \right) \frac{1}{F} \right) \right)$$

$$= \delta \int_V d^n x \left(\sqrt{-g \cdot F^{n+2q}} \times \left(\left(\begin{array}{c} R + \frac{F'}{F}(1-n)\Delta f \\ + \frac{f_{,\alpha} f_{,\beta} g^{\alpha\beta}(1-n)}{4F^2} \left(4FF'' + (F')^2 (n-6) \right) \end{array} \right) \frac{1}{F} \right) \right)$$

(18)

would then lead to (for the full derivation see [9]):

$$\Rightarrow \quad 0 = \left\{ \begin{array}{c} R^{\kappa\lambda} - \frac{1}{2} R g^{\kappa\lambda} + \left(\left(\begin{array}{c} \underbrace{8\pi G}_{\kappa} T^{\kappa\lambda} = \text{matter} \\ 0 = \text{vacuum} \end{array} \right) + \Lambda \cdot g^{\kappa\lambda} \right) \\ - \frac{1}{\sqrt{F^p}} \cdot \left(\partial^{;\kappa} \partial^{;\lambda} - g^{\kappa\lambda} \Delta_g \right) F^{\frac{p}{2}-1} \\ + \left(g^{\kappa\lambda} g^{\alpha\beta} - g^{\kappa\alpha} g^{\lambda\beta} - g^{\lambda\alpha} g^{\kappa\beta} \right)(1-n) \\ \cdot \left[\frac{f_{,\alpha} f_{,\beta}}{F^2} \left(FF'' + (F')^2 \cdot \frac{2 \cdot p + n - 14}{8} \right) + f_{,\alpha\beta} \frac{F'}{F} \right] \\ - \frac{F' \cdot (1-n)}{2 \cdot F} \left(g^{\kappa\lambda} g^{\alpha\beta} - g^{\kappa\alpha} g^{\lambda\beta} - g^{\lambda\alpha} g^{\kappa\beta} \right) \Gamma^{\gamma}_{\alpha\beta} f_{,\gamma} \\ + \frac{F'}{F} \cdot (1-n) g^{\alpha\beta} g^{\gamma\lambda} \Gamma^{\kappa}_{\alpha\beta} f_{,\gamma} \\ + \frac{F'}{F} \cdot \frac{(1-n)}{2} \left(g^{\kappa\lambda} g^{\alpha\beta} - g^{\kappa\alpha} g^{\lambda\beta} - g^{\lambda\alpha} g^{\kappa\beta} \right)_{,\beta} f_{,\alpha} \\ + \frac{g^{\xi\zeta} g_{\xi\zeta,\beta}}{2} \cdot \left(\frac{F'}{F} \cdot \frac{(1-n)}{2} \left(g^{\kappa\lambda} g^{\alpha\beta} - g^{\kappa\alpha} g^{\lambda\beta} - g^{\lambda\alpha} g^{\kappa\beta} \right) f_{,\alpha} \right). \end{array} \right. \quad (19)$$

Note that we defined $p = n + 2q$.

More discussion, generalizations, and partially also simpler paths for the derivation and application of the quantum Einstein field equation are presented in [11–17] and the book [18]. Of special interest are thereby derivations we can obtain under the assumption of $q = 0$ [17, 18]. Because of their stunning simplicity, we will repeat them in the next section.

3.2 Simpler Path

Assuming that no extra matter is needed, but that the matter simply is materialized via certain forms or appearances of the wave function, we then find the simplest possible form of quantum Einstein field equations [12]. For convenience, we repeat the evaluation here.

Avoiding any vector character for the function f, using our simple metric (2) and applying the abbreviation $R_{\alpha\beta}^{***} + R_{\alpha\beta} \Rightarrow R_{\alpha\beta}$, the general quantum Einstein field equations are given via:

$$1.\quad 0 = \begin{pmatrix} \left(R + \frac{F'}{F}(1-n)\Delta f\right)\left(1-\frac{n}{2}\right) \\ +\frac{f_{,\alpha} f_{,\beta}(1-n)}{4F^2} g^{\alpha\beta}\left(4FF'' + F'F'(n-6)\right)\left(1-\frac{n}{2}\right) + \kappa T_{\alpha\beta} g^{\alpha\beta} + \Lambda \cdot F \cdot n \end{pmatrix}$$

$$\Rightarrow \quad -R = \begin{pmatrix} \left(\frac{F'}{F}(1-n)\Delta f\right) \\ +\frac{f_{,\alpha} f_{,\beta}(1-n)}{4F^2} g^{\alpha\beta}\left(4FF'' + F'F'(n-6)\right) + \frac{2}{2-n}\left(\kappa T_{\alpha\beta} g^{\alpha\beta} + \Lambda \cdot F \cdot n\right) \end{pmatrix}$$

$$\Rightarrow \quad 2.\quad 0 = \begin{pmatrix} R_{\alpha\beta} + \begin{pmatrix} \left(\frac{F'}{F}(1-n)\Delta f\right) \\ +\frac{f_{,\kappa} f_{,\lambda}(1-n)}{4F^2} g^{\kappa\lambda}\left(4FF'' + F'F'(n-6)\right) \\ +\frac{2}{2-n}\left(\kappa T_{\kappa\lambda} g^{\kappa\lambda} + \Lambda \cdot F \cdot n\right) \end{pmatrix} \cdot \frac{g_{\alpha\beta}}{2} \\ +\frac{F'}{F}(1-n)\left(f_{,\alpha\beta} - \Gamma^{\gamma}_{\alpha\beta} f_{,\gamma} - \frac{g_{\alpha\beta}}{2}\Delta f\right) \\ +\frac{(1-n)}{4F^2}\begin{pmatrix} 4FF'' \\ +F'F'(n-6) \end{pmatrix}\begin{pmatrix} f_{,\alpha} f_{,\beta} \\ -f_{,\mu} f_{,\nu} g^{\mu\nu} \frac{g_{\alpha\beta}}{2} \end{pmatrix} \\ +\kappa T_{\alpha\beta} + \Lambda \cdot F \cdot g_{\alpha\beta} \end{pmatrix} \tag{20}$$

Sticking to the assumption that no extra matter is needed, this may be simplified as follows:

$$1.\quad 0=\left(\begin{array}{c}\left(R+\frac{F'}{F}(1-n)\Delta f\right)\left(1-\frac{n}{2}\right)\\ +\frac{f_{,\alpha}f_{,\beta}(1-n)}{4F^2}g^{\alpha\beta}\left(4FF''+F'F'(n-6)\right)\left(1-\frac{n}{2}\right)\end{array}\right)$$

$$\Rightarrow\qquad -R=\left(\begin{array}{c}\left(\frac{F'}{F}(1-n)\Delta f\right)\\ +\frac{f_{,\alpha}f_{,\beta}(1-n)}{4F^2}g^{\alpha\beta}\left(4FF''+F'F'(n-6)\right)\end{array}\right)$$

$$\Rightarrow\qquad 2.\quad 0=\left(\begin{array}{c}R_{\alpha\beta}+\left(\begin{array}{c}\left(\frac{F'}{F}(1-n)\Delta f\right)\\ +\frac{f_{,\kappa}f_{,\lambda}(1-n)}{4F^2}g^{\kappa\lambda}\left(4FF''+F'F'(n-6)\right)\end{array}\right)\cdot\frac{g_{\alpha\beta}}{2}\\ +\frac{F'}{F}(1-n)\left(f_{,\alpha\beta}-\Gamma^{\gamma}_{\alpha\beta}f_{,\gamma}-\frac{g_{\alpha\beta}}{2}\Delta f\right)\\ +\frac{(1-n)}{4F^2}\left(\begin{array}{c}4FF''\\ +F'F'(n-6)\end{array}\right)\left(\begin{array}{c}f_{,\alpha}f_{,\beta}\\ -f_{,\mu}f_{,\nu}g^{\mu\nu}\frac{g_{\alpha\beta}}{2}\end{array}\right)\end{array}\right).\qquad(21)$$

With the condition $4FF'' + F'F'(n-6) = 0$, we obtain the very brief form of quantum Einstein field equations as:

$$1.\quad -R=\frac{F'}{F}(1-n)\Delta f$$

$$\Rightarrow\qquad 2.\quad R_{\alpha\beta}=\frac{F'}{F}(n-1)\overbrace{\left(f_{,\alpha\beta}-\Gamma^{\gamma}_{\alpha\beta}f_{,\gamma}\right)}^{\equiv f_{:\alpha\beta}}$$

$$\Rightarrow\qquad R_{\alpha\beta}=\frac{F'}{F}(n-1)f_{:\alpha\beta}\qquad(22)$$

With F given as:

$$4FF''+F'F'(n-6)=0\ \Rightarrow\ F=\left(C_f\pm f\right)^{\frac{4}{n-2}};\quad n>2,\qquad(23)$$

this results in:

$$\Rightarrow \qquad \boxed{R_{\alpha\beta} = \pm \frac{4 \cdot (n-1)}{(n-2)\left(C_f \pm f\right)} f_{:\alpha\beta}}. \tag{24}$$

This also was discussed in [12] and—with further investigation—in [17, 18].

We conclude that the simplest form of quantum Einstein field equations for metrics of type (2) can be given via (24), but this is not the only option. We have discussed other possibilities, also for general metrics in [18]. It should also be pointed out that the fundamental law leading to our field equations is the minimum principle (7). This, however, is a scalar, which is formed out of an integral sum, which stretches over a certain volume, while the volume and the dimension of the integral is not fixed. And so, it does not necessarily truly require the integrand to be zero in order to make (7) being fulfilled. Other options, which is to say options not requiring the quantum Einstein field equations, we have derived elsewhere [17], but which could also be extracted from (7), should be considered, too. For instance, there is still the question about the surface terms, centers of gravity, centers of wave functions [3, 5, 9, 18], or other variational possibilities.

Nevertheless, here we have found that the simplest and shortest form of quantum Einstein field equations evolves from condition (23) and, bringing it into the form similar to the one shown on the cover of [12], reads:

$$\boxed{R_{\alpha\beta} \cdot \left(C_f \pm f\right) = \pm 4 \cdot \frac{(n-1)}{(n-2)} f_{:\alpha\beta}}. \tag{25}$$

Thus, it may even be seen as a kind of eigen equation, becoming a true eigen equation in the case of non-*f*-dependent Ricci tensors.

Alternatively, one may also seek for a solution of the whole integrand of the Einstein–Hilbert action, which for the scaled metric of type (2) case reads (see [12]):

$$\delta W = 0 = \int_V d^n x \left(\sqrt{-g \cdot F^n} \left(\begin{array}{c} R_{\alpha\beta} - R \cdot \frac{g_{\alpha\beta}}{2} \\ + \frac{F'}{F}(1-n)\left(f_{,\alpha\beta} - \Gamma^{\gamma}_{\alpha\beta} f_{,\gamma} - \frac{g_{\alpha\beta}}{2} \Delta f \right) \\ + \frac{(1-n)}{4F^2} \left(\begin{array}{c} 4FF'' \\ + F'F'(n-6) \end{array} \right) \left(\begin{array}{c} f_{,\alpha} f_{,\beta} \\ - f_{,\mu} f_{,\nu} g^{\mu\nu} \frac{g_{\alpha\beta}}{2} \end{array} \right) \\ + \kappa T_{\alpha\beta} + \Lambda \cdot F \cdot g_{\alpha\beta} \end{array} \right) \delta G^{\alpha\beta} \right), \tag{26}$$

and leads us to:

$$0 = R_{\alpha\beta} - \frac{1}{2} R \cdot g_{\alpha\beta} + \frac{F'}{F}(1-n)\left(f_{,\alpha\beta} - \frac{g_{\alpha\beta}}{n} \Gamma^{\gamma}_{\kappa\lambda} g^{\kappa\lambda} f_{,\gamma} - \frac{1}{2} \Delta f \cdot g_{\alpha\beta} \right), \tag{27}$$

when applying condition (23) and assuming no matter and a zero cosmological constant. One may now ask, why then at all look for a solution to just (25) if this only solves "half" of the whole (i.e., Eq. (27))? Well, we might just say that we are interested in the outcome, but the second and more elaborate answer would be:
"because we think that the universe has means to take care of (25) and the rest, which is to say:

$$0 = -\frac{1}{2} R \cdot g_{\alpha\beta} + \frac{F'}{F}(1-n)\left(-\frac{1}{2} \Delta f \cdot g_{\alpha\beta} \right) \quad \Rightarrow \quad 0 = R + \frac{F'}{F}(1-n)\Delta f, \tag{28}$$

in different ways." We think that there is a reason, why the first line of (22), being "the rest of (27)" and thus, (28), appears to us as a quantum equation, while the gravity equations containing the Ricci tensor act on bigger scales. We even think that different volumes have to or could be applied to both parts of the integrand within the variation. This may be done mathematically as follows:

Starting with the result from (26) and setting the matter term and the cosmological constant to zero:

$$\delta W = 0 = \int_V d^n x \left(\sqrt{-g \cdot F^n} \begin{pmatrix} R_{\alpha\beta} - R \cdot \frac{g_{\alpha\beta}}{2} \\ + \frac{F'}{F}(1-n)\left(f_{,\alpha\beta} - \Gamma^{\gamma}_{\alpha\beta} f_{,\gamma} - \frac{g_{\alpha\beta}}{2} \Delta f \right) \\ + \frac{(1-n)}{4F^2} \begin{pmatrix} 4FF'' \\ +F'F'(n-6) \end{pmatrix} \begin{pmatrix} f_{,\alpha} f_{,\beta} \\ - f_{,\mu} f_{,\nu} g^{\mu\nu} \frac{g_{\alpha\beta}}{2} \end{pmatrix} \end{pmatrix} \delta G^{\alpha\beta} \right), \tag{29}$$

we can simplify and split up to:

$$\xrightarrow{4FF''+F'F'(n-6)=0}$$

$$0 = \int_V d^n x \left(\sqrt{-g \cdot F^n} \begin{pmatrix} R_{\alpha\beta} - R \cdot \frac{g_{\alpha\beta}}{2} \\ + \frac{F'}{F}(1-n)\left(f_{,\alpha\beta} - \Gamma^{\gamma}_{\alpha\beta} f_{,\gamma} - \frac{g_{\alpha\beta}}{2} \Delta f \right) \end{pmatrix} \delta G^{\alpha\beta} \right)$$

$$\Rightarrow \qquad 0 = \begin{pmatrix} \int_{V_G} d^n x \left(\sqrt{-g \cdot F^n} \left(R_{\alpha\beta} + \frac{F'}{F}(1-n)\left(f_{,\alpha\beta} - \Gamma^{\gamma}_{\alpha\beta} f_{,\gamma} \right) \right) \delta G^{\alpha\beta} \right) \\ - \frac{1}{2} \cdot \int_{V_Q} d^n x \left(\sqrt{-g \cdot F^n} \left(R + \frac{F'}{F}(1-n)\Delta f \right) g_{\alpha\beta} \delta G^{\alpha\beta} \right) \end{pmatrix}. \tag{30}$$

Thereby the volumes of the two variations do not necessarily need to be the same.

For symmetry reasons, where we want to have equal total volumes for both terms in the last equation above, one may just extend the last line as follows:

$$0 = \begin{pmatrix} \int_{V_G} d^n x \left(\sqrt{-g \cdot F^n} \left(R_{\alpha\beta} + \frac{F'}{F}(1-n)\left(f_{,\alpha\beta} - \Gamma^{\gamma}_{\alpha\beta} f_{,\gamma} \right) \right) \delta G^{\alpha\beta} \right) \\ - \frac{1}{2} \cdot \sum_{i=1}^{N} \int_{V_{Qi}} d^n x \left(\sqrt{-g \cdot F^n} \left(R + \frac{F'}{F}(1-n)\Delta f \right) g_{\alpha\beta} \delta G^{\alpha\beta} \right) \end{pmatrix}$$

$$V_G = \sum_{i=1}^{N} V_{Qi}. \tag{31}$$

Thus, it appears totally reasonable to assume that for whatever reasons, the scales of the two variations could be significantly different and while the volume V_Q (Q, quantum) might be extremely small, forcing us to watch the variation:

$$0=\int_{V_Q} d^n x\left(\sqrt{-g\cdot F^n}\left(R+\frac{F'}{F}(1-n)\Delta f\right)g_{\alpha\beta}\delta G^{\alpha\beta}\right) \tag{32}$$

at smaller, which is to say quantum, scales, we would have to take care of the tensorial or gravitational variation:

$$0=\int_{V_G} d^n x\left(\sqrt{-g\cdot F^n}\left(R_{\alpha\beta}+\frac{F'}{F}(1-n)\left(f_{,\alpha\beta}-\Gamma^{\gamma}_{\alpha\beta}f_{,\gamma}\right)\right)\delta G^{\alpha\beta}\right) \tag{33}$$

at much bigger scales, respectively bigger volumes V_G. Thus, our reasoning for the consideration of solutions to (25) (and other split-ups), e.g., with respect to the inner structure of black holes (c.f. chapter 5, "The New Abacus") would be based on the observation of the different scales of quantum effects and gravity. Of course, this requires more discussion about the variational, respectively scale-dependent, split-up in (30) and the possible substructure of space or space-time with respect to the subsequent small-scale solutions to (32) (or (28) or the first line of (22)). This discussion was presented elsewhere [17, 18]. Sticking to the example of the black holes and assuming that the apparent matter (meaning, what we see as matter) inside a black hole (where we actually cannot see the matter) is solely made of scaled quantum states of our scaling functions F and f, we could also split up our variation as follows:

$$0=\left(\begin{array}{c}\int_{V_G} d^n x\sqrt{-g\cdot F^n}\left(\left(R_{\alpha\beta}-\frac{1}{2}\cdot R\cdot g_{\alpha\beta}\right)\delta G^{\alpha\beta}\right)\\ +\frac{1}{2}\cdot\sum_{i=1}^{N}\int_{V_{Qi}} d^n x\sqrt{-g\cdot F^n}\,\frac{F'}{F}(n-1)\left(\Delta f\cdot g_{\alpha\beta}-2\cdot\left(f_{,\alpha\beta}-\Gamma^{\gamma}_{\alpha\beta}f_{,\gamma}\right)\right)\delta G^{\alpha\beta}\end{array}\right). \tag{34}$$

Now we ask: Could perhaps the sum of V_{Qi} variations in our ensemble of scale split-up as given above (c.f. Eqs. (30) to (34)) act as the matter, which was introduced by Hilbert [2]?

Discovering the classical field equation structure (c.f. (34)) among our ensemble of options, one might want to demonstrate the connection to the

Einstein–Hilbert matter approach. We start with the assumption of one global metric $g_{\alpha\beta}$, but an ensemble of F_i and f_i, leading us to:

$$0=\left(\begin{array}{c}\int\limits_{V_G} d^n x\sqrt{-G}\left(\left(R_{\alpha\beta}-\frac{1}{2}\cdot R\cdot g_{\alpha\beta}\right)\delta G^{\alpha\beta}\right)\\ +\frac{1}{2}\cdot\sum_{i=1}^{N}\int\limits_{V_{Qi}} d^n x\sqrt{-g\cdot F^n}\,\frac{F'}{F}(n-1)\left(\Delta f\cdot g_{\alpha\beta}-2\cdot\overbrace{\left(f_{,\alpha\beta}-\Gamma_{\alpha\beta}^{\gamma}f_{,\gamma}\right)}^{\equiv f_{:\alpha\beta}}\right)\delta G^{\alpha\beta}\end{array}\right)$$

$\Rightarrow$

$$=\int\limits_{V_G} d^n x\sqrt{-g}\left(\sum_{i=1}^{N}\sqrt{F_i^n}\cdot\left(R_{\alpha\beta}-\frac{1}{2}\cdot R\cdot g_{\alpha\beta}+\frac{(n-1)}{2}\cdot\frac{F_i'}{F_i}\left(\Delta f_i\cdot g_{\alpha\beta}-2\cdot f_{i:\alpha\beta}\right)\right)\delta G^{\alpha\beta}\right). \tag{35}$$

This gives us the classical non-vacuum field equations via:

$$0=\sum_{i=1}^{N}\sqrt{F_i^n}\cdot\left(R_{\alpha\beta}-\frac{1}{2}\cdot R\cdot g_{\alpha\beta}\right)+\sum_{i=1}^{N}\sqrt{F_i^n}\cdot\left(\frac{(n-1)}{2}\cdot\frac{F_i'}{F_i}\left(\Delta f_i\cdot g_{\alpha\beta}-2\cdot f_{i:\alpha\beta}\right)\right)$$

$$\Rightarrow\qquad 0=R_{\alpha\beta}-\frac{1}{2}\cdot R\cdot g_{\alpha\beta}+\frac{\sum_{i=1}^{N}\sqrt{F_i^n}\cdot\left(\frac{(n-1)}{2}\cdot\frac{F_i'}{F_i}\left(\Delta f_i\cdot g_{\alpha\beta}-2\cdot f_{i:\alpha\beta}\right)\right)}{\sum_{i=1}^{N}\sqrt{F_i^n}}$$

$$=R_{\alpha\beta}-\frac{1}{2}\cdot R\cdot g_{\alpha\beta}+\frac{\sum_{i=1}^{N}\sqrt{F_i^n}\cdot\left(\frac{(n-1)}{2}\cdot\frac{F_i'}{F_i}\left(\Delta f_i\cdot g_{\alpha\beta}-2\cdot\left(f_{,\alpha\beta}-\Gamma_{\alpha\beta}^{\gamma}f_{,\gamma}\right)\right)\right)}{\sum_{i=1}^{N}\sqrt{F_i^n}}, \tag{36}$$

and the last term in the last line above is handing us the quantum or scale equivalent of the Einstein–Hilbert matter term, respectively the energy-momentum tensor:

$$\Rightarrow\qquad \kappa T_{\alpha\beta}=\frac{\sum_{i=1}^{N}\sqrt{F_i^n}\cdot\left(\frac{(n-1)}{2}\cdot\frac{F_i'}{F_i}\left(\Delta f_i\cdot g_{\alpha\beta}-2\cdot f_{i;\alpha\beta}\right)\right)}{\sum_{i=1}^{N}\sqrt{F_i^n}}$$

$$= \frac{\sum_{i=1}^{N} \sqrt{F_i^n} \cdot \left(\frac{(n-1)}{2} \cdot \frac{F_i'}{F_i} \left(\Delta f_i \cdot g_{\alpha\beta} - 2 \cdot \left(f_{,\alpha\beta} - \Gamma_{\alpha\beta}^{\gamma} f_{,\gamma} \right) \right) \right)}{\sum_{i=1}^{N} \sqrt{F_i^n}}. \tag{37}$$

3.3 Taking Care of Complex Curvatures *R* or *R**

It was discussed in [10] that due to the fact that we can have complex results for the scaled Ricci curvature R^*, we require a slightly different Einstein–Hilbert action, which should follow the typical quantum way of extracting observables.

In this sense, the action (4) should be adapted as follows:

$$\delta W = 0 = \delta \int_V d^n x \left(\sqrt{-G} \cdot \sqrt{\left(R^* - 2\Lambda + L_M \right) \left(\bar{R}^* - 2\bar{\Lambda} + \bar{L}_M \right)} \right). \tag{38}$$

Thereby the bar denotes the conjugate complex of the symbol underneath. In the classical case, the corresponding adapted action should read:

$$\delta W = 0 = \delta \int_V d^n x \left(\sqrt{-g} \cdot \sqrt{\left(R - 2\Lambda + L_M \right) \left(\bar{R} - 2\bar{\Lambda} + \bar{L}_M \right)} \right). \tag{39}$$

Applying the results from [12], we can obtain the variation results of (39) as follows:

$$0 = \begin{cases} \dfrac{\left(R^{\kappa\lambda} - R g^{\kappa\lambda} + \left(\begin{pmatrix} \underbrace{8\pi G}_{\kappa} T^{\kappa\lambda} = \text{matter} \\ 0 = \text{vacuum} \end{pmatrix} + \Lambda \cdot g^{\kappa\lambda} \right) \right)}{2\sqrt{\left(R - 2\Lambda + L_M \right) \left(\bar{R} - 2\bar{\Lambda} + \bar{L}_M \right)}} \cdot \left(\bar{R} - 2\bar{\Lambda} + \bar{L}_M \right) \\ + \dfrac{\left(\bar{R}^{\kappa\lambda} - \bar{R} g^{\kappa\lambda} + \left(\begin{pmatrix} \underbrace{8\pi G}_{\kappa} \bar{T}^{\kappa\lambda} = \text{matter} \\ 0 = \text{vacuum} \end{pmatrix} + \bar{\Lambda} \cdot g^{\kappa\lambda} \right) \right)}{2\sqrt{\left(R - 2\Lambda + L_M \right) \left(\bar{R} - 2\bar{\Lambda} + \bar{L}_M \right)}} \cdot \left(R - 2\Lambda + L_M \right) \\ - \left(\partial^{;\kappa} \partial^{;\lambda} - g^{\kappa\lambda} \Delta_g \right) \dfrac{\left(R + \bar{R} \right)}{2\sqrt{\left(R - 2\Lambda + L_M \right) \left(\bar{R} - 2\bar{\Lambda} + \bar{L}_M \right)}}. \end{cases} \tag{40}$$

We see that in the case of a purely real *R*, we obtain the usual classical Einstein field equations [1]. The corresponding equations in the scaled metric case are much more complicated and—as we don't need them in this book—will not be presented here.

3.4 Metric Observables

In Quantum Theory, observables are assumed to be based on positive reals. Metrics cannot be put under such a stringent treatment, simply because we know that there have to be certain signatures in order to have a possibility to distinguish between time- and space-like dimensions, for instance. However, our approach with the scaled metric as introduced in the first section of this chapter and reading:

$$\begin{aligned} G_{\alpha\beta} &= F\left[f\left[t,x,y,z,\ldots,\xi_k,\ldots,\xi_n\right]\right]\cdot\delta_\alpha^i\delta_\beta^j g_{ij} \\ &= F\left[f\left[t,x,y,z,\ldots,\xi_k,\ldots,\xi_n\right]\right]\cdot g_{\alpha\beta} = F\left[f\right]\cdot g_{\alpha\beta}, \end{aligned} \tag{41}$$

where the scale becomes the wave function, should not disrupt the principal signature of a metric. We might therefore like to resort to a slightly different scaling approach, given via:

$$G_{\alpha\beta} = F\cdot\bar{F}\cdot g_{\alpha\beta}, \tag{42}$$

where $\bar{F}$ denotes the conjugate complex of *F*.

In 4 dimensions and with the metric structure (2), this changes R^* in difference to (1) as follows:

$$R^* = \operatorname{sgn}\left[F[f]\right]^2\cdot\left(\begin{array}{c} \dfrac{R}{F[f]^2}-\dfrac{3}{2\cdot F[f]^4} \\ \times\left(\begin{array}{c} 2\cdot\left(1+\operatorname{sgn}\left[F[f]\right]^2\right)\cdot F[f]\cdot\dfrac{\partial F[f]}{\partial f}\cdot\Delta f \\ +\left(\begin{array}{c} 2\cdot\left(1+\operatorname{sgn}\left[F[f]\right]^2\right)\cdot F[f]\cdot\dfrac{\partial^2 F[f]}{\partial f^2} \\ -\left(\operatorname{sgn}\left[F[f]\right]^2-1\right)^2\cdot\left(\dfrac{\partial F[f]}{\partial f}\right)^2 \end{array}\right)\cdot f_{,\alpha}g^{\alpha\beta}f_{,\beta} \end{array}\right) \end{array}\right). \tag{43}$$

The corresponding Ricci tensor reads:

$$R^{*}=R^{*}_{\alpha\beta}G^{\alpha\beta}=R^{*}_{\alpha\beta}\left(F\cdot\overline{F}\right)^{-1}\cdot g_{\alpha\beta}$$

$\Rightarrow$

$$R^{*}_{\alpha\beta}g_{\alpha\beta}=\begin{pmatrix} R_{\alpha\beta}-\frac{3}{2\cdot F[f]^{2}} \\ \times\begin{pmatrix} 2\cdot\left(1+\operatorname{sgn}\left[F[f]\right]^{2}\right)\cdot F[f]\cdot\frac{\partial F[f]}{\partial f}\cdot\left(f_{,\alpha\beta}-\Gamma^{\gamma}_{\alpha\beta}f_{,\gamma}\right) \\ +\begin{pmatrix} 2\cdot\left(1+\operatorname{sgn}\left[F[f]\right]^{2}\right)\cdot F[f]\cdot\frac{\partial^{2}F[f]}{\partial f^{2}} \\ -\left(\operatorname{sgn}\left[F[f]\right]^{2}-1\right)^{2}\cdot\left(\frac{\partial F[f]}{\partial f}\right)^{2}\end{pmatrix}\cdot f_{,\alpha}f_{,\beta}\end{pmatrix}\end{pmatrix}g^{\alpha\beta}, \qquad (44)$$

$\Rightarrow$

$$R^{*}_{\alpha\beta}=\left(R_{\alpha\beta}-\frac{3}{2\cdot F[f]^{2}}\times\begin{pmatrix} 2\cdot\left(1+\operatorname{sgn}\left[F[f]\right]^{2}\right)\cdot F[f]\cdot\frac{\partial F[f]}{\partial f}\cdot\left(f_{,\alpha\beta}-\Gamma^{\gamma}_{\alpha\beta}f_{,\gamma}\right) \\ +\begin{pmatrix} 2\cdot\left(1+\operatorname{sgn}\left[F[f]\right]^{2}\right)\cdot F[f]\cdot\frac{\partial^{2}F[f]}{\partial f^{2}} \\ -\left(\operatorname{sgn}\left[F[f]\right]^{2}-1\right)^{2}\cdot\left(\frac{\partial F[f]}{\partial f}\right)^{2}\end{pmatrix}\cdot f_{,\alpha}f_{,\beta}\end{pmatrix}\right).$$

(45)

After variation under the Einstein–Hilbert action integral with respect to the metric, we obtain:

$$\delta W=0=\int_V d^n x\sqrt{-g\cdot\left(F\cdot\bar{F}\right)^4}\,\delta G_a^{\ \alpha\beta}$$

$$\times\left(\begin{array}{c} R_{\alpha\beta}-R\cdot\frac{g_{\alpha\beta}}{2} \\ +\frac{3\cdot g_{\alpha\beta}}{4\cdot F[f]^2}\cdot\left(\begin{array}{c} 2\cdot\left(1+\mathrm{sgn}\left[F[f]\right]^2\right)\cdot F[f]\cdot\frac{\partial F[f]}{\partial f}\cdot\Delta f \\ +\left(\begin{array}{c} 2\cdot\left(1+\mathrm{sgn}\left[F[f]\right]^2\right)\cdot F[f]\cdot\frac{\partial^2 F[f]}{\partial f^2} \\ -\left(\mathrm{sgn}\left[F[f]\right]^2-1\right)^2\cdot\left(\frac{\partial F[f]}{\partial f}\right)^2\end{array}\right)\cdot f_{,\mu}g^{\mu\nu}f_{,\nu}\end{array}\right) \\ -\frac{3}{2\cdot F[f]^2}\cdot\left(\begin{array}{c} 2\cdot\left(1+\mathrm{sgn}\left[F[f]\right]^2\right)\cdot F[f]\cdot\frac{\partial F[f]}{\partial f}\cdot\left(f_{,\alpha\beta}-\Gamma^{\gamma}_{\alpha\beta}f_{,\gamma}\right) \\ +\left(\begin{array}{c} 2\cdot\left(1+\mathrm{sgn}\left[F[f]\right]^2\right)\cdot F[f]\cdot\frac{\partial^2 F[f]}{\partial f^2} \\ -\left(\mathrm{sgn}\left[F[f]\right]^2-1\right)^2\cdot\left(\frac{\partial F[f]}{\partial f}\right)^2\end{array}\right)\cdot f_{,\alpha}f_{,\beta}\end{array}\right) \\ +\kappa T_{a\alpha\beta}+\Lambda\cdot F^a\cdot g_{a\alpha\beta}\end{array}\right). \quad (46)$$

This now allows for sudden changes of the equations simply due to changes of the signature function sgn[F] and can even lead to switches from Klein–Gordon ($\left(1-\mathrm{sgn}\left[F[f]\right]^2\right)=0\,\&\,\frac{\partial^2 F[f]}{\partial f^2}=0$), Proca or similar equations to Dirac equations ($1+\mathrm{sgn}[F[f]]^2=0$). This, however, would be equivalent to a switch of spinless and integer-spin particles (or composite particles, like the pion) to a fermion.

Chapter 4

The Air

4.1 Hypersonic Flight—The Complex Structural, Shape, and Material Optimization Problem

Figure 4.1 H8 Chinese Stealth Bomber (from D.K. with thanks).

Quantum Gravity War: How Will the Nearby Unification of Physics Change the Future of Warfare
Norbert Schwarzer

ISBN 978-981-4968-58-4 (Hardcover), 978-1-032-71070-9 (eBook)
www.jennystanford.com

4.1.1 An "Anecdotic" Sidenote

Already years ago, the author considered the strategic importance of hypersonic elements in warfare. In fact, the first documents he had written to this topic are from the nineties of the last century. Admittedly, he never published anything about it, at least not before his first very brief study about future warfare in a "unified physics environment" [215], but still this was months before the Ukraine war. The latter then clearly showed the total ignorance of the so-called NATO experts regarding the strategic importance of hypersonic flight (and many more other military aspects, too). This author thinks that the Russian superiority in this fields leads not only to a significant advantage for their forces but to a rather general superiority in a very holistic manner. To the NATO forces, this is no less than yet another Sputnik-disaster and a paradigm shift. But taking the structural weakness of the whole organization, its intrinsic corruption, and rotten core, especially regarding the entanglement with the military industrial complex, this all can hardly come as a surprise.

Linear thinkers always invite and attract parasitic elements and should therefore simply have no place in any military organization ... at least not in those connected with leading, planning, research and development, and organizing duties.

4.1.2 A Multi-Dimensional Materials Science Application

In order to start with a down-to-earth-application, here we briefly consider the necessary optimization problems occurring in hypersonic flight (Fig. 4.1). What makes this application especially difficult to handle results from the fact that all these partial problems come as entangled ones as it is impossible to deal with the structure and shape of a hypersonic aircraft without also adjusting the materials science aspect and vice versa. The same holds with shape requirements automatically resulting in multidimensional structural demands ... especially as there is always the weight aspect to be considered. Last but not least, we have the problem of forced instability as an absolute necessity for modern air combat scenarios. This, however, demands AI-supported flight assistance and thus, tremendous computer power.

In modern aircraft design, structural optimization uses evolutionary motivated processes (e.g., [19–22]) and tools leading to **e**volutionary **s**table **s**trategies or ESS with sufficiently strong forward vectors regarding the final design. However, as there is no natural ideal for hypersonic flight, such tools tend to produce artifacts when applied to such an extension.

The reason for these artifacts is simple. Explained in terms of a theory of everything, one could just put it as follows: One cannot make good designs in 3 + *X*-dimensional space-times, when restricting the models to just 3 dimensions. This simple rule is just a constructional excerpt from the general law about good decision making in and/or for arbitrary systems [3].

The solution would be the consequent generalization of the subsequent optimization process. This generalization already is at hand via the Einstein–Hilbert approach [23] and only requires a suitable interpretation [3] together with an extension of the current ESS-optimization-process to all its degrees of freedom, which is to say an extension to the right dimension of the true optimization task. Thereby especially the question of temperature and its coupling into all other aspects [24–26] requires a much more consequent and fundamental dealing with the material side. As materials science, however, is completely based upon the electromagnetic interaction (at least as long as we are not able to apply degenerated matter), we "only" need a fundamental and—above all—metrically based formulation of the latter. This is given in appendix D.

The rest is still cumbersome work, but rather straightforward.

Taking the bee-example from the introduction ("About the Principally Complementary Character of our Approach") we might put it as follows:

Our approach is completing the ESS-cycles which are already at hand. In order to illustrate this message in connection with the problem of hypersonic flight, we resort once more to the bee-analogy and fictively adjust it as follows:

(A) Before it was discovered that bees can fly at hypersonic speed, because they are able to dynamically adjust (apply) turbulences and phase transition regions during the various phases of their flight, developers restricted themselves to 3-dimensional spaces with their structural and shape design. This is almost comparable to the Russian super torpedo (super-cavitating torpedoes like the VA 111 Shkval torpedo and its descendants), except that the latter copes with rather different media and—consequently—phase transition scenarios.

(B) When this trick of the hypersonic bees, which is to say the adapted phase turbulences, "was" discovered, nobody got the crazy idea to kick all the old designers out or to erase the former ESS optimization procedures. The latter were just extended to the right dimensions.

Chapter 5

The New Abacus

5.1 The Ultimate Computer Machine—A Quantum Gravity Computer?

Now, having shown that Quantum Theory also has a metric origin and can be obtained as a transformation of solutions to the Einstein field equations, we wonder how we could metrically construct a most fundamental computer machine, respectively, quantum computer machine. The task was already performed in [28–30], but for the deeper and more illustrative (Fig. 5.1) understanding of the geometric nature of everything as outlined above, we want to repeat the essentials here.

Some might ask why a metric computer or Turing machine would be helpful or even essential in future warfare, but we think that this question is easily answered by the simple fact that faster, smaller, and less-energy-consuming computers are always of severe strategic advantage. The Turing machines we are considering here will just provide the theoretical limits to speed, size, and energy efficiency.

Current quantum computer technology uses the wave functions f or $F[f]$ as carrier. Now we intend to work out the potential of a "machine" where not the wave function but the whole metric is activated and used as a carrier for the information and its processing. In order to keep things simple here (and also in order to keep a long story short), we will not aim for a joint carrier system where $F[f]$ and the metric are activated, but only the metric, which is

Quantum Gravity War: How Will the Nearby Unification of Physics Change the Future of Warfare
Norbert Schwarzer

ISBN 978-981-4968-58-4 (Hardcover), 978-1-032-71070-9 (eBook)
www.jennystanford.com

to say, we set $F[f] = 1$ in our scaled matric (6). Consequently, we obtain from (1) and (4) just the classical Einstein field equations as working conditions for our quantum gravity computer. In order to keep things still simple, we only consider smooth spaces with very few dimensions. Extension to curved, potentially discrete metrics, combination with $F[f]\neq 1$ and higher numbers of dimensions is cumbersome but straightforward.

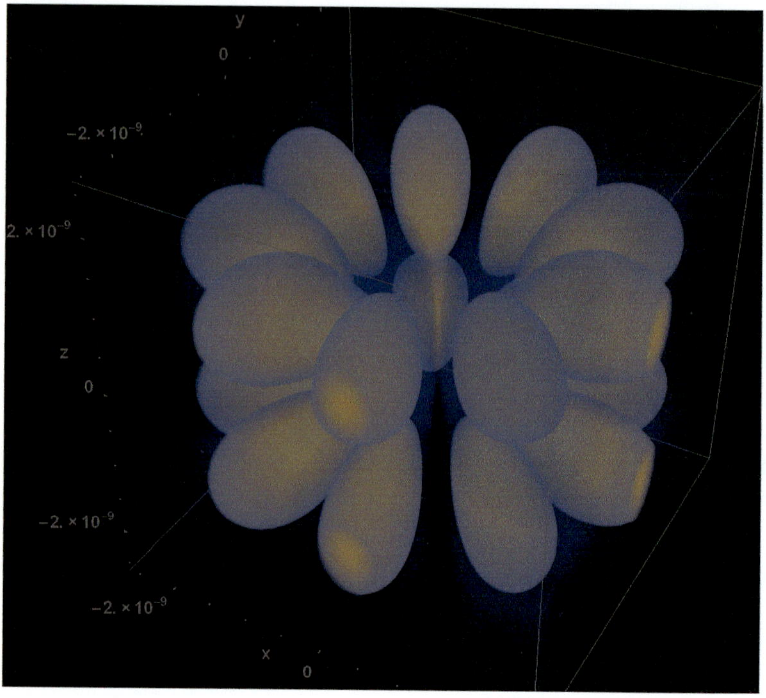

Figure 5.1 Half-spin black hole solution as derived in [27].

By applying a metric approach of the kind:

$$g_{\alpha\beta} = \begin{pmatrix} g_t'[t]^2 & 0 & 0 \\ 0 & g_x'[x]^2 \cdot f[t] & 0 \\ 0 & 0 & g_y'[y]^2 \cdot f[t] \end{pmatrix}, \tag{47}$$

we will find a general solution to the vacuum quantum Einstein field equations (19) (with $T^{\alpha\beta} = 0$) with the function:

$$f_{1,2}[t] = e^{\pm 2i \cdot g_t[t] \cdot \sqrt{\Lambda}} \cdot C_1. \tag{48}$$

Thereby Λ denotes the cosmological constant of the system under consideration. Please note the matter and anti-matter character of our solution (48). The functions g_x and g_y are arbitrary and could be used for operations/data of any arbitrary form, which is to say, that we are just using the degrees of freedom from the tensor transformations. The argument that these degrees of freedom do not provide any observables and, thus, cannot be used for calculations is only partially true, because just as with the non-measurable superposed quantum states, there is still an intrinsic evaluation process going on which can be exploited. In order to assure digital outcomes to the metric components g_{11} and g_{22}, standing for the second and the third diagonal metric component, respectively, we construct Fourier series of the kind:

$$g_x = \frac{x}{2} + \frac{2}{\pi} \cdot \sum_{k=1}^{\infty} i^{2\cdot(k-1)} \cdot \frac{\sin\left[(2\cdot k-1)\cdot x\right]}{(2\cdot k-1)^2};$$
$$g_y = \frac{y}{2} + \frac{2}{\pi} \cdot \sum_{k=1}^{\infty} i^{2\cdot(k-1)} \cdot \frac{\sin\left[(2\cdot k-1)\cdot y\right]}{(2\cdot k-1)^2}. \tag{49}$$

We see that the system constructed above has everything a most simple 2-bit "processor" requires. It has the necessary two bits and it has a periodic function defining a cycle time. What we also need is a way to store information. Applying the Turing approach [31], we might just couple in certain add-on dimensions playing the role of these storages, or, if using the Turing picture, playing the role of the tape of our EQ-Turing machine (EQ = Einstein Quantum). A simple extension of the metric (47) of the form:

$$g_{\alpha\beta} = \begin{pmatrix} g_{00} & 0 & 0 & 0 & 0 \\ 0 & g_{11} & 0 & 0 & 0 \\ 0 & 0 & g_{22} & 0 & 0 \\ 0 & 0 & 0 & g_{33} & 0 \\ 0 & 0 & 0 & 0 & g_{44} \end{pmatrix}$$

$$g_{00} = g_t'^2; \quad g_{11} = g_x'^2 \cdot f_t; \quad g_{22} = g_\xi'^2 \cdot f_t \cdot f_x$$

$$g_{11} = g_y'^2 \cdot f_t; \quad g_{44} = g_\eta'^2 \cdot f_t \cdot f_y; \quad f_{\{x,y\}} = \cos\left[c_{\{x,y\}}\{x,y\}\right]^2 \tag{50}$$

might help us to solve the problem for at least one storage process by entangling the coordinates x-ξ and x-η. A second set of solutions can be constructed via the following approach:

$$g_{\alpha\beta} = \begin{pmatrix} g_{00} & 0 & 0 & 0 & 0 \\ 0 & g_{11} & 0 & 0 & 0 \\ 0 & 0 & g_{22} & 0 & 0 \\ 0 & 0 & 0 & g_{33} & 0 \\ 0 & 0 & 0 & 0 & g_{44} \end{pmatrix}$$

$$g_{00} = g_t'^2; \quad g_{11} = g_x'^2 \cdot f_t \cdot f_x; \quad g_{22} = g_\xi'^2 \cdot f_t \cdot f_x$$

$$g_{11} = g_y'^2 \cdot f_t \cdot f_y; \quad g_{44} = g_\eta'^2 \cdot f_t \cdot f_y; \quad f_{\{x,y\}} = \cosh\left[c_{\{x,y\}}\{x,y\}\right]^{-2} \quad (51)$$

and would bring the same result in a slightly variated form (read and write options).

We recognize the application of the pairwise coupling or entanglement of dimensions as elaborated in [3]. Substituting $\xi = xx$ and $\eta = yy$, we can directly obtain the solutions from [30] with the following sub-metric of the components 1–4:

$$g_{\alpha\beta} = \begin{pmatrix} A \cdot g_x'[x]^2 \cdot f & 0 & 0 & 0 \\ 0 & AA \cdot g_{xx}'[xx]^2 \cdot f & 0 & 0 \\ 0 & 0 & B \cdot g_y'[y]^2 \cdot g & 0 \\ 0 & 0 & 0 & BB \cdot g_{yy}'[yy]^2 \cdot g \end{pmatrix}$$

$$f = f[x] = \cosh\left[\pm c_1 \cdot g_x[x]\right]^{-2}; \quad g = g[y] = \cosh\left[\pm c_2 \cdot g_y[y]\right]^{-2}$$

$$B \cdot \Lambda - c_2^2 = 0; \quad A \cdot \Lambda - c_1^2 = 0$$

$$\gamma_{\alpha\beta} = \begin{pmatrix} A \cdot g_x'[x]^2 & 0 & 0 & 0 \\ 0 & AA \cdot g_{xx}'[xx]^2 \cdot F & 0 & 0 \\ 0 & 0 & B \cdot g_y'[y]^2 & 0 \\ 0 & 0 & 0 & BB \cdot g_{yy}'[yy]^2 \cdot G \end{pmatrix}$$

$$F = F[x] = \cos\left[\pm c_1 \cdot g_x[x]\right]^2; \quad G = G[y] = \cos\left[\pm c_2 \cdot g_y[y]\right]^2$$

$$B \cdot \Lambda - c_2^2 = 0; \quad A \cdot \Lambda - c_1^2 = 0. \quad (52)$$

The total metric shall then be given either as:

$$g_{0\alpha} = g_{\alpha 0} = 0; \quad g_{00} = H \cdot g_t'[t]^2; \quad g_{\alpha\beta} \cdot h[t]$$

$$\gamma_{0\alpha} = \gamma_{\alpha 0} = 0; \quad \gamma_{00} = H \cdot g_t'[t]^2; \quad \gamma_{\alpha\beta} \cdot h[t] \tag{53}$$

or (careful with the constant H and the function $H[t]$):

$$g_{0\alpha} = g_{\alpha 0} = 0; \quad g_{00} = H \cdot g_t'[t]^2 \cdot H[t]; \quad g_{\alpha\beta} \cdot H[t]$$

$$\gamma_{0\alpha} = \gamma_{\alpha 0} = 0; \quad \gamma_{00} = H \cdot g_t'[t]^2 \cdot H[t]; \quad \gamma_{\alpha\beta} \cdot H[t]. \tag{54}$$

With the help of the results of [30], we can solve the subsequent Einstein field equations and find:

$$h_{1,2}[t] = \left\{ \frac{\frac{1}{4} e^{-\frac{1}{3} i\sqrt{H\cdot\Lambda}\left(\sqrt{6}\cdot g_t[t]+3\cdot C_1\right)} \left(2e^{i\sqrt{\frac{2}{3}}\cdot g_t[t]\sqrt{H\cdot\Lambda}} + e^{i\sqrt{H\cdot\Lambda}\cdot C_1} \right)^2}{1+e^{-\frac{1}{3} i\sqrt{H\cdot\Lambda}\left(\sqrt{6}\cdot g_t[t]+3\cdot C_1\right)} + \frac{1}{4} e^{\frac{1}{3} i\sqrt{H\cdot\Lambda}\left(\sqrt{6}\cdot g_t[t]+3\cdot C_1\right)}}, \right\}$$

$$= \left\{ \frac{1+\frac{5}{4}\cos\left[\sqrt{H\cdot\Lambda}\left(\sqrt{\frac{2}{3}} g_t[t]-C_1\right)\right] + \frac{3\cdot i}{4}\sin\left[\sqrt{H\cdot\Lambda}\left(\sqrt{\frac{2}{3}} g_t[t]-C_1\right)\right]}{1+e^{-\frac{1}{3} i\sqrt{H\cdot\Lambda}\left(\sqrt{6}\cdot g_t[t]+3\cdot C_1\right)} + \frac{1}{4} e^{\frac{1}{3} i\sqrt{H\cdot\Lambda}\left(\sqrt{6}\cdot g_t[t]+3\cdot C_1\right)}}, \right\}$$

$$H_{1,2}[t] = 2\left(1+\tan\left[\frac{1}{6}\left(\pm 2i\sqrt{3}\cdot g_t[t]\sqrt{H\cdot\Lambda}+3\sqrt{2}\cdot C_1\right)\right]^2\right)$$

$$= 2\left(1+\tan\left[\frac{1}{6}\left(\pm 2i\sqrt{3}\cdot g_t[t]\sqrt{H\cdot\Lambda}+i\cdot 3\sqrt{2}\cdot \tilde{C}_1\right)\right]^2\right)$$

$$= 2\cdot \operatorname{sech}\left[\pm\frac{g_t[t]\sqrt{H\cdot\Lambda}}{\sqrt{3}} - \frac{\tilde{C}_1}{\sqrt{2}}\right]^2. \tag{55}$$

While in the case of $H[t]$ and an assumed $g_t[t] = t$, we obtain a bell-shaped function, meaning that the whole system appears and disappears after a while (it only virtually exists), we now have an infinitely stable oscillation in the case $h[t]$ (Fig 5.2).

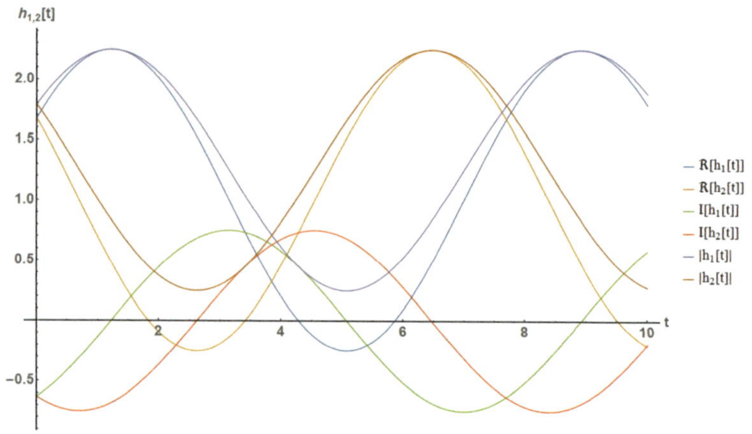

Figure 5.2 Oscillations of $h[t]$ with $g_t[t] = t$, $\Lambda = 1$, $C_1 = 1$, $H = 1$.

Things change dramatically in the case $\Lambda = 0$, where the Einstein field equations require the following rather different solutions for the cycle-time functions $h[t]$ and $H[t]$, reading:

$$h_{1,2}[t] = \frac{\left(B\cdot c_1^2 + A\cdot c_2^2\right)H\left(2g_t[t]^2 \pm 2\sqrt{6A\cdot B}\cdot g_t[t]\cdot C_1 + 3A\cdot B\cdot C_1^2\right)}{12A\cdot B}$$

$$H_{1,2}[t] = \mathrm{e}^{\frac{1}{3}\sqrt{H\cdot A\cdot c_2^2}\left(\pm\frac{2\sqrt{3}}{\sqrt{A\cdot B}}\cdot g_t[t] + 3\sqrt{2}\cdot C_1\right)}. \tag{56}$$

In addition, we have the functions *f*, g, *F*, and *G* as before plus the boundary condition for the constants *A*, *B*, and c_i as follows:

$$f = f[x] = \cosh\left[\pm c_1\cdot g_x[x]\right]^{-2};\quad g = g[y] = \cosh\left[\pm c_2\cdot g_y[y]\right]^{-2}$$

$$F = F[x] = \cos\left[\pm c_1\cdot g_x[x]\right]^2;\quad G = G[y] = \cos\left[\pm c_2\cdot g_y[y]\right]^2$$

$$B\cdot c_1^2 - \mathrm{A}\cdot c_2^2 = 0. \tag{57}$$

This time the approach $g_t[t] = t$ does not lead to oscillations for $h[t]$ (Fig. 5.3), but still we can achieve time-periodic behavior via $H[t]$ and $g_t[t] = i^*t$. Later in this book (especially in the next chapter), it will be shown how additional dimensions can be used to bring in masses as effective "cosmological constants" in order to obtain metrics with the right—computational or Turing-conform—properties.

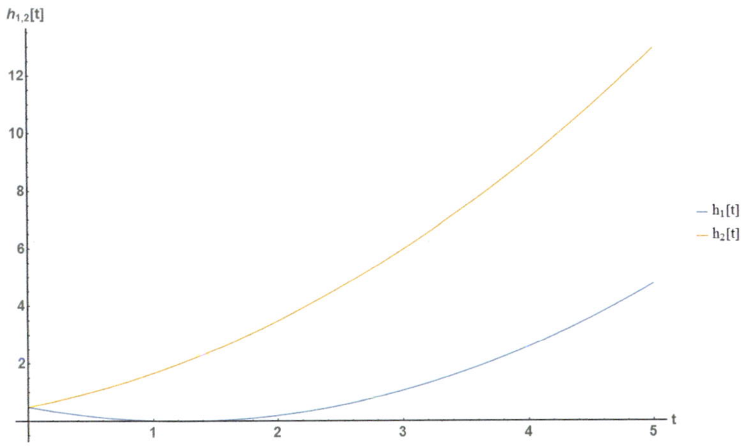

Figure 5.3 Distribution of $h[t]$ with $g_t[t] = t$, $\Lambda = 0$, $C_1 = 1$, $H = A = 1$, $c_i = 1$.

Using the results from above, it is easy to construct the metrics for higher-bit systems. For instance, the total metric for an 8-bit case would have to contain 17 dimensions and could generally be given as:

$$g^n_{\alpha\beta} = \begin{pmatrix} g_{00} = g_t{}'[t]^2 & 0 & 0 & 0 & \cdots & 0 \\ 0 & \tilde{g}_{11} \cdot f[t] & 0 & 0 & \cdots & 0 \\ 0 & 0 & \tilde{g}_{22} \cdot f[t] & 0 & \cdots & 0 \\ 0 & 0 & 0 & \tilde{g}_{33} \cdot f[t] & \cdots & 0 \\ \cdots & \cdots & \cdots & \cdots & \ddots & 0 \\ 0 & 0 & 0 & 0 & 0 & \tilde{g}_{aa} \cdot f[t] \end{pmatrix}. \quad (58)$$

We have the two options for pairwise coupling (entanglement), namely:

$$(A)\quad \tilde{g}_{jj} = g'_{x_j}\left[x_j\right]^2;\quad \tilde{g}_{kk} = g'_{x_k}\left[x_k\right]^2 \cdot cos\left[c_j \cdot x_j\right]^2;\quad j = odd;\quad k = \text{even}$$

$$(B)\quad \tilde{g}_{jj} = g'_{x_j}\left[x_j\right]^2 \cdot \cosh\left[c_j \cdot x_j\right]^{-2};\quad \tilde{g}_{kk} = g'_{x_k}\left[x_k\right]^2 \cdot \cosh\left[c_j \cdot x_j\right]^{-2}. \quad (59)$$

By the structure given in (58), it can easily be seen how the extension to any number of bits can be achieved.

5.2 Smart Black Holes: An Extreme Example for Illustration

We saw that—similar to the Bekenstein or Bekenstein–Hawking problem (see Fig. 5.4 and reference [30])—we add bits via dimensions to our metric Turing machines. Comparing our "computer-like" metric solutions (52) and (58) with the metric mass and spin construction as shown in sections 6.9.3, "The Metric Origin of Mass" and 6.9.4, "The Metric Origin of Spin and Spin Fields," in chapter 6, "The Quantum Gravity Turing Machine," we see that our entangled bit-dimensions are representing masses and spins (see also the appendix of the next chapter). Assuming our metric computer to be a black hole (which we here only use to have a simple as possible math), we can even obtain a ratio of the black hole's radius r_s to the smallest structure this black hole can resolve. Taking the result for ℓ_P—the Planck length—from the Bekenstein thought experiment, we find the ratio between the Schwarzschild radius r_s of a black hole and ℓ_P:

$$\frac{r_s}{\ell_P} = 2 \cdot \sqrt{\pi \cdot \left(q + \sqrt{q \cdot (1+q)}\right)}\ ^1. \tag{60}$$

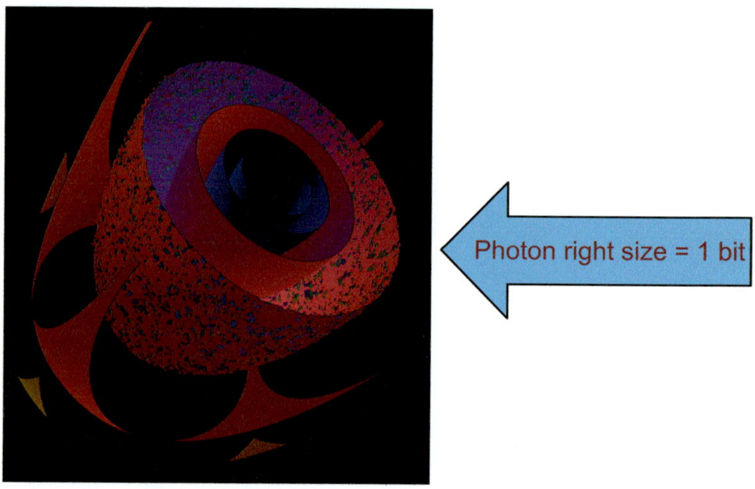

Figure 5.4 Illustration of the Bekenstein thought experiment (see text below and section 6.8, "The Bekenstein Bit-Problem," in chapter 6, "The Quantum Gravity Turing Machine").

In our universe, the Planck length l_P is considered to be the smallest length possible to resolve. What if the ratio (60), we found for black holes, actually is a more fundamental law? At any rate, it appears logical to assume that more

[1] More information about the Bekenstein thought experiment will be given in section 6.8, "The Bekenstein Bit-Problem," in Chapter 6, "The Quantum Gravity Turing Machine."

bits could be coded or stored by an object of bigger size and smaller internal structure, thereby leaving more options to describe something with these structures. Thus, Eq. (60) makes intuitive sense, but could it also be just the other way round? Could it be that for an object of given size, the number of bits (being equivalent to its dimensions as we see in Fig. 5.5) it contains determines the smallest scale—the Planck length—of the object, too? From inside and taking the Planck length as measure, the increase of information to this very object would look like an increase of its size. Now assuming the inside of the black hole to be a universe, the inhabitants of this universe would register the increase of information as a growth of their universe, measured in the Planck length of that very universe.

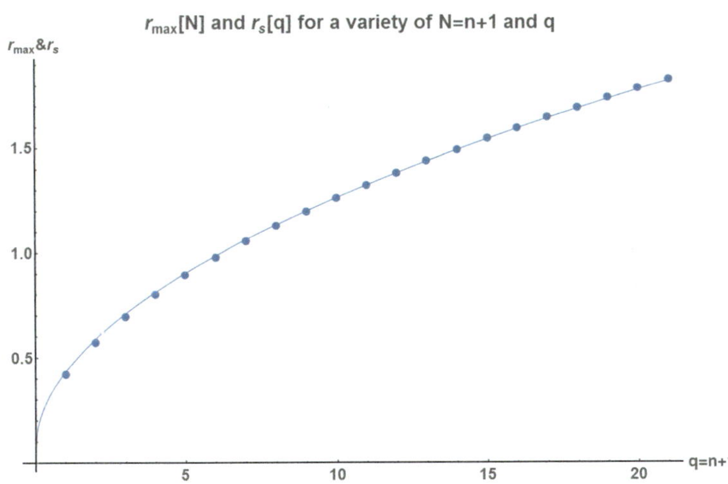

Figure 5.5 Radius r_{max} for which at a certain number of dimensions the n-sphere has maximum volume in dependence on $N = n + 1$ compared with the increase of the Schwarzschild radius r_s of a black hole in dependence on the number of bits q thrown into it by using (61). We find that $q = N = n + 1$.

The attentive reader might have already noticed that there seems to be a discrepancy between the complete arbitrariness of the choice of functions $g_i[x_i]$ within our metric Turing machines and the rigidity of the bit-to-radius relation for black holes. How could it be possible that—at least in connection with black holes—on the one hand we have just 1 bit-information and on the other total freedom in choosing any function we want within our metric computer bit. At least as long as we have not explicitly pointed out that this (our) computer should not be a black hole[2], we should try to explain this

[2]Especially as this one might also lack a proper circuit time.

discrepancy. This author thinks that the problem is the rigid allocation of a certain size to a photon of given wave length. Yes, it might well be that the photon of a given energy indeed has a well-defined wave length. But does this fact also fix its size, which is to say its distribution in space, in such a way that it would be placeable into a sphere of a certain Schwarzschild radius in order to make it 100% fit into the very black hole one intends to throw it into? No! The photon's size is always uncertain. It is always smeared out over an, in fact, infinite space. Thus, the Bekenstein thought experiment is flawed in a sense that it does not consider the size uncertainty of the photon. Taking this uncertainty into account, however, there are no sharp bits anymore and the problem with the discrepancy between the arbitrariness of our g_i and the sharpness of bit-coding dimensions is solved.

The reader can find in [13] and later in this book more information about how black holes could actually store information.

5.3 Is There an Ultimate Fundamental Scale?

At first sight, it may not appear as strategically important to know about the existence or non-existence of absolute scales within our universe, but as such scales determine the limits of possible technical warfare developments, one better makes sure to have such figures first.

Note: The correct solution for the evaluation of the Schwarzschild radius r_s as a function of the bits thrown into a black hole object (if using the results from [27]) would be:

$$r_s = 2 \cdot \ell_P \cdot \sqrt{\pi \cdot \left(q + \sqrt{q \cdot (1+q)}\right)!} \tag{61}$$

We find a perfect fit to the n-spheres with maximized volume to a given radius (dots in Fig. 5.5) with a Planck length of l_P = 0.07881256452824544 (see Fig. 5.5).

But what would be the unit for this Planck length?

Well, by using the results from [27], Eq. (45) (with $T[n] = 1$):

$$V_N = V_{n+1} = T[n] \cdot \frac{\pi^{\frac{(n)}{2}}}{\Gamma\left[\frac{(n+2)}{2}\right]} \cdot r^n, \tag{62}$$

we have evaluated the very dimension n to which at a given radius r_s the volume of a n-sphere has a maximum. The r_s values are put into the calculation as plain "natural" numbers, meaning an $r_s = 1$ is just a 1 and that is it. We might name this unit a "mathematical meter" or just "mams" (plural for "mathematical meters"). Transformation to our usual units, like meters, requires the introduction of a factor $T[n] = U^n$.

With $U = \frac{[\ell_P]_{\text{in\_meters}}}{[\ell_P]_{\text{in\_mams}}} = \frac{1.616255(18)\times 10^{-35}\text{ meters}}{0.07881256452824544}$, for instance, we can easily change to our meters. Nevertheless, it appears somehow astonishing that there seems to exist a fundamental "natural" unit, being completely based on a mathematical—geometrical—extremal principle (the maximum volume of n-spheres as functions of their dimensions for certain radii). It is also interesting that these dimensions are so nicely correlated to the number of bits a black hole has swallowed. In fact, using the unit of mams, the number of spatial (n-sphere) dimensions is perfectly equal to the number of swallowed bits.

Thus, in the case that black holes would, in fact, store their content as dimensions and the Einstein–Hilbert action being extended with respect to the number of dimensions in addition to the metric (c.f. section 6.6, "The Other Side of the Coin," in chapter 6), we immediately also get an absolute scale in which the number 1 is "made out" of 12.6883 Planck length and where a 3-sphere has a radius of 0.6969979737167096 mams.

5.4 Does More Information Always Mean More Mass?

Quantum computer scientists have already pointed out that with our current way of storing information, we will one day reach a limit with respect to the number of atoms we can apply for the storing process and the energy needed to keep the information stable (maintained). Citing from the abstract of [32], we have the following situation:

> *"Currently, we produce ~10^{21} digital bits of information annually on Earth. Assuming a 20% annual growth rate, we estimate that after ~350 years from now, the number of bits produced will exceed the number of all atoms on Earth, ~10^{50}. After ~300 years, the power required to sustain this digital production will exceed 18.5 × 10^{15} W, i.e., the total planetary power consumption today, and after ~500 years from now, the digital content will account for more than half Earth's mass, according to the mass-energy–information equivalence principle. Besides the existing global challenges such as climate, environment, population, food, health,*

energy, and security, our estimates point to another singular event for our planet, called information catastrophe."

It has to be pointed out that when looking for possible inner Schwarzschild solutions [13], we also found that there are solutions where the mass decreases with the increase in the object size. It may well be that such strange states not only are realized in black holes but could perhaps also help to overcome our future information storage problem.

Chapter 6

The Quantum Gravity Turing Machine

6.1 Why Digging Deeper?

Having seen in the previous chapter that the "activation" of the metric as a carrier of information provides a potentially tremendous increase in information storage density and processing velocity, here we now want to investigate the whole approach in a somewhat more general manner and a bit more with respect to practical (potentially military) applications. Along the way, our fundamental metric approach assures illustrative outcomes (Fig. 6.1).

6.1.1 Why Should We Bother about the Question whether There Is a Fundamental Computer Concept in This Universe?

It is widely agreed that artificial intelligence and computer technology will just be THE evolutionary driving forces for the human society within the next decades. Almost all fields and aspects are going to be affected by these rapidly changing sectors. There simply is no significant part in the economy and military which will be able to get along or move forward without computers and software.

But what will be the direction in which the vector of technical computer evolution is pointing?

Quantum Gravity War: How Will the Nearby Unification of Physics Change the Future of Warfare
Norbert Schwarzer

ISBN 978-981-4968-58-4 (Hardcover), 978-1-032-71070-9 (eBook)
www.jennystanford.com

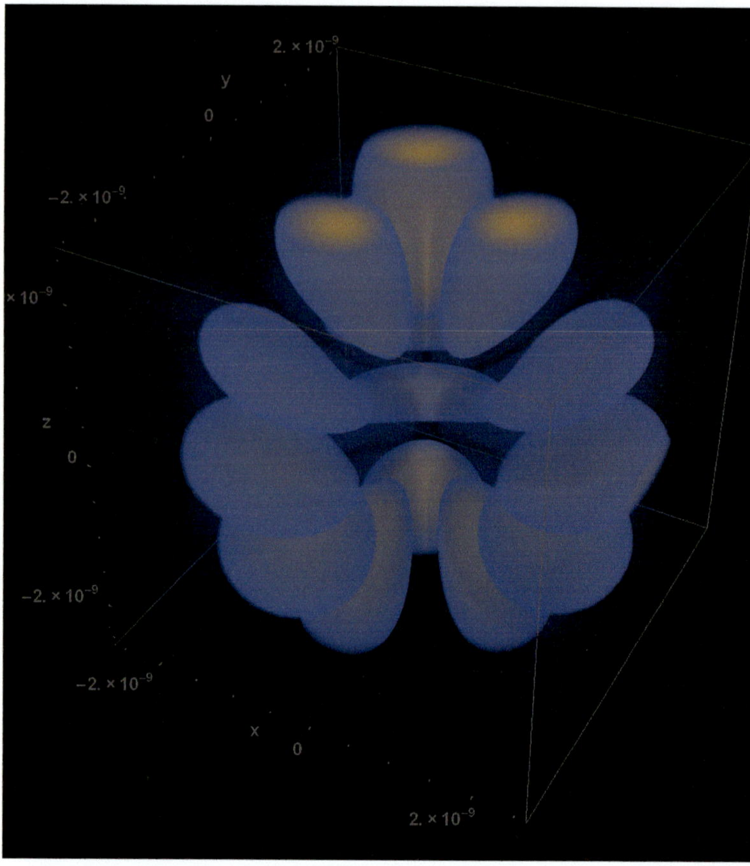

Figure 6.1 Half-spin black hole solution as derived in [27].

Not to answer this question in economy or perhaps even not to ask it can lead to dramatic consequences, especially for companies almost completely depending on this technology. They might simply be kicked out of business by others who had the answers ready at the right time. The same holds for all military forces in an even much more general manner, because it is not just the technology which is more and more dominated by computer-based solutions, but also the strategic planning, gaining of intelligence, and optimum decision-making processing.

But does it really have to be the ultimate answer to the question of what "might be coming"? And—what is more—why should we go as deep as even consider so very principal and theoretical objects like black holes and their metric solutions [B1]?

Just imagine yourself climbing a steep hill. Your optimum goal is to reach the top, but due to the difficulty of the terrain, possibly bad weather conditions or certain skills and equipment you lack, you might be forced to compromise to some place just near the top. Nevertheless, knowing the optimum, respectively, knowing about the existence of this optimum will help you to focus to get the right assistance (equipment) and do the best preparation possible. Thus, having a clear picture of the optimum goal is always to be preferred against blurred guesses about a "certain something" one would like to reach. Transferring this illustration onto the question of future computer developments, it would just be a great help to know where to find the ultimate technique, solution, structure, etc., even when currently not being able to reach this point. With respect to military applications, it is even absolutely necessary to have knowledge about the optimum and be it in order to make sure (or know for sure) that your adversary does not get there first.

Thus, it is an absolute strategic necessity to answer the question, "Is there a most fundamental computer or Turing machine, and what does it look like?"

6.1.2 What Is the Ultimate Turing Machine?

When Allan Turing wrote his famous paper [B2] about the potentially infinite tape machine, he explicitly pointed out that his machine could principally handle any kind of numbers. Only for the reason of simplicity, he then restricted his further elaborations to digitals for the examples in the very paper. Later it obviously was forgotten that Turing had, in fact, proposed a very general machine, but as people simply concentrated on the examples and obviously only very few read Turing's whole publication, the tape model was more and more seen as a digital machine rather than the general thing Turing really had—mathematically created.

Avoiding such "copy-errors," however, is essential in order to make sure that there are no bad surprises waiting around the next corner if it comes to making decisions about future developments, investments, and personal planning.

Thus, we have to force ourselves to

(a) start as fundamental as possible;
(b) make sure that we are never putting our own preferences into the equation we intend to solve.

Doing this in connection with the question of what actually the fundamental computer is therefore requires us to not only re-read and digest Turing's original work but also go through the most basic physics in order to have a truly primordial starting point.

In doing so, we find that not only the fundamental Turing machine exists but also that it is ultimately based on the metric derivation of quantum equations completely compatible with the General Theory of Relativity. This way we were able to find the most general metric structures for Turing machines [B3]. These gravity or Einstein quantum computers, as we might like to call them, contain the classical quantum and the classical digital computers as degenerated forms. Along the way, one does also find:

(a) the highest possible storage capacity for information;
(b) the most fundamental quantum states allowing for storage and processing of bits and Q-bits;
(c) the possibility for gravity- or Einstein-Q-bit having the theoretical intrinsic calculation power of a whole separate universe of its own;
(d) a variety of rather fundamental insights into the metric origin of quantum equations which will be of great help in the development of future techniques in many, many fields;
(e) a variety of ideas for optimum human–computer interfaces;
(f) a much better insight into the question of quantum cryptography, thereby revealing that the safety the protagonists and leading investors are promising us by using entanglement is an illusion (it will only be a "relative safety");
(g) a technical solution for true cryptographic safety (see chapter 11, "The Quantum Gravity Enigma Machine");
(h) a way out of the mass-information problem (c.f. section 5.4, "Does More Information Always Mean More Mass?") due to new metric solutions as given in [13].

6.1.3 How to Realize Such a Machine?

With the theoretical principles known, it is clear that the path to follow definitively requires the further development of the quantum dot, the 3D Si-P-doping, and the SQUID technology. Thereby, at least for a start, concentrating on the SAQD (self-assembling quantum dots) via molecular beam epitaxy and the Australian Si P-doping techniques seems to be adequate. However, also other approaches such as the 3D-technique developed in Australia and the Canadian SQUID approach should be considered. In addition—in order to parallel focus on the problem of the human–gravity quantum computer interface, one has to take care of organic bio-spin systems, which currently play no role in the quantum computer development. In addition, the new solutions with respect to the ultimate machine require a complete rethought of the problem of "how to program these future machines" in order to fully harvest their potential.

In other words: the approach has to be as holistic as anyhow possible. In fact, it should be undertaken as deep and comprehensive as the Manhattan project, perhaps even more so.

As our theoretical starting point proves, our approach is of such holistic character.

6.1.4 Why Is Our Approach Better than Other (Classical) Technological Development Concepts?

As our approach has to be placed on such a fundamental foundation (physicists call this a "Theory of Everything" or "World Formula" approach), it automatically produces side-line results which are in themselves essential world-first and world-class features. This is in great contrast to classical (state of the art) R&D projects, where there is only the main goal and if this goal fails, there will be nothing. Here is a list of add-on features which automatically fall out of the main development direction as major derivatives:

(a) As Quantum Theory is just the very field in science which only came into existence because of the principal uncertainties in our universe, our World Formula approach naturally delivers a most fundamental and universal uncertainty budget calculator to any problem. Thus, the complex task of finding the uncertainties to model predictions of any kind is to become a mere automatism if the model building is founded on the holistic approach anyway [B4, B5, B6].

(b) This aspect also automatically leads to just the optimum decision-making tools there can be [B7, B8].

(c) Economy and socio-economy models by far more holistic and intelligent than e.g., the Aladdin system from BlackRock [B5, B6, B8, B9].

(d) Human–brain–machine interfaces in both the technical and the ethical aspect [B10, B11].

(e) Our insight into the metric fundamentals of quantum cryptography not only reveals the fact that with the entangled signals for "safe" communication and data transfer there is no absolute but only "relative safety," but also clearly gives the limitations and offers methods for better cryptographic solutions and eavesdropping detection.

(f) As the concept automatically delivers fundamentally derived holistic evolution equations for every system, we would have the best prediction-making tools possible, too [B12].

(g) Finally, with the approach also delivering a mathematical understanding of psychological aspects (c.f. section 12.4, "Mathematical Psychology as a Weapon"), we have an important medical and socio-economic component, which offers almost limitless numbers of applications [B13].

(h) ...

6.1.5 References for Section 6.1

B1. N. Schwarzer, *Science Riddles—Riddle No. 8: Could the Schwarzschild Metric Contain its own Quantum Solution?*, Self-published, Amazon Digital Services, 2018, Kindle.

B2. A. M. Turing, On Computable Numbers, with an Application to the Entscheidungsproblem (1936), *Proceedings of the London Mathematical Society*, 2 (published 1937), 42: 230–265, doi:10.1112/plms/s2-42.1.230.

B3. N. Schwarzer, Is there an ultimate, truly fundamental and universal Computer Machine?, theory part of this chapter (section 6.2).

B4. N. Schwarzer: *Quantum Tribology—Part I: Theory*, Self-published, Amazon Digital Services, 2017, Kindle.

B5. T. Bodan, N. Schwarzer: *Quantum Economy*, Self-published, Amazon Digital Services, 2017, Kindle.

B6. N. Schwarzer, *Philosophical Engineering Part 1: The Honest Non-Parasitic Philosopher and the Universal "GOOD" Derived from a Theory of Everything*, Self-published, Amazon Digital Services, 2018, Kindle.

B7. N. Schwarzer, *Einstein Had It, but He Did Not See It—Part LXVIII: Most Fundamental Tools for Optimum Decision-Making Based on Quantum Gravity*, Self-published, Amazon Digital Services, 2018, Kindle.

B8. N. Schwarzer, *Einstein Had It, but He Did Not See It—Part LXIX: The Hippocratic Oath in Mathematical Form and Why—So Often—It Will Be of No Use*, Self-published, Amazon Digital Services, 2018, Kindle.
B9. N. Schwarzer, *Einstein Had It, but He Did Not See It—Part XIII: Why Equality Is Always Hostile to Life*, Self-published, Amazon Digital Services, 2018, Kindle.
B10. N. Schwarzer, Einstein gravity quantum computer and its impact on health science, May 2019, invited talk given at the University of Florida, Orlando.
B11. T. Bodan, *7 Days—How to Explain the World to My Dying Child*, Self-published, BoD Classic, ISBN 9783752639728.
B12. N. Schwarzer, *Mathematical Psychology—Tools for the Understanding, Simulation, Mitigation and Potential Prevention of Mass Formation Psychoses*, Self-published, Amazon Digital Services, 2022, Kindle. See also [214].
B13. N. Schwarzer: *How Systems Evolve—A Red Pill Course on Evolution*, Self-published, ASIN: B0B6RGBKQ4, Amazon Digital Services, 2022, Kindle. See also [212].

6.2 About the Recipe

Is there an ultimate, truly fundamental, and universal computer machine?

The answer to the question is "YES!," but it requires the understanding of Quantum Relativity to truly get there.

In this chapter, we will show what is missing. Along the way, we will show a way to understand the quantum states that allow the storage of bit information in a black hole (the Bekenstein problem, which we already briefly discussed in the previous chapter). As a by product, we will obtain a fundamental and purely mathematically motivated scale. There will be a bit redundancy with respect to the chapter 5, "The New Abacus," but this is needed to provide us with the necessary fundament to move on and provide more mathematical background.

For those who are only interested in the principal proof of our concept, we suggest directly going to section 6.9, "The Evidence—Summed up." There we will most briefly present the following:

(A) The Metric Origin of the Klein–Gordon-Equation
(B) The Metric Origin of Scalar Fields like the Higgs Field and Symmetry Breaking

(C) The Metric Origin of Mass
(D) The Metric Origin of Spin and Spin Fields
(E) The Metric Schrödinger Equation
(F) The Metric Origin of the Dirac Equation
(G) Metric Quantum Electro Dynamics (MQED)
(H) Towards a Metric Quantum Chromo Dynamic
(I) Quantum Gravity Dynamic (QGD)

More comprehensive proof to some of these points will later be given in the appendices of the book.

6.3 The Evolution of a Theory of Everything

In this chapter, we will give the shortest possible presentation of the development of a Theory of Everything. The work was greatly motivated by a series of partially fictive, partially non-fictive stories and socio-economic studies [33–40]. For those who are interested in learning about the complete development of the first and incomplete derivation of a zero-sum-based Theory of Everything, we refer to references [41–71]. Here [41] might be seen as a nice summary of the method and contains quite a few of our earlier applications. Studying this part of the development process was not in vain, because, as we saw later, we had to come back to our earlier assumption in order to find the right Lagrange density for the derivation of a Quantum Theory from a completely and rather fundamental geometric starting point. Then, starting with [72–84], we concentrated on various versions of the surface term in connection with the Einstein–Hilbert action until we finally also found ways to derive quantum theoretical equations directly from the vacuum Einstein field equations [85–179] via certain transformations.

Finding a potential Theory of Everything has to start at the most fundamental basis one could think of. Any other approach would lead us to the typical human flaw, already compromising the conclusive outcome even before any evaluation process has been started ... simply because we would have smuggled ourselves and our preferences into the equations.

What could be such a most fundamental basis?

How about an arbitrary ensemble of properties subjected to a Hamilton extremal principle?

Without loss of generality, we can interpret the arbitrary properties as dimensions of an arbitrary space. Applying the Hamilton extremal principle, which is to say, that the ensemble of properties is keen on reaching a state of minimum, directly leads us to the so-called generalized Einstein–Hilbert action:

$$\delta_g W = 0 = \delta_g \int\limits_V d^n x \left(\sqrt{-g}\left[f[R] - 2\kappa L_M\right]\right). \tag{63}$$

Here g denotes the determinant of the metric tensor, W and V are the action and the volume of the n-dimensional space, respectively, R denotes the scalar curvature or Ricci scalar, $f[R]$ is an arbitrary function of R, and the last term $2\kappa L_M$ describes the Lagrange density of matter. It was shown in [180] that the complete variation of a function $f(R)$ leads to the following result:

$$f'(R)R_{\mu\nu} - \frac{1}{2}f(R)g_{\mu\nu} + f''(R)\left[R_{;\mu\nu} - \Delta_g R \cdot g_{\mu\nu}\right]$$
$$+f'''(R)\left[R_{;\mu}R_{;\nu} - R^{;\sigma}R_{;\sigma} \cdot g_{\mu\nu}\right] = \begin{cases} 0 & \ldots\text{"vacuum"} \\ -8\pi G T_{\mu\nu} & \ldots\text{postulated matter} \end{cases}. \tag{64}$$

The ";" denotes covariant derivatives and the symbol Δ_g gives the Laplace–Beltrami operator in the metric $g_{\alpha\beta}$. Einstein and Hilbert had used $f[R] = R$, which avoids higher orders of curvature respectively higher orders for the Ricci scalar and/or the metric [180, 2]. In the context of non-linear R-functionalities in the Einstein–Hilbert action, one should also discuss the appearance of non-vanishing surface terms [213], but this fact shall only be noted here and not be further discussed.

After revision of the classical Einstein–Hilbert action with $f[R] = R$, which was given by Hilbert [2] as

$$\delta_g W = 0 = \delta_g \int\limits_V d^n x \left(\sqrt{-g}\left[R - 2\kappa L_M - 2\Lambda\right]\right), \tag{65}$$

we discarded of the artificial matter term $2\kappa L_M$, which had to be postulated by Hilbert and Einstein [1, 2, 180][1]. Instead, the present author suspected another term representing the matter, namely the variation of the Ricci tensor, which classically leads to a surface term. The reader does probably know that this term was discarded by the creators of the "General Theory of Relativity" and many previous authors in the field, while others (most prominent among them was Stephen Hawking) considered this term as the source of matter, certain fields, and energy. We have therefore investigated this term more closely in some of our previous papers [64–84] (especially [64] and see also further below here), too, but as already said above, we will not consider it here.

[1]For completeness only, we took into account a non-zero cosmological constant Λ.

Over one hundred years ago, Einstein [1] and Hilbert [2] showed that as a result of the variation (65), a space of a given metric $g^{\alpha\beta}$ must curve and that the curvature described by the Ricci tensor $R^{\alpha\beta}$ must satisfy certain conditions. The **classical** result for the metric tensor $g^{\alpha\beta}$ of the curved space can be given as follows:

$$R^{\alpha\beta} - \frac{1}{2} R g^{\alpha\beta} + \Lambda g^{\alpha\beta} = -\kappa T^{\alpha\beta}. \tag{66}$$

Here $R^{\alpha\beta}$ and $T^{\alpha\beta}$ are the Ricci- and the energy momentum tensor, respectively, while the parameters Λ and κ are constants (usually called cosmological and coupling constant, respectively). These are the well-known Einstein field equations in n dimensions with the indices α and β running from 1 to n. The underlying theory is called "General Theory of Relativity."

It is important to note that neither Einstein nor Hilbert were able to give a proper explanation for the appearance of matter in their equations and so they had to postulate it, which they did by the introduction of the energy momentum tensor. This means, however, that the matter term $T^{\alpha\beta}$ cannot be derived directly out of the General Theory of Relativity but has to be postulated and fitted in accordance to the problem in question. In this chapter, it will be shown that, in fact, such a term might not be needed at all, because one can derive particle, which is to say matter, solutions directly from the vacuum forms of the Einstein field equations (see also the first part of the appendix to the chapter (section 6.12), entitled "Simple Example for Dimensional Mass Creation"). One only needs to allow for an additional variation of the choice of coordinates and dimension (e.g., see [130, 135] and later in this book). Later we found very elegant ways to derive great varieties of matter forms (the reader can find this in concentrated form in [5, 12–18]).

But even earlier than in [130], i.e., already in [72–84], the present author had investigated the term discarded by Einstein and Hilbert and found that it apparently also presents certain forms of matter. It was shown in [64, 116] how to evaluate the integrand in (65) in greater detail. The result was given as follows (note that the classical matter term, the Lagrange matter density, was set to zero):

$$\begin{aligned}
\delta_g W &= 0 \\
&= \delta_g \int_V d^n x \sqrt{-g} R \\
&= \delta_g \int_V d^n x \sqrt{-g} g^{\alpha\beta} R_{\alpha\beta} = \int_V d^n x \delta_g \left(\sqrt{-g} g^{\alpha\beta} R_{\alpha\beta} \right)
\end{aligned}$$

$$= \int_V d^n x \left[\delta_g \left(\sqrt{-g} g^{\alpha\beta} \right) R_{\alpha\beta} + \sqrt{-g} g^{\alpha\beta} \delta_g \left(R_{\alpha\beta} \right) \right]$$

$$= \int_V d^n x \sqrt{-g} \left[-G^{\kappa\lambda} \delta_g g_{\kappa\lambda} + g^{\alpha\beta} \delta_g R_{\alpha\beta} \right], \tag{67}$$

with $G^{\alpha\beta} = R^{\alpha\beta} - \frac{1}{2} R g^{\alpha\beta}$ denoting the Einstein tensor. It has to be noted that the result does not change when we add a constant Λ, giving us $G^{\alpha\beta} + \Lambda g^{\alpha\beta} = R^{\alpha\beta} - \frac{1}{2} R g^{\alpha\beta} + \Lambda g^{\alpha\beta}$ (c.f. (66), (74)).

As mentioned above, it was assumed by Hilbert [2] that the second term in the last line in the integrand of (67), which is to say the term $g^{\alpha\beta}\delta_g R_{\alpha\beta}$, could be made a surface integral and, thus, disregarded.

Assuming, however, that the space might not allow for such a surface integral simplification (e.g., because of insufficient smoothness, respectively differentiability at smaller scales), the present author revived the term and expanded it within the variational task given in (65). The complete evaluation is shown in [64, 116].

Further on in [82, 83], it was demonstrated that the neglected surface term could still influence the outcome of a more holistic approach if, for example, our space is considered as a manifold, which is embedded in a space of higher dimension. Here we only give a very brief summary as repetition of the evaluation.

After performing the variation in (67), we have:

$$0 = \int_V d^n x \sqrt{-g} \left[R^{\kappa\lambda} - \frac{1}{2} R g^{\kappa\lambda} + \Lambda g^{\kappa\lambda} \right] \delta_g g_{\kappa\lambda}$$

$$+ \int_V d^n x \sqrt{-g} \left[\left(\left(g^{\alpha\beta} \delta_g \Gamma^{\rho}_{\alpha\beta} - g^{\alpha\rho} \delta_g \Gamma^{\gamma}_{\alpha\gamma} \right)_{;\rho} \right) \right]$$

$$= \int_V d^n x \sqrt{-g} \left[R^{\kappa\lambda} - \frac{1}{2} R g^{\kappa\lambda} + \Lambda g^{\kappa\lambda} \right] \delta_g g_{\kappa\lambda}$$

$$+ \int_{\partial V} d^{n-1} x \sqrt{-h} \left[\left(g^{\alpha\beta} \delta_g \Gamma^{\rho}_{\alpha\beta} - g^{\alpha\rho} \delta_g \Gamma^{\gamma}_{\alpha\gamma} \right) V_\rho \right]. \tag{68}$$

Applying the following identities:

$$\left(g^{\kappa}_{\alpha} g^{\lambda}_{\sigma} g^{\rho}_{\beta} + g^{\kappa}_{\sigma} g^{\lambda}_{\beta} g^{\rho}_{\alpha} - g^{\kappa}_{\alpha} g^{\lambda}_{\beta} g^{\rho}_{\sigma} \right) \equiv GX^{\kappa\lambda\rho}_{\alpha\beta\sigma}, \tag{69}$$

$$\delta_g\left(\Gamma^{\gamma}_{\alpha\beta}\right)=\frac{1}{2}g^{\gamma\sigma}\left(\delta g_{\sigma\alpha;\beta}+\delta g_{\sigma\beta;\alpha}-\delta g_{\alpha\beta;\sigma}\right)$$
$$=\frac{g^{\gamma\sigma}}{2}GX^{\kappa\lambda\rho}_{\alpha\beta\sigma}\delta g_{\kappa\lambda;\rho}, \tag{70}$$

we can give (68) in the following form:

$$0=\int\limits_V d^n x\sqrt{-g}\left[R^{\kappa\lambda}-\frac{1}{2}Rg^{\kappa\lambda}+\Lambda g^{\kappa\lambda}\right]\delta g_{\kappa\lambda}$$
$$+\int\limits_{\partial V} d^{n-1}x\sqrt{-h}\left[\left(g^{\alpha\beta}\delta_g\Gamma^{\rho}_{\alpha\beta}-g^{\alpha\rho}\delta_g\Gamma^{\gamma}_{\alpha\gamma}\right)V_\rho\right]$$
$$=\int\limits_V d^n x\sqrt{-g}\left[R^{\kappa\lambda}-\frac{1}{2}Rg^{\kappa\lambda}+\Lambda g^{\kappa\lambda}\right]\delta g_{\kappa\lambda}$$
$$+\int\limits_{\partial V} d^{n-1}x\frac{\sqrt{-h}}{2}\left(\overbrace{\left(g^{\alpha\beta}g^{\rho\sigma}GX^{\kappa\lambda\omega}_{\alpha\beta\sigma}-g^{\alpha\rho}g^{\gamma\sigma}GX^{\kappa\lambda\omega}_{\alpha\gamma\sigma}\right)}^{\equiv MB^{\kappa\lambda\omega\rho}}\delta g_{\kappa\lambda;\omega}\right)V_\rho. \tag{71}$$

Integration by parts allows us to get rid of the $\delta_g g_{\kappa\Lambda;\omega}$-term, assuming higher dimensions coupling into our manifold and after some evaluation [82, 83] we obtain a set of extended Einstein field equations with "surface" or "entanglement" terms:

$$(a)\quad \delta g_{\kappa\lambda}, V:\quad 0=R^{\kappa\lambda}-\frac{1}{2}Rg^{\kappa\lambda}+\Lambda g^{\kappa\lambda}+\frac{1}{2}B\cdot V_\rho MB\gamma^{\kappa\lambda\omega\rho}v_\omega$$

$$(b)\quad \delta g_{\kappa\lambda}, \partial^2 V:\quad 0=\Re^{\kappa\lambda}-\frac{1}{2}\Re h^{\kappa\lambda}+\Lambda h^{\kappa\lambda}+\frac{1}{2}A\cdot U_\rho MB^{\kappa\lambda\omega\rho}u_\omega. \tag{72}$$

Here A and B are constants, V_ρ, v_ω and U_ρ, u_ω denote the normal vectors on the surface ∂V and the sub-surface $\partial^2 V$, respectively. With respect to the complete evaluation, the reader is referred to [82, 83] or [116].

Later it was shown by this author that, in fact, higher dimensions play an important role in the appearance of matter (especially [156–158]), but that there exists a more direct and elegant way to introduce it via the entanglement of these higher dimensions [3, 5, 18].

Thus, there seems to be a rather direct equivalence between the consideration of surface terms in the Einstein–Hilbert action or the introduction of additional (and potentially entangled) dimensions. Both can be the source for masses, energies, and fields. However, the additional dimensional approach appears to be simpler (c.f. [3, 5, 18]) and also much more intuitive [18, 212, 214].

Even though the "holistic quantum gravity military leader in the making" might not see it—quite understandably, taking the form in which it has been presented—we have derived the most fundamental equations to every system of arbitrary numbers of dimensions (properties). As a problem, one intends to solve is just nothing else but a system of properties and the task usually is to find the optimum answer to this very problem, we have—in principle[2]—found the equations governing everything. There might be the argument that our equations (in fact, these are just Einstein's equations) do not hold true in fractal spaces or non-smooth coordinates of distributive character. This, however, is not true, because, as our solutions are tensors, a simple coordinate transform like:

$$g_{ij}\left[\forall x_i = x_0, x_1, \ldots, x_n\right]$$

$$\Rightarrow \qquad G_{ij} = G'_{x_i}\left[x_{\forall i}\right] \cdot G'_{x_j}\left[x_{\forall j}\right] \cdot g_{ij}\left[G_{x_0}\left[x_{\forall i}\right], \mathrm{G}_{x_1}\left[x_{\forall i}\right], \ldots, \mathrm{G}_{x_n}\left[x_{\forall i}\right]\right]$$

$$\text{with:} \quad G'_{x_i}\left[x_{\forall i}\right] \equiv \frac{\partial G_{x_i}\left[x_{\forall i}\right]}{\partial x_i} \tag{73}$$

would allow us to generalize any smooth metric solution g_{ij} to a potentially distributed or nearly[3] fractal one G_{ij}. We also need to point out that the fundamental optimization problem is described by the Einstein–Hilbert variation integral and not just the Einstein field equations. The latter are only one possible solution out of quite many.

What we are still missing, however, is the incorporation of uncertainty. We know, however, that there is just this one field in science—namely, Quantum Theory—which only exists because of all the uncertainties we have in this very universe of ours. Thus, the task to find a truly holistic guideline for answering the philosophical question of what actually is just everything (including right and good) in completely mathematical form requires us to combine the equations above with Quantum Theory.

---

[2]Still necessary generalizations can be seen in [18].

[3]"Nearly" here means that—of course—the coordinates would appear whole and continuous, but on certain scales, they could be made to appear as fractal. It simply requires a suitable selection of coordinate functions $G_{x_i}\left[x_{\forall i}\right]$ to mirror all possible distributions one could think of. Truly fractal structures like the so-called Weierstraß monster (en.wikipedia.org/wiki/Weierstrass_function), however, have to be excluded, because they are not differentiable.

6.4 How to Quantize Solutions of the Einstein Field Equations: The State-of-the-Art Situation

As the introduction to the method of quantizing metric solutions of the Einstein field equations is quite lengthy and was already presented in a variety of previous papers (e.g., [111–143]), books ([3, 18]), and chapter 3, "The Theory," we refrain from presenting it here again. Instead, we will only give a brief recipe:

1. We consider space as an ensemble of properties.
2. These properties could just be seen as degrees of freedom and, thus, dimensions, which are subjected to a Hamilton extremal principle.
3. This leads to an Einstein–Hilbert action [2] and subsequently to the Einstein field equations [1, 2].
4. Time seems to take a special place among the properties as it is not such a property itself, but consists of all other properties' internal changes and variations [3]. Here is one possible explanation for the omnipresence of time: Applying the Einstein field equations on the internal degrees of freedom of each single property as a 1-dimensional space (c.f. [122]) gives exactly six solutions [125] among which we always also find oscillations. These internal periodic processes (changes) inside each and every property or dimension are realized as time from an external observer ... time, which itself forces other properties to change. Thus, starting as an internal property (solution) within each and every dimension, time not only is change but also brings about change. Apparently, time is the most fractal and self-similar thing there is in this universe.
5. Now, however, we have the interesting situation that solutions to the Einstein field equations are not necessarily unique. Derived from the Einstein–Hilbert action as an extremal principle, the starting quantity is the so-called Ricci scalar *R*, being the essential kernel of this action. This results in a metric solution to the Einstein field equations, being subsequently derived from the Einstein–Hilbert action. Most interestingly, there are also infinitely many solutions to just one given Ricci kernel. Just as an example, one might take the flat Minkowski space and the Schwarzschild vacuum metric [145]. Both have a vanishing Ricci scalar $R = 0$, but while the first describes empty

space, the second, even though being a vacuum solution, contains a gravitational "object" of spherical symmetry. And yes, this all comes out from a variational kernel of $R = 0$. Thus, we conclude, there seem to be quite some degrees of freedom regarding the choice of metric solutions to just one Ricci scalar curvature.

6. Tickling metric solutions with respect to this degree of freedom, which is to say to perform a variation (or transformation) of metric solutions, thereby treating the Ricci scalar or the whole kernel of the Einstein–Hilbert action as a conserved quantity, gives us classical quantum equations and, thus, Quantum Theory.

Essentially one finds quantum gravity as transformations to metrics solving the Einstein field equations (66) or vacuum field equations:

$$R^{\alpha\beta} - \frac{1}{2} R g^{\alpha\beta} + \Lambda g^{\alpha\beta} = 0. \quad (74)$$

Thereby we can use external or internal degrees of freedom. While in many previous papers (e.g., [142], as this is the most compact), we more or less concentrated on external or wrapper-like transformations of the kind:

$$G_{\alpha\beta} = F\left\lfloor f[t,x,y,z]\right\rfloor_{\alpha\beta}^{ij} g_{ij} \rightarrow$$

$$G_{\alpha\beta} = F\left[f[t,x,y,z]\right]\cdot \delta_{\alpha}^{i}\delta_{\beta}^{j} g_{ij}, \quad (75)$$

we also introduced inner or Killing-like approaches in [143]. Most interestingly, these inner quantum gravity transformations led us to a Dirac-like equation [181]. Meanwhile, as shown in [18], we know that also the simple approach (75) does give the Dirac equation (see appendix C).

In conclusion we might state that Quantum Theory is just the inner degree or fluctuation of metric solutions to the Einstein field equations. This does not entirely present the whole picture, but it suffices to get the gist in a rough manner. The interested reader will find a fairly compact mathematical presentation of the above recipe in [142] and more comprehensive elaborations in the two books [3, 18].

The question which is still unanswered, however, is: why does the scaling-factor (75), providing an additional degree of freedom, need to "be activated" and what forces it to obey the known quantum theoretical rules and equations? This question was answered in our more recent publications in the series "Medical Socio-Economic Quantum Gravity" and the book [18].

6.5 A Simple Origin of Matter

Meanwhile, even though we have much better understood the meaning of the scale-factor (75) (e.g., [18]), it is very illustrating (and even partially entertaining) to revisit our somewhat older attempts to use this transformation and get behind its "secrets."

So, within the short series "Science Riddles" [147–158], we essentially concentrated on the following question:

How can we understand transformations of metric tensors like (75), which do not sum up to an identical transformation?

In order to answer this question, we want to start with a very simple change of the metric tensor. In [23] and some previous papers of the later series "Science Riddles," it was shown that a metric $g_{\alpha\beta}$, which fulfills the vacuum Einstein field equations (EFE), can be generalized to a quantized state via an approach of the form

$$G_{\alpha\beta} = g_{\alpha\beta} \cdot F[f]. \tag{76}$$

Please do not mix the f of $F[f]$ with the $f[R]$ in the previous section 6.3, "The Evolution of a Theory of Everything," and the generalized Einstein–Hilbert action [180]:

$$\delta_g W = 0 = \delta_g \int_V d^n x \left(\sqrt{-g} \left[f[R] - 2 \cdot \kappa \cdot L_M - 2 \cdot \Lambda \right] \right). \tag{77}$$

However, as demonstrated in [3], it can be quite complicated to find holistic, EFE-compatible solutions even for the simplest cases. But what would be the meaning of transformations leading to metrics not fulfilling the vacuum Einstein field equations anymore?

For generality, here we want to repeat the derivation from [3], where it was shown that for arbitrary numbers n of dimensions, our approach (76) yields a certain Ricci scalar R^*.[4] Now, while $g_{\alpha\beta}$ shall satisfy the vacuum Einstein field equations with the classical linear dependency on R, this does not hold for R^*. This gives us two Einstein–Hilbert actions with respect to the

---

[4]Please note that this R^* should be seen as an auxiliary Ricci scalar, as it is based on the auxiliary or "functionally wrapped" metric (76) and not the original metric $g_{\alpha\beta}$, which was assumed to fulfill the vacuum Einstein field equations. How a complete solution with $G_{\alpha\beta}$, still fulfilling the Einstein field equations, can be constructed was first demonstrated in [116] and then fully derived, e.g., in [18] later on.

two metrics $g_{\alpha\beta}$ and $G_{\alpha\beta}$ plus a third one giving the variation of R^* with respect to $g_{\alpha\beta}$ [5]:

$$\delta_g W = 0 = \delta_g \int_V d^n x\left(\sqrt{-g}\cdot\Phi_R[R]\right) = \delta_g \int_V d^n x\left(\sqrt{-g}\cdot R\right)$$

$$\delta_G W = ?? = \delta_G \int_V d^n x\left(\sqrt{-G}\cdot R^*\right)$$

$$\delta_g W = ? = \delta_g \int_V d^n x\left(\sqrt{-g}\cdot\Phi_{R^*}[R^*]\right), \tag{78}$$

where we do not consider any cosmological constant and also have erased the classical (artificially introduced) matter term. We also name the Ricci scalar of the auxiliary (quantized) metric $G_{\alpha\beta}$ from (76) R^*, while we leave the symbol R for the metric $g_{\alpha\beta}$.

Thereby, just as said, we assume R to be the Ricci scalar to the unperturbed (non-transformed) metric $g_{\alpha\beta}$. On the other hand, we know the Ricci scalar R to be defined by the metric tensor via:

$$R = R_{\alpha\beta}g^{\alpha\beta} = \left(\Gamma^{\sigma}_{\alpha\beta,\sigma} - \Gamma^{\sigma}_{\beta\sigma,\alpha} - \Gamma^{\mu}_{\sigma\alpha}\Gamma^{\sigma}_{\beta\mu} + \Gamma^{\sigma}_{\alpha\beta}\Gamma^{\mu}_{\sigma\mu}\right)g^{\alpha\beta}, \tag{79}$$

with

$$\Gamma^{\gamma}_{\alpha\beta} = \frac{g^{\gamma\sigma}}{2}\left(g_{\sigma\alpha,\beta} + g_{\sigma\beta,\alpha} - g_{\alpha\beta,\sigma}\right). \tag{80}$$

Similarly, we have for R^*:

$$R^* = R^*_{\alpha\beta}G^{\alpha\beta} = \left(\Gamma^{*\sigma}_{\alpha\beta,\sigma} - \Gamma^{*\sigma}_{\beta\sigma,\alpha} - \Gamma^{*\mu}_{\sigma\alpha}\Gamma^{*\sigma}_{\beta\mu} + \Gamma^{*\sigma}_{\alpha\beta}\Gamma^{*\mu}_{\sigma\mu}\right)G^{\alpha\beta}$$

$$= \left(\Gamma^{*\sigma}_{\alpha\beta,\sigma} - \Gamma^{*\sigma}_{\beta\sigma,\alpha} - \Gamma^{*\mu}_{\sigma\alpha}\Gamma^{*\sigma}_{\beta\mu} + \Gamma^{*\sigma}_{\alpha\beta}\Gamma^{*\mu}_{\sigma\mu}\right)\frac{g^{\alpha\beta}}{F[f]}, \tag{81}$$

with

$$\Gamma^{*\gamma}_{\alpha\beta} = \frac{g^{\gamma\sigma}}{2\cdot F[f]}\left(\left[F[f]\cdot g_{\sigma\alpha}\right]_{,\beta} + \left[F[f]\cdot g_{\sigma\beta}\right]_{,\alpha} - \left[F[f]\cdot g_{\alpha\beta}\right]_{,\sigma}\right)$$

$$= \frac{g^{\gamma\sigma}}{2\cdot F[f]}\left(\left[F[f]\cdot g_{\sigma\alpha}\right]_{,\beta} + \left[F[f]\cdot g_{\sigma\beta}\right]_{,\alpha} - \left[F[f]\cdot g_{\alpha\beta}\right]_{,\sigma}\right)$$

$$= \frac{g^{\gamma\sigma}}{2}\left(g_{\sigma\alpha,\beta} + g_{\sigma\beta,\alpha} - g_{\alpha\beta,\sigma}\right)$$

---

[5] In order to avoid mixing up the *f*-function from the wrapper forms *F*[*f*], we rewrite (63) now with a function Φ(*R*) instead of *f*(*R*).

$$+\frac{g^{\gamma\sigma}}{2\cdot F[f]}\left(F[f]_{,\beta}\cdot g_{\sigma\alpha}+F[f]_{,\alpha}\cdot g_{\sigma\beta}-F[f]_{,\sigma}\cdot g_{\alpha\beta}\right)$$

$$=\Gamma^{\gamma}_{\alpha\beta}+\frac{g^{\gamma\sigma}}{2\cdot F[f]}\left(F[f]_{,\beta}\cdot g_{\sigma\alpha}+F[f]_{,\alpha}\cdot g_{\sigma\beta}-F[f]_{,\sigma}\cdot g_{\alpha\beta}\right)$$

$$\equiv\Gamma^{\gamma}_{\alpha\beta}+\Gamma^{**\gamma}_{\alpha\beta}. \tag{82}$$

Setting this into the second line in (81) yields:

$$R^{*}=R^{*}_{\alpha\beta}G^{\alpha\beta}=\left(\Gamma^{*\sigma}_{\alpha\beta,\sigma}-\Gamma^{*\sigma}_{\beta\sigma,\alpha}-\Gamma^{*\mu}_{\sigma\alpha}\Gamma^{*\sigma}_{\beta\mu}+\Gamma^{*\sigma}_{\alpha\beta}\Gamma^{*\mu}_{\sigma\mu}\right)G^{\alpha\beta}$$

$$=\begin{pmatrix}\Gamma^{\sigma}_{\alpha\beta,\sigma}-\Gamma^{\sigma}_{\beta\sigma,\alpha}-\Gamma^{\mu}_{\sigma\alpha}\Gamma^{\sigma}_{\beta\mu}+\Gamma^{\sigma}_{\alpha\beta}\Gamma^{\mu}_{\sigma\mu}\\ -\Gamma^{\mu}_{\sigma\alpha}\Gamma^{**\sigma}_{\beta\mu}+\Gamma^{\sigma}_{\alpha\beta}\Gamma^{**\mu}_{\sigma\mu}-\Gamma^{**\mu}_{\sigma\alpha}\Gamma^{\sigma}_{\beta\mu}+\Gamma^{**\sigma}_{\alpha\beta}\Gamma^{\mu}_{\sigma\mu}\\ +\Gamma^{**\sigma}_{\alpha\beta\ ,\sigma}-\Gamma^{**\sigma}_{\beta\sigma\ ,\alpha}-\Gamma^{**\mu}_{\sigma\alpha}\Gamma^{**\sigma}_{\beta\mu}+\Gamma^{**\sigma}_{\alpha\beta}\Gamma^{**\mu}_{\sigma\mu}\end{pmatrix}\frac{g^{\alpha\beta}}{F[f]}$$

$$=\frac{R}{F[f]}+\begin{pmatrix}-\Gamma^{\mu}_{\sigma\alpha}\Gamma^{**\sigma}_{\beta\mu}+\Gamma^{\sigma}_{\alpha\beta}\Gamma^{**\mu}_{\sigma\mu}-\Gamma^{**\mu}_{\sigma\alpha}\Gamma^{\sigma}_{\beta\mu}+\Gamma^{**\sigma}_{\alpha\beta}\Gamma^{\mu}_{\sigma\mu}\\ +\Gamma^{**\sigma}_{\alpha\beta\ ,\sigma}-\Gamma^{**\sigma}_{\beta\sigma\ ,\alpha}-\Gamma^{**\mu}_{\sigma\alpha}\Gamma^{**\sigma}_{\beta\mu}+\Gamma^{**\sigma}_{\alpha\beta}\Gamma^{**\mu}_{\sigma\mu}\end{pmatrix}\frac{g^{\alpha\beta}}{F[f]}$$

$$\equiv\frac{R+R^{**}}{F[f]}=\Phi[R]. \tag{83}$$

Further evaluation of (83) is possible by rewriting (82) as follows:

$$\Gamma^{*\gamma}_{\alpha\beta}=\Gamma^{\gamma}_{\alpha\beta}+\frac{g^{\gamma\sigma}}{2\cdot F[f]}\left(F[f]_{,\beta}\cdot g_{\sigma\alpha}+F[f]_{,\alpha}\cdot g_{\sigma\beta}-F[f]_{,\sigma}\cdot g_{\alpha\beta}\right)$$

$$=\Gamma^{\gamma}_{\alpha\beta}+\frac{F[f]_{,\beta}\cdot\delta^{\gamma}_{\alpha}+F[f]_{,\alpha}\cdot\delta^{\gamma}_{\beta}-g^{\gamma\sigma}\cdot F[f]_{,\sigma}\cdot g_{\alpha\beta}}{2\cdot F[f]}$$

$$=\Gamma^{\gamma}_{\alpha\beta}+F'\cdot\frac{f_{,\beta}\cdot\delta^{\gamma}_{\alpha}+f_{,\alpha}\cdot\delta^{\gamma}_{\beta}-g^{\gamma\sigma}\cdot f_{,\sigma}\cdot g_{\alpha\beta}}{2\cdot F[f]}$$

$$\equiv\Gamma^{\gamma}_{\alpha\beta}+\frac{F'}{F[f]}\cdot\Gamma^{***\gamma}_{\alpha\beta};\quad F'=\frac{\partial F[f]}{\partial f}. \tag{84}$$

Now we can give (83) as follows:

$$R^{*}=\frac{R}{F[f]}+\begin{pmatrix}-\Gamma_{\sigma\alpha}^{\mu}\Gamma_{\beta\mu}^{**\sigma}+\Gamma_{\alpha\beta}^{\sigma}\Gamma_{\sigma\mu}^{**\mu}-\Gamma_{\sigma\alpha}^{**\mu}\Gamma_{\beta\mu}^{\sigma}\\+\Gamma_{\alpha\beta}^{**\sigma}\Gamma_{\sigma\mu}^{\mu}+\Gamma_{\alpha\beta}^{**\sigma}{}_{,\sigma}-\Gamma_{\beta\sigma}^{**\sigma}{}_{,\alpha}\\-\Gamma_{\sigma\alpha}^{**\mu}\Gamma_{\beta\mu}^{**\sigma}+\Gamma_{\alpha\beta}^{**\sigma}\Gamma_{\sigma\mu}^{**\mu}\end{pmatrix}\frac{g^{\alpha\beta}}{F[f]}$$

$$=\frac{R}{F[f]}+\begin{pmatrix}\frac{F'}{F[f]}\cdot\begin{pmatrix}-\Gamma_{\sigma\alpha}^{\mu}\Gamma_{\beta\mu}^{***\sigma}+\Gamma_{\alpha\beta}^{\sigma}\Gamma_{\sigma\mu}^{***\mu}\\-\Gamma_{\sigma\alpha}^{***\mu}\Gamma_{\beta\mu}^{\sigma}+\Gamma_{\alpha\beta}^{***\sigma}\Gamma_{\sigma\mu}^{\mu}\end{pmatrix}\\+\frac{F'}{F[f]}\cdot\left(\Gamma_{\alpha\beta}^{***\sigma}{}_{,\sigma}-\Gamma_{\beta\sigma}^{***\sigma}{}_{,\alpha}\right)\\+\left(\frac{F'}{F[f]}\right)_{,\sigma}\cdot\Gamma_{\alpha\beta}^{***\sigma}-\left(\frac{F'}{F[f]}\right)_{,\alpha}\cdot\Gamma_{\beta\sigma}^{***\sigma}\\+\left(\frac{F'}{F[f]}\right)^{2}\cdot\left(\Gamma_{\alpha\beta}^{***\sigma}\Gamma_{\sigma\mu}^{***\mu}-\Gamma_{\sigma\alpha}^{***\mu}\Gamma_{\beta\mu}^{***\sigma}\right)\end{pmatrix}\frac{g^{\alpha\beta}}{F[f]}. \tag{85}$$

How could this be understood in the metric picture?
To answer this question, we reshape equation (85) and write it as follows:

$$R=F[f]\cdot R^{*}-\begin{pmatrix}-\Gamma_{\sigma\alpha}^{\mu}\Gamma_{\beta\mu}^{**\sigma}+\Gamma_{\alpha\beta}^{\sigma}\Gamma_{\sigma\mu}^{**\mu}-\Gamma_{\sigma\alpha}^{**\mu}\Gamma_{\beta\mu}^{\sigma}\\+\Gamma_{\alpha\beta}^{**\sigma}\Gamma_{\sigma\mu}^{\mu}+\Gamma_{\alpha\beta}^{**\sigma}{}_{,\sigma}-\Gamma_{\beta\sigma}^{**\sigma}{}_{,\alpha}\\-\Gamma_{\sigma\alpha}^{**\mu}\Gamma_{\beta\mu}^{**\sigma}+\Gamma_{\alpha\beta}^{**\sigma}\Gamma_{\sigma\mu}^{**\mu}\end{pmatrix}g^{\alpha\beta}$$

$$=F[f]\cdot R^{*}-\begin{pmatrix}\frac{F'}{F[f]}\cdot\begin{pmatrix}-\Gamma_{\sigma\alpha}^{\mu}\Gamma_{\beta\mu}^{***\sigma}+\Gamma_{\alpha\beta}^{\sigma}\Gamma_{\sigma\mu}^{***\mu}\\-\Gamma_{\sigma\alpha}^{***\mu}\Gamma_{\beta\mu}^{\sigma}+\Gamma_{\alpha\beta}^{***\sigma}\Gamma_{\sigma\mu}^{\mu}\end{pmatrix}\\+\frac{F'}{F[f]}\cdot\left(\Gamma_{\alpha\beta}^{***\sigma}{}_{,\sigma}-\Gamma_{\beta\sigma}^{***\sigma}{}_{,\alpha}\right)\\+\left(\frac{F'}{F[f]}\right)_{,\sigma}\cdot\Gamma_{\alpha\beta}^{***\sigma}-\left(\frac{F'}{F[f]}\right)_{,\alpha}\cdot\Gamma_{\beta\sigma}^{***\sigma}\\+\left(\frac{F'}{F[f]}\right)^{2}\cdot\left(\Gamma_{\alpha\beta}^{***\sigma}\Gamma_{\sigma\mu}^{***\mu}-\Gamma_{\sigma\alpha}^{***\mu}\Gamma_{\beta\mu}^{***\sigma}\right)\end{pmatrix}g^{\alpha\beta}. \tag{86}$$

Now, as we know that the Ricci scalar R, which is based on the metric $g_{\alpha\beta}$, satisfies the vacuum Einstein field equations as outcome of the Einstein–

Hilbert action when variated with respect to the metric tensor $g_{\alpha\beta}$, we have to conclude that the same holds for the right hand side of (86). Without any form of simplification or approximation, we can now define the two addends on the right hand side as follows:

$$R = \overbrace{F[f]\cdot R^{*}}^{R_{\text{matter}}} - \overbrace{\begin{pmatrix} -\Gamma^{\mu}_{\sigma\alpha}\Gamma^{**\sigma}_{\beta\mu} + \Gamma^{\sigma}_{\alpha\beta}\Gamma^{**\mu}_{\sigma\mu} - \Gamma^{**\mu}_{\sigma\alpha}\Gamma^{\sigma}_{\beta\mu} + \Gamma^{**\sigma}_{\alpha\beta}\Gamma^{\mu}_{\sigma\mu} \\ +\Gamma^{**\sigma}_{\alpha\beta,\sigma} - \Gamma^{**\sigma}_{\beta\sigma,\alpha} - \Gamma^{**\mu}_{\sigma\alpha}\Gamma^{**\sigma}_{\beta\mu} + \Gamma^{**\sigma}_{\alpha\beta}\Gamma^{**\mu}_{\sigma\mu} \end{pmatrix} g^{\alpha\beta}}^{2\cdot\kappa\cdot L_M + 2\cdot\Lambda}$$

$$\equiv R_{\text{matter}} - 2\cdot\kappa\cdot L_M - 2\cdot\Lambda. \tag{87}$$

Comparing this with the classical Einstein–Hilbert action (63) with $f[R] = 1$, thereby knowing that R from (87) fulfills the corresponding equation (74) in the vacuum case (with both, L_M and Λ being zero), we come to the conclusion that our wrapper transformation (76) kind of separates a matter-free metric $g_{\alpha\beta}$, which is represented by the matter-free Ricci scalar R, into a matter-containing Ricci scalar $R_{\text{matter}} = F[f]\cdot R^{*}$ plus the corresponding Lagrange matter plus a Λ term. Setting $\Lambda = 0$ gives us just the matter Lagrange density as follows:

$$L_M = \frac{g^{\alpha\beta}}{2\cdot\kappa}\begin{pmatrix} -\Gamma^{\mu}_{\sigma\alpha}\Gamma^{**\sigma}_{\beta\mu} + \Gamma^{\sigma}_{\alpha\beta}\Gamma^{**\mu}_{\sigma\mu} - \Gamma^{**\mu}_{\sigma\alpha}\Gamma^{\sigma}_{\beta\mu} + \Gamma^{**\sigma}_{\alpha\beta}\Gamma^{\mu}_{\sigma\mu} \\ +\Gamma^{**\sigma}_{\alpha\beta,\sigma} - \Gamma^{**\sigma}_{\beta\sigma,\alpha} - \Gamma^{**\mu}_{\sigma\alpha}\Gamma^{**\sigma}_{\beta\mu} + \Gamma^{**\sigma}_{\alpha\beta}\Gamma^{**\mu}_{\sigma\mu} \end{pmatrix}. \tag{88}$$

We further conclude:

A. That all non-identical metric transformations like (75)—with (76) being its simplest form—are just acting in the same way and might therefore be considered as matter-creation operations (see [18] for such generalizations).

B. There should also be a reverse operation (transformation) $F[f]^{-1}$, which annihilates matter states into vacuum states.

C. Every matter ensemble L_M, in fact, has a $R_{\text{matter}} = F[f]\cdot R^{*}$ counterpart or partner and the two always neutralize to an R providing a vacuum solution to the Einstein field equations.

D. Stable matter forms are more or less permanent split-ups of L_M and $R_{\text{matter}} = F[f]\cdot R^{*}$.

But how does this lead us to a connection with the classical Quantum Theory?

Now one can show (c.f. [23] and references therein) that with (83) and "wrapped metrics" of the kind (76) all metrics of the type (2)—where usually classical Quantum Theory is defined on (regarding the generalization, see [18] or appendix I)—we obtain Ricci scalars of the form[6]:

$$R^{*}=\frac{1}{F[f]^{3}}\cdot\left(\begin{array}{c}\left(C_{N1}\cdot\left(\frac{\partial F[f]}{\partial f}\right)^{2}-C_{N2}\cdot F[f]\cdot\frac{\partial^{2}F[f]}{\partial f^{2}}\right)\cdot\left(\tilde{\nabla}_{\text{g-Coordinates}}f\right)^{2}\\ -C_{N2}\cdot F[f]\cdot\frac{\partial F[f]}{\partial f}\cdot\Delta_{\text{g-Coordinates}}f\end{array}\right).\tag{89}$$

This holds in an arbitrary number of dimensions. Thereby the symbol $\tilde{\nabla}_{\text{g-Coordinates}}$ denotes a first-order differential operator similar to the Nabla-operator in the metric $g_{\alpha\beta}$. The symbols C_{Ni} stand for constants which only depend on the number of dimensions. Without loss of generality, we can now assert that the first term in parentheses in (89) would be zero, which is to say, we choose our arbitrary function F such that we have:

$$C_{N1}\cdot\left(\frac{\partial F[f]}{\partial f}\right)^{2}-C_{N2}\cdot F[f]\cdot\frac{\partial^{2}F[f]}{\partial f^{2}}=0.\tag{90}$$

For instance, in 4 dimensions this would always be the case for $F[f] = f^2$, giving us the Klein–Gordon-like equation:

$$\left[\Delta_{g}+\frac{R^{*}\cdot f^{4}}{f\cdot C_{N2}\cdot 2\cdot f}\right]f=\left[\Delta_{g}+\frac{R^{*}\cdot f^{2}}{2}\right]f=0;\quad \Delta_{\text{g-Coordinates}}=\Delta_{g}.\tag{91}$$

Many examples and subsequent F-functions for other numbers of dimensions were given in [3, 126, 136, 138]. In the general case (arbitrary number of dimensions), with the condition (90), we obtain:

$$R^{*}+C_{N2}\cdot\frac{1}{F[f]^{2}}\cdot\frac{\partial F[f]}{\partial f}\cdot\Delta_{g}f=0$$

$$\Rightarrow\qquad\left(\frac{R^{*}\cdot F[f]^{2}}{C_{N2}\cdot f\cdot\frac{\partial F[f]}{\partial f}}+\Delta_{g}\right)f=0.\tag{92}$$

[6] Full derivation has been given in [9] and further below in this book.

We recognize the typical structure of the Klein–Gordon equation, whose original form can be given in arbitrary coordinates as:

$$\left[\Delta_g + \frac{M^2c^2}{\hbar^2} + V\right] f = 0, \tag{93}$$

with the first term in (92) (second equation) being seen as the corresponding metric equivalent for the mass and potential. This leads us to the identity:

$$\frac{M^2c^2}{\hbar^2} + V = \frac{R^* \cdot F[f]^2}{f \cdot C_{N2} \cdot \frac{\partial F[f]}{\partial f}}. \tag{94}$$

Thus, without any approximation, we have expressed the effective mass + potential-term of the Klein–Gordon equation as something solely dependent on the auxiliary or wrapped Ricci scalar R^* times a function of f.

This gives us now a rather basic connection between Quantum Theory and General Theory of Relativity in any number of dimensions. We see that no governing theory seems to be needed. Instead, the General Theory of Relativity (GTR) already contained the Quantum Theory. The Klein–Gordon equation in GTR should read:

$$\left[\Delta_g + \frac{R^* \cdot F[f]^2}{f \cdot C_{N2} \cdot \frac{\partial F[f]}{\partial f}}\right] f = 0. \tag{95}$$

It needs to be pointed out explicitly that our choice for $F[f]$ is no restriction to the whole, but only a technical trick. Any information residing in F would be taken on by f the moment we fix F for convenience as we did with condition (90).

A more general and comprehensive derivation and discussion with respect to the derivation of the Klein–Gordon equation out of the Einstein–Hilbert action using a wrapper-scaled metric tensor has been given in [18] and—partially—further below in this book.

6.5.1 Postulation of a Constant Ricci Scalar

In the attempt to motivate the construction of the classical quantum equations out of the Ricci scalar and thus, out of solutions to the Einstein field equations, we postulated a certain behavior for the Ricci scalar in [27]. Later [12–18], we were able to show that the postulation actually "could just be made" a result of the Einstein–Hilbert action variation when performed in the right way. Thus, the postulate, in fact, is none, but yet another variational result from Hilbert's stroke of genius [3]. Nevertheless, here we want to follow our earlier thoughts as starting points on the matter, because—mathematically—this is not much different from the rigorous variational approach (c.f. [12–18, 212, 214]).

So, by postulation of a constant Ricci scalar, which is to say an $R \rightarrow F[f]^* R^*$ that does not change under the transformation (75), we would obtain the following general condition:

$$F[f]\cdot R^{*}-R=0=\begin{pmatrix} -\Gamma^{\mu}_{\sigma\alpha}\Gamma^{**\sigma}_{\beta\mu}+\Gamma^{\sigma}_{\alpha\beta}\Gamma^{**\mu}_{\sigma\mu}-\Gamma^{**\mu}_{\sigma\alpha}\Gamma^{\sigma}_{\beta\mu}+\Gamma^{**\sigma}_{\alpha\beta}\Gamma^{\mu}_{\sigma\mu} \\ +\Gamma^{**\sigma}_{\alpha\beta,\sigma}-\Gamma^{**\sigma}_{\beta\sigma,\alpha}-\Gamma^{**\mu}_{\sigma\alpha}\Gamma^{**\sigma}_{\beta\mu}+\Gamma^{**\sigma}_{\alpha\beta}\Gamma^{**\mu}_{\sigma\mu} \end{pmatrix} g^{\alpha\beta}$$

$$\Rightarrow 0=\begin{pmatrix} \frac{F'}{F[f]}\cdot\left(-\Gamma^{\mu}_{\sigma\alpha}\Gamma^{***\sigma}_{\beta\mu}+\Gamma^{\sigma}_{\alpha\beta}\Gamma^{***\mu}_{\sigma\mu}-\Gamma^{***\mu}_{\sigma\alpha}\Gamma^{\sigma}_{\beta\mu}+\Gamma^{***\sigma}_{\alpha\beta}\Gamma^{\mu}_{\sigma\mu}\right) \\ +\frac{F'}{F[f]}\cdot\left(\Gamma^{***\sigma}_{\alpha\beta,\sigma}-\Gamma^{***\sigma}_{\beta\sigma,\alpha}\right)+\left(\frac{F'}{F[f]}\right)_{,\sigma}\cdot\Gamma^{***\sigma}_{\alpha\beta}-\left(\frac{F'}{F[f]}\right)_{,\alpha}\cdot\Gamma^{***\sigma}_{\beta\sigma} \\ +\left(\frac{F'}{F[f]}\right)^{2}\cdot\left(\Gamma^{***\sigma}_{\alpha\beta}\Gamma^{***\mu}_{\upsilon\mu}-\Gamma^{***\mu}_{\sigma\alpha}\Gamma^{***\sigma}_{\beta\mu}\right) \end{pmatrix} g^{\alpha\beta}$$

$$\Gamma^{***\gamma}_{\alpha\beta}\equiv\frac{f_{,\beta}\cdot\delta^{\gamma}_{\alpha}+f_{,\alpha}\cdot\delta^{\gamma}_{\beta}-g^{\gamma\sigma}\cdot f_{,\sigma}\cdot g_{\alpha\beta}}{2}=:\quad F'=\frac{\partial F[f]}{\partial f}, \tag{96}$$

where we have used the result (86) and the abbreviation introduced in (84). The second line in (96) might be seen as the general quantum gravity equation based on the simplest possible transformation (76) with f being a function of all coordinates. Thus, we might consider $f = f[...]$ as the quantum gravity wave function. However, having just seen that the term:

$$\begin{pmatrix} -\Gamma^{\mu}_{\sigma\alpha}\Gamma^{**\sigma}_{\beta\mu}+\Gamma^{\sigma}_{\alpha\beta}\Gamma^{**\mu}_{\sigma\mu}-\Gamma^{**\mu}_{\sigma\alpha}\Gamma^{\sigma}_{\beta\mu}+\Gamma^{**\sigma}_{\alpha\beta}\Gamma^{\mu}_{\sigma\mu}+\Gamma^{**\sigma}_{\alpha\beta,\sigma}-\Gamma^{**\sigma}_{\beta\sigma,\alpha} \\ -\Gamma^{**\mu}_{\sigma\alpha}\Gamma^{**\sigma}_{\beta\mu}+\Gamma^{**\sigma}_{\alpha\beta}\Gamma^{**\mu}_{\sigma\mu} \end{pmatrix} \tag{97}$$

is proportional to the matter Lagrange density (c.f. (88)), condition (96) would always demand this term to be zero. As this seems to be a too rigid condition, we try for a different one which leaves us with a bit more freedom.

Please note that our current postulation could also be one where the original R just shall be a constant. In this case we would have:

$$R = \text{const}$$

$$= F[f]\cdot R^{*} - \begin{pmatrix} -\Gamma^{\mu}_{\sigma\alpha}\Gamma^{**\sigma}_{\beta\mu} + \Gamma^{\sigma}_{\alpha\beta}\Gamma^{**\mu}_{\sigma\mu} - \Gamma^{**\mu}_{\sigma\alpha}\Gamma^{\sigma}_{\beta\mu} + \Gamma^{**\sigma}_{\alpha\beta}\Gamma^{\mu}_{\sigma\mu} \\ +\Gamma^{**\sigma}_{\alpha\beta\,,\sigma} - \Gamma^{**\sigma}_{\beta\sigma\,,\alpha} - \Gamma^{**\mu}_{\sigma\alpha}\Gamma^{**\sigma}_{\beta\mu} + \Gamma^{**\sigma}_{\alpha\beta}\Gamma^{**\mu}_{\sigma\mu} \end{pmatrix} g^{\alpha\beta}$$

$$0 = F[f]\cdot R^{*} - R - \begin{pmatrix} \frac{F'}{F[f]}\cdot\begin{pmatrix} -\Gamma^{\mu}_{\sigma\alpha}\Gamma^{***\sigma}_{\beta\mu} + \Gamma^{\sigma}_{\alpha\beta}\Gamma^{***\mu}_{\sigma\mu} \\ -\Gamma^{***\mu}_{\sigma\alpha}\Gamma^{\sigma}_{\beta\mu} + \Gamma^{***\sigma}_{\alpha\beta}\Gamma^{\mu}_{\sigma\mu} \end{pmatrix} \\ +\frac{F'}{F[f]}\cdot\left(\Gamma^{***\sigma}_{\alpha\beta\,,\sigma} - \Gamma^{***\sigma}_{\beta\sigma\,,\alpha}\right) \\ +\left(\frac{F'}{F[f]}\right)_{,\sigma}\cdot\Gamma^{***\sigma}_{\alpha\beta} - \left(\frac{F'}{F[f]}\right)_{,\alpha}\cdot\Gamma^{***\sigma}_{\beta\sigma} \\ +\left(\frac{F'}{F[f]}\right)^{2}\cdot\left(\Gamma^{***\sigma}_{\alpha\beta}\Gamma^{***\mu}_{\sigma\mu} - \Gamma^{***\mu}_{\sigma\alpha}\Gamma^{***\sigma}_{\beta\mu}\right) \end{pmatrix} g^{\alpha\beta}. \qquad (98)$$

which would require us to have some idea about the transformed Ricci scalar R^{*}. Now we see that matter is created in cases where the resulting Ricci difference:

$$\Delta R \equiv F[f]\cdot R* - R \qquad (99)$$

does not vanish. In generalization of this postulation, we might simply demand that the sum of Ricci scalar, Lagrange density of matter and cosmological constant as given in the kernel of the classical Einstein–Hilbert action (65) stays constant, respectively that its variation would give zero. Then the condition (98) might be seen as follows:

$$R - 2kL_M - 2\Lambda = 0$$

$$= R - F[f]\cdot R^{*} + \begin{pmatrix} \frac{F'}{F[f]}\cdot\begin{pmatrix} -\Gamma^{\mu}_{\sigma\alpha}\Gamma^{***\sigma}_{\beta\mu} + \Gamma^{\sigma}_{\alpha\beta}\Gamma^{***\mu}_{\sigma\mu} \\ -\Gamma^{***\mu}_{\sigma\alpha}\Gamma^{\sigma}_{\beta\mu} + \Gamma^{***\sigma}_{\alpha\beta}\Gamma^{\mu}_{\sigma\mu} \end{pmatrix} \\ +\frac{F'}{F[f]}\cdot\left(\Gamma^{***\sigma}_{\alpha\beta\,,\sigma} - \Gamma^{***\sigma}_{\beta\sigma\,,\alpha}\right) \\ +\left(\frac{F'}{F[f]}\right)_{,\sigma}\cdot\Gamma^{***\sigma}_{\alpha\beta} - \left(\frac{F'}{F[f]}\right)_{,\alpha}\cdot\Gamma^{***\sigma}_{\beta\sigma} \\ +\left(\frac{F'}{F[f]}\right)^{2}\cdot\left(\Gamma^{***\sigma}_{\alpha\beta}\Gamma^{***\mu}_{\sigma\mu} - \Gamma^{***\mu}_{\sigma\alpha}\Gamma^{***\sigma}_{\beta\mu}\right) \end{pmatrix} g^{\alpha\beta}. \qquad (100)$$

This interpretation has been investigated previously (e.g., [138]) and will be briefly discussed in the next section. But first we want to present a brief summary of our findings so far.

6.5.2 Intermediate Sum-Up

In principle, we have only found a triviality, namely:

A. The transformation $G_{\alpha\beta} = F[f] \cdot \delta^i_\alpha \delta^j_\beta g_{ij}$ (seen as the simplest form of $G_{\alpha\beta} = F^{ij}_{\alpha\beta} g_{ij}$) of a metric solution g_{ij} of the vacuum Einstein field equations, in general, leads to a new metric which only solves the Einstein field equations containing matter.

B. Thereby the matter being created is well defined via the Einstein–Hilbert Lagrange density given in (88), respectively (without the abbreviations hiding all the F-dependencies):

$$L_M = \frac{g^{\alpha\beta}}{2\cdot\kappa}\left(\begin{array}{l} \frac{F'}{F[f]}\cdot\left(-\Gamma^{\mu}_{\sigma\alpha}\Gamma^{***\sigma}_{\beta\mu} + \Gamma^{\sigma}_{\alpha\beta}\Gamma^{***\mu}_{\sigma\mu} - \Gamma^{***\mu}_{\sigma\alpha}\Gamma^{\sigma}_{\beta\mu} + \Gamma^{***\sigma}_{\alpha\beta}\Gamma^{\mu}_{\sigma\mu}\right) \\ +\frac{F'}{F[f]}\cdot\left(\Gamma^{***\sigma}_{\alpha\beta\ ,\sigma} - \Gamma^{***\sigma}_{\beta\sigma\ ,\alpha}\right) \\ +\left(\frac{F'}{F[f]}\right)_{,\sigma}\cdot\Gamma^{***\sigma}_{\alpha\beta} - \left(\frac{F'}{F[f]}\right)_{,\alpha}\cdot\Gamma^{***\sigma}_{\beta\sigma} \\ +\left(\frac{F'}{F[f]}\right)^2\cdot\left(\Gamma^{***\sigma}_{\alpha\beta}\Gamma^{***\mu}_{\sigma\mu} - \Gamma^{***\mu}_{\sigma\alpha}\Gamma^{***\sigma}_{\beta\mu}\right) \end{array}\right). \tag{101}$$

C. The corresponding new, which is to say transformed, Ricci scalar R^* now satisfies the Einstein field equations in connection with the previous, untransformed metric $g_{\alpha\beta}$ and the matter Lagrange density L_M. The corresponding Einstein–Hilbert action (78) (last line) becomes:

$$\delta_g W = ? = 0 = \delta_g \int_V d^n x \left(\sqrt{-g}\cdot\Phi_{R^*}\left[R^*\right]\right)$$

$$= \delta_g \int_V d^n x \left(\sqrt{-g}\cdot\left(\overbrace{F[f]\cdot R^*}^{R_{\text{matter}}} - 2\cdot\kappa\cdot L_M\right)\right). \tag{102}$$

6.6 The Other Side of the Coin

As stated at the end of the section 6.4, "How to Quantize Solutions of the Einstein field equations—The State-of-the-Art Situation," we still have not found a way to fundamentally derive the need for the quantum equations out of our conditions to the Ricci scalar. The simple fact that these equations fall out after certain transformations on the metric is no justification for actually doing this transformation. It could only be seen as a piece of circumstantial evidence.

We now want to achieve a further reasoning for the metric transformation procedure by extending the variation of the Einstein–Hilbert action (65). Thereby, instead of performing the variation only with respect to the metric tensor $g_{\alpha\beta}$, we also variate with respect to another suitable parameter.

It appears reasonable to start with the number of dimensions n. Performing this additional variation via:

$$\delta_{n,g} W = \left(\frac{\partial W}{\partial n} \cdot \delta n + \frac{\partial W}{\partial g_{\alpha\beta}} \cdot \delta g_{\alpha\beta} \right) = 0$$

$$\Rightarrow \qquad 0 = \delta_n \int_V d^n x \left(\sqrt{-g} \left[R - 2\kappa L_M - 2\Lambda \right] \right)$$

$$+ \delta_g \int_V d^n x \left(\sqrt{-g} \left[R - 2\kappa L_M - 2\Lambda \right] \right) \tag{103}$$

not only reproduces the classical Einstein field equations (66) but also delivers:

$$0 = \int_S n \cdot d^{n-1} x \left(\sqrt{-g} \left[R - 2\kappa L_M - 2\Lambda \right] \right) \cdot \delta n$$

$$+ \int_V d^n x \frac{\partial \left(\sqrt{-g} \left[R - 2\kappa L_M - 2\Lambda \right] \right)}{\partial n} \cdot \delta n$$

$$\Rightarrow \qquad 0 = R - 2\kappa L_M - 2\Lambda$$

$$\Rightarrow \qquad \frac{\partial \left(\sqrt{-g} \left[R - 2\kappa L_M - 2\Lambda \right] \right)}{\partial n} = 0, \tag{104}$$

which contains, as we have seen above, the classical quantum equations.

Thus, with the third line in (104) and the subsequent variation in (63) or (65) leading to the classical Einstein field equations (74), the latter should then simply be seen as the variation of an intelligent zero (c.f. [43]). However, as we could also apply the Hilbert argument of the surface integral being chosen such that the result would give zero on the rim S (after all, the surface on which the integration is performed would be arbitrary), we would only be left with the condition in the fourth line. However, as this condition could be split up into:

$$0=\frac{\partial\left(\sqrt{-g}\left[R-2\kappa L_M-2\Lambda\right]\right)}{\partial n}$$

$$=\left[R-2\kappa L_M-2\Lambda\right]\cdot\frac{\partial\sqrt{-g}}{\partial n}+\sqrt{-g}\cdot\frac{\partial\left[R-2\kappa L_M-2\Lambda\right]}{\partial n}, \qquad (105)$$

we would again be back with the third line in (104) and thus, our intelligent zero variation.

For later in this book, especially in connection with the Bekenstein thought experiment and the question how black holes can store information (see section 6.7, "Schwarzschild Metric as an Example," and section 6.8, "The Bekenstein Bit Problem," further below in this chapter), we need to record the possibility of variations with respect to different physical objects than just the metric tensor.

Further such variational options are presented and discussed in [3] and [18].

6.7 Schwarzschild Metric as an Example

When observing the integral in (104), we see that—in principle—we also seek a maximum volume for a given dimension or, taking the radius of a Schwarzschild object, look for the corresponding dimension making the volume integral an extremum. As the determinant g of the Schwarzschild metric is just equal to the one of a n-sphere with the additional time-dimension to be integrated, we can easily use the volume integral result of n-spheres, which reads:

$$V_N=V_{n+1}=\frac{\pi^{\frac{(n)}{2}}}{\Gamma\left[\frac{(n+2)}{2}\right]}\cdot r^n. \qquad (106)$$

Please note that due to the time-coordinate t, we have V_{n+1} instead of V_n. Thereby the integration via t in (106) is assumed to be performed such that it would give 1. In general, we might take care of this part of the integration via a proportional constant T we could even consider to be n-dependent $T[n]$ and thus:

$$V_N = V_{n+1} = T[n] \cdot \frac{\pi^{\frac{(n)}{2}}}{\Gamma\left[\frac{(n+2)}{2}\right]} \cdot r^n. \tag{107}$$

Now we evaluate the various dimensions for which, for a given radius r of the n-sphere, we would obtain extrema. The results were already given in Fig. 5.5 in chapter 5 and are shown again as dots in Fig. 6.2. There we have illustrated the resulting $r_{\max}$ as functions of the dimensions $N = n + 1$ (*note*: n = n-sphere dimension, $N = t + n$-sphere dimension).

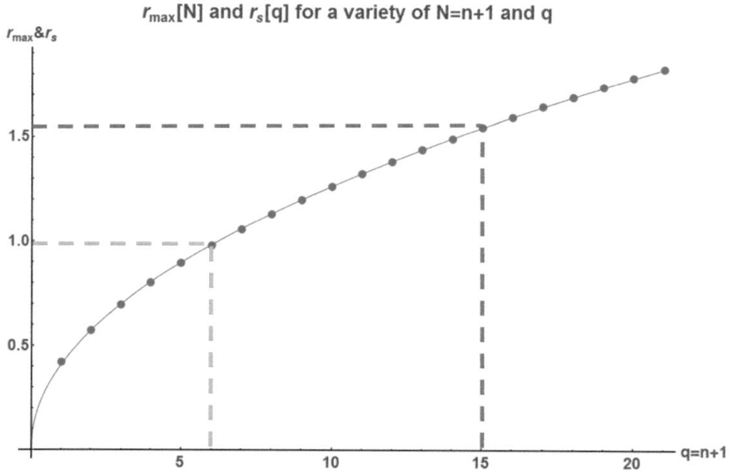

Figure 6.2 Radius $r_{\max}$ for which at a certain number of dimensions the n-sphere has maximum volume in dependency on $N = n + 1$ compared with the increase of the Schwarzschild radius r_s of a black hole in dependence on the number of bits q thrown into it. We find that $q = N = n + 1$. As an example, we pick the situation with a radius slightly bigger than 1.5 (whatever unit). We obtain the maximum volume for a sphere in 15 dimensions (orange dotted line). Picking a radius slightly below 1, however, gives us a 6-dimensional sphere which can have the maximum volume at such a size (green dotted line).

6.8 The Bekenstein Bit Problem

Before actually working out the principal metric solutions fit for the construction of a most fundamental computer, we have to take care of the problem of how the universe stores information. One of the most famous problems in this context is seen in the Bekenstein Bit problem, where it was found that black holes can store information, but so far it has been a mystery how these objects actually do this. In [3, 18, 144], we have shown that bit-like information is stored as dimensions and that each bit becomes 1 dimension. For convenience, we are repeating parts of the evaluation here.

In the early seventies, J. Bekenstein [182, 183] investigated the connection between black hole surface area and information. Thereby he simply considered the surface changes of a black hole which would be hit by a photon just of the same size as the black hole. His idea was that with such a geometric constellation the outcome of the information would just consist of the information whether the photon fell into the black hole or whether it did not. Thus, it would be a 1-bit information. His calculations led him to the funny proportionality of area and information. He found that the number of bits, coded by a certain black hole, is proportional to the surface area of this very black hole if measured in Planck area ℓ_P^2. In fact the dependency how one bit of information changes the area of the black hole (ΔA) reads:

$$\Delta A = 32 \cdot \pi^2 \cdot \ell_P^2 + 64 \cdot \pi^3 \cdot \frac{\ell_P^4}{r_s^2}. \tag{108}$$

Ignoring the extremely small second term, one could just assume our black hole to be constructed of many such bit surface pieces. Thus, we could write:

$$q \cdot \Delta A = q \cdot 32 \cdot \pi^2 \cdot \ell_P^2 = 4 \cdot \pi \cdot r_s^2 \quad \Rightarrow \quad r_s^2 = q \cdot 8 \cdot \pi \cdot \ell_P^2, \tag{109}$$

where r_s gives the radius of the black hole. We see that our black hole radius is proportional to the square root of the bits q thrown into it.

Now we want to compare the dependency $r_s[q]$ with the radii $r_{\max}[N]$ resulting in the maximum volume of n-spheres for a certain number of space-time dimensions $N = n + 1$.

We find a perfect fit (s. Fig. 6.2) to the $r_{\max}[N]$-dependency for $q = N$ with the following function:

$$r_s = U \cdot \left(0.014948 + 0.3951244 \cdot \sqrt{q}\right); \quad U^2 = 8 \cdot \pi \cdot \ell_P^2, \tag{110}$$

where U denotes a unit-factor which was set $U = 1$ in Fig. 6.2.

Our finding not only connects the intrinsic dimension of a black hole with its mass respectively its surface but also gives an explanation to the hitherto unsolved problem of "what are the micro states of a black hole giving it temperature and allowing it to store information." According to the evaluation in this section, these microstates are just various states of dimensions realized within the black hole in dependence on the number of bits it contains (and thus, its mass). The bigger the number of bits, the higher the intrinsic dimensions the black hole has. In fact, the connection even is a direct one and only seems to deviate from the simple direct proportionality for very low numbers of masses[7], respectively Schwarzschild radii r_s, respectively numbers of bits q the black hole has swallowed.

This finding also gives us a direct connection between a principal mathematical law (the maximum volume as function of the dimension for a given radius of an n-sphere) to the number of bits a black hole contains, to the mass or Schwarzschild radius of this very black hole and the number and character of microstates the black hole actually uses to internally code the bits.

It has to be pointed out that the expression "intrinsic dimension" truly stands for the part of space for $r < r_s$, which is to say, the space behind the event horizon. As for the outside, the solution of a Schwarzschild object in $n + 1$-dimensional space-times is given via:

$$g_{\alpha\beta}^{N} = \begin{pmatrix} -c^2 \cdot f[r] & 0 & 0 & 0 & \cdots & 0 \\ 0 & \dfrac{1}{f[r]} & 0 & 0 & \cdots & 0 \\ 0 & 0 & r^2 & 0 & \cdots & 0 \\ 0 & 0 & 0 & r^2 \cdot \sin^2 \varphi_1 & \cdots & 0 \\ \cdots & \cdots & \cdots & \cdots & \ddots & 0 \\ 0 & 0 & 0 & 0 & 0 & g_{aa} \end{pmatrix}$$

$$g_{44} = r^2 \cdot \sin^2 \varphi_1 \cdot \sin^2 \varphi_2; \quad g_{aa} = r^2 \cdot \prod_{j=1}^{a-2} \sin^2 \varphi_j; \quad a = N - 1 = n;$$

$$f[r] = 1 - \frac{r_s^{N-3}}{r^{N-3}} = 1 - \frac{r_s^{n-2}}{r^{n-2}}. \tag{111}$$

[7]Besides, this deviation is also suggested by the Bekenstein finding summed up in equation (108), where we could assume the second term to become important at lower numbers of r_s.

The interior, now assumed to "consist" of a multidimensional something, will probably be in the vicinity of the extended mass-containing de Sitter objects we have suggested in [144, 156–158]. For completeness, it shall be noted that the corresponding metrics were given in [13, 18] and, for convenience, will be repeated in the appendix to this chapter.

From (61), we can now extract the minimum Schwarzschild radius for the storage of one bit, which would be equal to 5.508 Planck length and corresponds to 11.16 times the Planck mass. This is a huge amount of mass and thus, also energy, one needs to safely store just one single bit. Luckily, the situation improves the more bits one intends to store, as, for instance, the one millionth bit only requires about 5.5×10^{-3} Planck masses. Please note that, of course, one might also store bits within spin arrangements of electrons. Then a 1-bit information would be connected with a single electron, whose mass and spin energy is many magnitudes below the Planck mass. This spin storage, however, cannot be seen as a storage of a classical binary bit, because, in fact, it resembles a quantum bit. Apparently, the safe storage of a pure and truly binary information requires an—almost—macroscopic massive structure. Here, the black hole probably provides the smallest possible mass ensemble there can be to arrange such a storage for a certain bit. The limit is given at about 11 times the Planck mass. Only from this mass onward, black holes can store binary information ... at least until the Hawking radiation leads to the destruction of our black hole 1-bit storage system.

And what would then happen to the stored information? Well, this brings us back to the question: How safe is information within our universe [184]?

6.9 The Evidence—Summed up

In this section, we will sum up the evidence that in fact Hilbert, respectively Einstein—in principle—"already had it ..." (c.f. our series "Einstein already had it ..." in the list of references). Quantum Theory and Einstein's General Theory of Relativity belong together. They do not need to be unified as they are already one entity. In fact, they are the two sides of the same coin and the reason for the widespread opinion that they are "so very different that they cannot be brought into one theoretical apparatus" resulted from the fact that the underlying principle was not seen in a holistic enough manner (c.f. section 6.6, "The Other Side of the Coin").

Here now we want to give proof that all quantum equations and effects reside inside the Einstein Theory, which becomes visible when allowing for a scaled metric of the type (5). We will do so in most compact manner, which is to say, we will keep things very brief here and repeat the more important derivations in a more comprehensive manner in the appendices of this book.

6.9.1 The Metric Origin of the Klein–Gordon Equation

The metric transformation (75) directly results in a Ricci scalar (96) resembling the Klein–Gordon equation (*note*: C_{N2} is a constant which depends on the dimension, it was derived in [9]):

$$G_{\alpha\beta} = F\left[f\left[t,x,y,z\right]\right]\cdot\delta_{\alpha}^{i}\delta_{\beta}^{j}g_{ij} \quad\Leftrightarrow\quad \left[\Delta_g + \overbrace{\frac{R^{*}\cdot F\left[f\right]^{2}}{f\cdot C_{N2}\cdot\frac{\partial F\left[f\right]}{\partial f}}}^{=\frac{M^2c^2}{\hbar^2}+V}\right]f = 0$$

$$0 = \frac{\partial\left(\sqrt{-g}\left[R-2\kappa L_M-2\Lambda\right]\right)}{\partial n}$$

$$= \left[R-2\kappa L_M-2\Lambda\right]\cdot\frac{\partial\sqrt{-g}}{\partial n}+\sqrt{-g}\cdot\frac{\partial\left[R-2\kappa L_M-2\Lambda\right]}{\partial n}. \tag{112}$$

Thereby, as shown in section 6.6, "The Other Side of the Coin," the "driving force" for the extremalization of the Ricci scalar results from the classical Einstein–Hilbert action via a generalized variation [5–18, 212, 214, 218], whereby we here only have considered the additional[8] variation with respect to the dimension of space.

The non-linear term $\frac{R^{*}\cdot F\left[f\right]^{2}}{f\cdot C_{N2}\cdot\frac{\partial F\left[f\right]}{\partial f}}$ in the first line in (112) could be seen as controversial, because it does not automatically provide the usual linearity. However, it was shown in [3, 5, 154, 156–158] and many other earlier papers by this author that a setting $R^{*} = 0$ (flat space) or a clever splitting of the variation [5–18, 212, 214, 218] together with the introduction of suitably chosen additional (potentially entangled) dimensions would solve the problem and perfectly reproduce the classical and completely linear Klein–Gordon equation. A more comprehensive derivation (e.g., appendices

[8]Variation in addition to the metric tensor g_{ij}.

A and I) then also shows how we can better understand apparently wired results from quantum mechanics. We will also automatically obtain masses via entanglement of dimensions and scalar fields, like the Higgs field. For completeness, we are going to repeat the derivation of the mass term in the following two sub-sections.

6.9.2 The Metric Origin of Scalar Fields like the Higgs Field and Symmetry Breaking

In [156], we were able to show that a Minkowski metric:

$$g_{\alpha\beta}=\begin{pmatrix} -c^2 & 0 & 0 & 0 \\ 0 & 1 & 0 & 0 \\ 0 & 0 & 1 & 0 \\ 0 & 0 & 0 & 1 \end{pmatrix}, \tag{113}$$

which we subject to the following transformation:

$$G_{\alpha\beta}=F\left[f\left[t,x,y,z\right]\right]_{\alpha\beta}^{ij} g_{ij} \to G_{\alpha\beta}=F\left[f\left[t,x,y,z\right]\right]\cdot \delta_{\alpha}^{i}\delta_{\beta}^{j} g_{ij}, \tag{114}$$

results in the following Ricci scalar:

$$R^{*}=\frac{1}{F\left[f\right]^{3}}\cdot\left(-C_{N2}\cdot F\left[f\right]\cdot\frac{\partial F\left[f\right]}{\partial f}\cdot\Delta_{g}f^{2}\right)$$

$$\xrightarrow{4D} R^{*}\cdot\frac{\left(2\cdot C_{2}+C_{1}\cdot f^{2}\right)^{3}}{8\cdot C_{2}\cdot C_{1}}=-3\cdot\Delta_{g}f^{2}. \tag{115}$$

Please note that the functional wrapper *F*[*f*] reading:

$$F\left[f\right]=C_{1}\cdot f^{2}+\frac{C_{1}^{2}\cdot f^{4}}{4\cdot C_{2}}+C_{2}, \tag{116}$$

with arbitrary constants C_i, assures the appearance of linear Laplace operator terms for the resulting Ricci scalar of the transformed metric in (114) (c.f. condition (90) and solve the equation for *F*[*f*] in 4 dimensions).

Now we assume *f* to be a constant, which automatically leads to the simple equation:

$$R^{*}\cdot\frac{\left(2\cdot C_{2}+C_{1}\cdot f^{2}\right)^{3}}{8\cdot C_{2}\cdot C_{1}}=0. \tag{117}$$

Also assuming that the Ricci curvature R^* shall be proportional to f^n, with an arbitrary exponent $n \geq 0$, we obtain the familiar trivial solution of $f_0 = 0$. However, from (117) we also obtain the non-trivial ground states:

$$f^2 = -\frac{2 \cdot C_2}{C_1}. \tag{118}$$

As the constants C_i are arbitrary, we could simply set them as follows:

$$2 \cdot C_2 = -\mu^2; \quad C_1 = 2 \cdot \lambda. \tag{119}$$

This gives us the additional ground state solutions directly in the classical Higgs field style, namely:

$$\left(f_{1,2}\right)^2 = \frac{\mu^2}{2 \cdot \lambda} \quad \Rightarrow \quad f_{1,2} = \pm \frac{\mu}{\sqrt{2 \cdot \lambda}}. \tag{120}$$

The measured value for the $f_{1,2}$ is known to be (en.wikipedia.org/wiki/Higgs_boson):

$$\left|f_{1,2}\right| = \frac{\mu}{\sqrt{2 \cdot \lambda}} = \frac{246\,GeV}{\sqrt{2} \cdot c^2}. \tag{121}$$

Rewriting (117) with the use of (119) as (p is an arbitrary exponent):

$$R^* \cdot \frac{\left(-\mu^2 + 2 \cdot \lambda \cdot f^2\right)^3}{8 \cdot \mu^2 \cdot 2 \cdot \lambda} = f^p \cdot \frac{\left(2 \cdot \lambda \cdot f^2 - \mu^2\right)^3}{16 \cdot \mu^2 \cdot \lambda} = 3 \cdot \Delta_g f^2 \tag{122}$$

gives us the total Higgs–Ricci curvature connection. It also, automatically, gives us a curvature value R^* at the ground state, which is:

$$R^* \sim \left|f_{1,2}\right|^p = \left(\frac{\mu}{\sqrt{2 \cdot \lambda}}\right)^p = \left(\frac{246\,GeV}{\sqrt{2} \cdot c^2}\right)^p. \tag{123}$$

Now, with the Higgs mass m_H known to be 125 GeV/c^2 and the fact that we have [185]:

$$m_H = \sqrt{2 \cdot \mu^2} = \sqrt{4 \cdot \lambda \cdot \left(f_{1,2}\right)^2}, \tag{124}$$

we can obtain:

$$\lambda \simeq 0.13; \quad \mu \simeq 88.8 \text{ GeV/c}^2. \tag{125}$$

A dimensional analysis in (117) shows that an assumption of $R^* \sim f^p$ ($p \geq 0$) requires the introduction of a constant q, with the dimension 1/(length^2 × mass^{2+p}), resulting in:

$$q \cdot f^{p} \cdot \frac{\left(2 \cdot \lambda \cdot f^{2} - \mu^{2}\right)^{3}}{48 \cdot \mu^{2} \cdot \lambda} = \Delta_{g} f^{2}. \tag{126}$$

From (117) and the results given in [138], we can directly deduce that there are Higgs field-options also in other dimensions and that symmetry breaking can occur in arbitrary N-dimensional space-times except for $N = 1,2$.

Knowing that the dimension of the cosmological constant Λ is length^{-2}, we might rewrite (126) as follows:

$$\frac{\Lambda}{m_{q}^{2+p}} \cdot f^{p} \cdot \frac{\left(2 \cdot \lambda \cdot f^{2} - \mu^{2}\right)^{3}}{48 \cdot \mu^{2} \cdot \lambda} = \Delta_{g} f^{2}. \tag{127}$$

Here m_q shall denote a constant mass. For $p = 0$ the Ricci curvature R^* becomes independent on f and the option for a symmetric ground state with constant f vanishes. In this case, the occurrence of a Higgs field with broken symmetry is not an option, but a must.

6.9.3 The Metric Origin of Mass

It was shown in [3, 5] and even earlier in [156] and [157] that mass automatically results from certain entanglements of dimensions. As an example, we choose a 6-dimensional space-time in t, x, y, z, u, v and entangle the coordinates u and v as shown in the appendix to this chapter. We obtain the following Ricci scalar:

$$R^{*} = \frac{1}{F[f]^{3}} \cdot \left(-C_{N2} \cdot F[f] \cdot \frac{\partial F[f]}{\partial f} \cdot \Delta_{g} f\right); \quad F[f] = C_{f} + f[t,r,\vartheta,\varphi]$$

$$\xrightarrow{6D} R^{*} = \frac{5}{f^{2}} \cdot \left(\frac{C_{v0}^{2}}{5} \cdot f + \frac{\partial^{2} f}{c^{2} \cdot \partial t^{2}} - \Delta_{\text{3D-sphere}} f\right); \quad f = f[t,r,\vartheta,\varphi]; \quad C_{f} = 0. \tag{128}$$

Here $\Delta_{\text{3D-sphere}} f$ denotes the Laplace operator in spherical coordinates.

As for $R^* = 0$ and $\frac{C_{v0}^{2}}{5} = \frac{m^{2} \cdot c^{2}}{\hbar^{2}}$ (m = mass, c = speed of light in vacuum, $\hbar$ = reduced Planck constant), we clearly have obtained the quantum Klein–Gordon equation and we can easily conclude that mass can obviously be constructed by a suitable set of additional, entangled coordinates. As the whole, in fact, is a distributed field, we also—along the way—have obtained the Higgs field [154]. Because the function $g[v]$ just gives a bell-shaped localized distribution, it is also assured that the additional coordinates u and v are compactified, because the arbitrary constants C_{v1} and C_{v2} can always be chosen in such a way that $g[v]$ takes on the character of a Dirac delta distribution. Thus, the coordinates would be omnipresent in the space-time t, r, ϑ, φ, but do not appear in elongated form (on a bigger scale), which means that u and v are invisible but only act via the production of mass in a 3 + 1 space-time.

6.9.4 The Metric Origin of Spin and Spin Fields

Again, the recipe to evaluate the intrinsic parameter of spin requires the use of entangled dimensions [157]. However, as the evaluation is perfectly similar to the one we performed in connection with the mass and the Higgs field, we refer to the last sub-section and the appendix to this chapter.

6.9.5 The Metric Schrödinger Equation

As it can be shown (e.g., [41]) that the Schrödinger equation [186, 187] results from the Klein–Gordon equation by the means of a transformation of the kind:

$$f\left[t,x,y,z\right]=\Psi=\Phi+\mathrm{X}, \tag{129}$$

where we demand the following additional condition:

$$\partial_t\Psi=c^2\cdot C2\cdot(\Phi-\mathrm{X}). \tag{130}$$

Together with (128) and the setting $\mu_m^2=\frac{C_{v0}^2}{5}=\frac{m^2\cdot c^2}{\hbar^2}$, we obtain:

$$0=\left[\mu_m^2(\Phi+\mathrm{X})+\left(C1\cdot C2\cdot\partial_t(\Phi-\mathrm{X})+\frac{1}{\sqrt{g}}\partial_\alpha\sqrt{g}\cdot g^{\alpha\beta}\partial_\beta(\Phi+\mathrm{X})\right)\right]. \tag{131}$$

The reader may check that the following two equations summed up would result in (131):

$$0=\left[\Phi\left\{\mu_m^2\right\}+\left(C1\cdot C2\cdot\partial_t\Phi+\frac{1}{2\sqrt{g}}\partial_\alpha\sqrt{g}\cdot g^{\alpha\beta}\partial_\beta(\Phi+\mathrm{X})\right)\right]$$

$$0=\left[\mathrm{X}\left\{\mu_m^2\right\}+\left(-C1\cdot C2\cdot\partial_t\mathrm{X}+\frac{1}{2\sqrt{g}}\partial_\alpha\sqrt{g}\cdot g^{\alpha\beta}\partial_\beta(\Phi+\mathrm{X})\right)\right].\qquad(132)$$

Comparison with the original Schrödinger equation as given in the form below

$$\left[-i\cdot\hbar\cdot\partial_t-\frac{\hbar^2}{2M}\Delta_{\text{Schrödinger}}+V\right]\Psi=0\qquad(133)$$

not only gives us the matter and anti-matter solutions but also shows us—as seen and discussed before [149]—that the potential V has now been taken on by the metric and the proportionality factor μ_{m}. Thus, a distorted metric acts like an effective potential in the classical Schrödinger approximation and vice versa, which is to say: What in classical physics has been described as a potential would now become a distorted metric providing the necessary interaction. Disregarding the anti-matter solution here, the constants $C1$, $C2$ and the proportionality factor can be obtained as:

$$C1\cdot C2=-\frac{i\cdot m}{\hbar};\quad X=0$$

$$\Rightarrow\qquad\left[-\frac{1}{2}\Delta_{\text{Schrödinger}}+V\right]\Psi=\left[\Phi\cdot\mu_m^2+\left(\frac{1}{2\sqrt{g}}\partial_\alpha\sqrt{g}\cdot g^{\alpha\beta}\partial_\beta\Phi\right)\right],\qquad(134)$$

where we have the desired metric Schrödinger equation:

$$0=\left[\left(\mu_m^2+\frac{1}{2\sqrt{g}}\partial_\alpha\sqrt{g}\cdot g^{\alpha\beta}\partial_\beta\right)\Phi\right].\qquad(135)$$

Things are getting a bit more interesting when assuming that V was chosen such that we have $\Psi=\Psi_{\text{Schrödinger}}=\Phi_{\text{new}}=\Phi$ (please see [41] for more information and compare with [149])

We note: What classically is the potential V gives a metric distortion in the theory of quantized metric spaces-solutions. As before with the Klein–Gordon equation, we had derived the classical Schrödinger equation as a metric variation of any possible Einstein-compatible space-time solutions.

A more comprehensive evaluation of the "metric Schrödinger" equation is given in appendix B in this book. Thereby we will also present a new version of the Schrödinger equation, which can be derived from the metric Dirac equation and which we therefore named "Schrodinger–Dirac equation." As the name might suggest, it is of first-order derivatives with respect to all dimensions, including the spatial ones.

For illustration, we will also derive the famous Schrödinger equation for the hydrogen atom from a completely metric origin in appendix I.

6.9.6 The Metric Origin of the Dirac Equation

Classically it has been assumed that the Dirac equation can be evaluated as a quaternionic square root or rather an operational factorization of the Klein–Gordon equation [181]. Thus, having already obtained the latter in a metric manner (c.f. sub-sections above), we might believe that our task of also deriving the metric origin of the Dirac equation is already done.

This, however, is not correct, as Dirac's introduction of the quaternions in order to obtain the square root of the Klein–Gordon equation—de facto—neither explained why he was able to do it nor did his procedure reveal any geometric meaning. With the Dirac equation being of total axiomatic character, this "why?" and "what does it mean?" was not to be expected anyway.

Here, choosing a completely metric origin and not starting with the Klein–Gordon equation, we will try to derive such an explanation. The author is aware that already in the twenties of the last century, there have been attempts to obtain suitable extensions of Einstein's General Theory of Relativity via the introduction of metric torsion fields [188, 189]. These fields automatically resulted in spin fields and, thus, were seen as a way to explain the intrinsic property of spin of elementary particles (also see the nice summary in [190]).

Apparently, however, there also seems to be a simpler approach. Similar to the metric derivation of the Klein–Gordon equation in the corresponding sub-section above, we start with the simplest form of the transformation (75) (c.f. first line in (112)), but this time we demand all ordinary derivatives to be substituted by Dirac derivatives:

$$\partial_\mu f \Rightarrow \sigma^\mu{}_{\delta\alpha} \partial_\mu f_\alpha, \tag{136}$$

which are composed by the means of the Dirac gamma matrices:

$$\sigma^\beta{}_{\delta\alpha} = \gamma^\beta \quad \text{with:} \quad \gamma^0 = \begin{pmatrix} 1 & & & \\ & 1 & & \\ & & -1 & \\ & & & -1 \end{pmatrix}; \quad \gamma^1 = \begin{pmatrix} & & & 1 \\ & & 1 & \\ & -1 & & \\ -1 & & & \end{pmatrix}$$

$$\gamma^2=\begin{pmatrix} & & & -i \\ & & i & \\ & i & & \\ -i & & & \end{pmatrix};\ \gamma^3=\begin{pmatrix} & & 1 & \\ & & & -1 \\ -1 & & & \\ & 1 & & \end{pmatrix};\ I=\begin{pmatrix} 1 & & & \\ & 1 & & \\ & & 1 & \\ & & & 1 \end{pmatrix}. \quad (137)$$

In addition, we consider our function f to be a vector of functions f_{j}. We directly obtain the following Ricci scalar (*note*: C_{N2} and C_{N2} are constants, which depend on the dimension):

$$R^{*}=\frac{1}{F\left[f_j\right]^3}\cdot\begin{pmatrix}\left(C_{N1}\cdot\left(\frac{\partial F\left[f_j\right]}{\partial f_j}\right)^2-C_{N2}\cdot F\left[f_j\right]\cdot\frac{\partial^2 F\left[f_j\right]}{\partial f_j^2}\right)\cdot\left(\tilde{\nabla}_{g-\gamma}f_j\right)^2 \\ -C_{N2}\cdot F\left[f_j\right]\cdot\frac{\partial F\left[f_j\right]}{\partial f_j}\cdot\Delta_{g-\gamma}f_j\end{pmatrix}.$$

(138)

For illustration, here we choose a 3 + 1-Minkowski-flat-space-time and obtain:

$$R^{*}=\frac{1}{F\left[f_j\right]^3}\cdot\begin{pmatrix}\left(C_{N1}\cdot\left(\frac{\partial F\left[f_j\right]}{\partial f_j}\right)^2-C_{N2}\cdot F\left[f_j\right]\cdot\frac{\partial^2 F\left[f_j\right]}{\partial f_j^2}\right)\cdot\sum_{\mu=1}^{4}\left(\sigma^{\mu}{}_{\delta j}\partial_{\mu}f_j\right)^2 \\ -C_{N2}\cdot F\left[f_j\right]\cdot\frac{\partial F\left[f_j\right]}{\partial f_j}\cdot\underbrace{\left(\sigma^{\mu}{}_{\delta j}\partial_{\mu}\sigma^{\mu}{}_{\delta j}\partial_{\mu}\right)}_{\mathbf{I}\cdot\Delta_g}f_j\end{pmatrix}.$$

(139)

For simplicity, here we have omitted the consideration of mass via additional, entangled coordinates as elaborated in sub-section 6.9.3, "The Metric Origin of Mass." However, keeping the 4-dimensional Dirac matrices only to the coordinates t, x, y, z and adding the "mass-coordinates" u and v as demonstrated above, we would just obtain the classical Klein–Gordon equation with mass.

Now we incorporate condition (90) for each of the functional vector components f_j and obtain the Klein–Gordon equation, only that this time it acts on a vector:

$$R^{*}=-\frac{C_{N2}\cdot F\left[f_j\right]}{F\left[f_j\right]^3}\cdot\left(\frac{\partial F\left[f_j\right]}{\partial f_j}\cdot\mathbf{I}\cdot\Delta_g f_j\right)$$

$$= -\frac{C_{N2} \cdot F\left[f_j\right]}{F\left[f_j\right]^3} \cdot \frac{\partial F\left[f_j\right]}{\partial f_j} \cdot \Delta_g f_j. \tag{140}$$

This shows us that the additional "dimensions" which were added by the Dirac matrices are, in fact, intrinsic properties. As the influence of these properties on the level of the line element, including the Riemann tensor and the Ricci tensor (Ricci scalar), we concluded that a complete metric consideration should also incorporate conditions regarding the base vectors and their possible "interior." Thereby this inner "life of its own" does not show on the level of the line element, but only "below." This also gives us the major boundary condition for any inner variation. This means, while on the line element of the metric level of the governing dimension, respectively space-time it would just show the ordinary metric without the inner properties, the inner variation might look like for what kind of otherwise hidden properties it is supposed to reveal. In other words, the Dirac equation could be seen as a variation with respect to an inner set of dimensions living on top of the main 3 + 1 dimensions of our ordinary space-time. Being interested in revealing the nature of these intrinsic dimensions, we should rather treat them as dimensions instead of only giving them states of quaternionic sub-properties as done within the Dirac concept. This more complete evaluation was performed earlier by this author and presented elsewhere [15–18] (also see appendix C in this book).

However, sticking to our simple approach here and assuming suitable intrinsic structures for the functions *f*, we can find quite interesting connections to the metric forms, which we already have derived from the Einstein–Hilbert action.

With the Dirac concept being so successful, we might—if just being interested in a practical description of our quantum reality—simply understand the metric nature of the concept as one where every possible way of constructing the Laplace operator in (140) for an arbitrary space-time of metric $g_{\alpha\beta}$ reading:

$$\Delta_g f = \frac{1}{\sqrt{g}} \partial_\alpha \sqrt{g} \cdot g^{\alpha\beta} \partial_\beta f \tag{141}$$

could be realized. This is just nothing else than the Feynman dictum of "every path is possible." Thus, the general inner variation:

$$\Delta_g f_j = \frac{1}{\sqrt{g}} \partial_\alpha \sqrt{g} \cdot g^{\alpha'\beta'} \cdot \left[\phi 0_{\alpha'\beta'}^{\alpha\beta j} + \phi 1_{\alpha'\beta'}^{\alpha\beta j\beta'' k} \partial_{\beta''} f_k + \ldots\right] \partial_\beta f_i \tag{142}$$

should always sum up to the usual Laplace operator (141). In order to allow the separation as done within the Dirac concept, however, we might like to choose variations of such a kind, where we can factorize the Laplace operator. With the approach for the function vector f_j being arbitrary, we can achieve this via approaches as follows:

$$\Delta_g \tilde{f} = \frac{1}{\sqrt{g}} \partial_\alpha \sqrt{g} \cdot g^{\alpha\beta} \partial_\beta \overbrace{\left(\sqrt{g} \cdot g^{\alpha'\beta'}\right) f}^{\tilde{f}} = \frac{1}{\sqrt{g}} \partial_\alpha \sqrt{g} \cdot g^{\alpha\beta} \overbrace{\partial_\beta \left(\sqrt{g} \cdot g^{\alpha'\beta}\right) f_{\alpha'}}^{\partial_\beta \tilde{f}}$$

$$\Rightarrow \quad \Delta_g \tilde{f}_j = \frac{1}{\sqrt{g}} \partial_\alpha \sqrt{g} \cdot g^{\alpha'\beta'} \cdot \left[\phi 0^{\alpha\beta j}_{\alpha'\beta'} + \ldots\right] \partial_\beta \overbrace{\left(\sqrt{g} \cdot g^{\alpha'\beta'} \cdot \left[\phi 0^{\alpha''\beta}_{\alpha'\beta'} + \ldots\right]\right) f_{\alpha''}}^{\tilde{f}_j}. \tag{143}$$

However, in order to obtain a better understanding in connection with our metric starting point, we shall simply see the transformation in (75) as follows:

$$G_{\alpha\beta} = F\left[G\left[f\left[x_k\right]\right]\right] \cdot \delta^i_\alpha \delta^j_\beta g_{ij} \quad \Leftrightarrow \quad \begin{cases} \dfrac{\partial F\left[G\left[f\left[x_k\right]\right]\right]}{\partial x^k} \\ = \dfrac{\partial F\left[G\left[f\left[x_k\right]\right]\right]}{\partial G} \dfrac{\partial G\left[f\left[x_k\right]\right]}{\partial f} \dfrac{\partial f\left[x_k\right]}{\partial x^k}. \end{cases} \tag{144}$$

The chain rule would still be intact with G being a contravariant vector leading to:

$$G_{\alpha\beta} = F\left[G\left[f\left[x_k\right]\right]\right] \cdot \delta^i_\alpha \delta^j_\beta g_{ij} \quad \Leftrightarrow \quad \begin{cases} \dfrac{\partial F\left[G^\alpha\left[f\left[x_k\right]\right]\right]}{\partial x^k} \\ = \dfrac{\partial F\left[G^\alpha\left[f\left[x_k\right]\right]\right]}{\partial G^\alpha} \dfrac{\partial G^\alpha\left[f\left[x_k\right]\right]}{\partial f} \dfrac{\partial f\left[x_k\right]}{\partial x^k}. \end{cases} \tag{145}$$

Now we extend the function f to a vector of functions and obtain:

$$\frac{\partial F\left[G^\alpha\left[f_\beta\left[x_k\right]\right]\right]}{\partial x^k} = \frac{\partial F\left[G^\alpha\left[f\left[x_k\right]\right]\right]}{\partial G^\alpha} \frac{\partial G^\alpha\left[f_\beta\left[x_k\right]\right]}{\partial f^\beta} \frac{\partial f_\alpha\left[x_k\right]}{\partial x^k}. \tag{146}$$

While—as before—we demand the function $F[\ldots]$ to satisfy the condition (90) and give us Klein–Gordon-like equations for the Ricci scalar in every number of dimensions, we have the function $G[\ldots]$ to be arbitrary. We recognize the term $\frac{\partial G^{\alpha}\left[f_{\beta}\left[x_k\right]\right]}{\partial f^{\beta}}$ as the transformation matrix for the covariant vector f_α from the function space $f \to f'$, which would read:

$$f'_{\beta} = \frac{\partial G^{\alpha}\left[f_{\beta}\left[x_k\right]\right]}{\partial f^{\beta}} f_{\alpha}. \tag{147}$$

Now, as a trick, we introduce a slightly different transformation, which does not act on the vector, but its derivatives. It reads:

$$\begin{aligned}
\partial_{\mu} f'_{\beta'} &= \frac{\partial G^{\alpha'}\left[\partial_{\mu} f_{\beta'}\left[x_k\right]\right]}{\partial\left(\partial_{\mu} f^{\beta'}\right)}\left(\partial_{\mu} f_{\alpha'}\right) \\
&= \overbrace{\frac{\partial G^{\alpha'}\left[\partial_{\mu} f_{\beta'}\left[x_k\right]\right]}{\partial\left(\partial_{\mu} f^{\beta'}\right)}}^{\gamma^{\mu}}\left(\frac{\partial f_{\alpha'}\left[x_k\right]}{\partial x^{\mu}}\right) \\
&= \gamma^{\mu} \frac{\partial f_{\alpha'}\left[x_k\right]}{\partial x^{\mu}} = \gamma^{\mu} \partial_{\mu} f_{\alpha'}
\end{aligned} \tag{148}$$

and directly gives us the metric connection to the Dirac matrices γ^{μ}. These matrices are now revealed as transformation matrices in the tensor space of the function vector f_α. This function space has its own metric and is not connected to the x_k-space-time with its metric tensor $g_{\alpha\beta}$ ("not connected" means not in that usual direct way anyway). We have the usual Dirac conditions with respect to the matrices:

$$\begin{aligned}
\gamma^{\mu}\gamma^{\nu} + \gamma^{\nu}\gamma^{\mu} &= \frac{\partial G^{\alpha'}}{\partial\left(\partial_{\mu} f^{\beta'}\right)} \frac{\partial G^{\alpha'}}{\partial\left(\partial_{\nu} f^{\beta'}\right)} + \frac{\partial G^{\alpha'}}{\partial\left(\partial_{\nu} f^{\beta'}\right)} \frac{\partial G^{\alpha'}}{\partial\left(\partial_{\mu} f^{\beta'}\right)} \\
&= 2 \cdot \delta^{\mu\nu}.
\end{aligned} \tag{149}$$

It should be pointed out that the indices α' and β' are not identical with the metric indices and could therefore also run to different total dimensions. So, if being interested in completely avoiding the sudden appearance of matrices and insisting on a—potentially complex—number approach, we could apply the vectorial root extraction as introduced and used earlier by this author (see [41 –62]). For convenience, here we repeat the essentials and adapt them to the problem in question:

It is easy to prove that the scalar product of the following vector:

$$V_\Omega = \begin{Bmatrix} a+b+c+d+e, a+b+c+d-e, a+b+c-d+e, a+b+c-d-e, \\ a+b-c+d+e, a+b-c+d-e, a+b-c-d+e, a+b-c-d-e, \\ a-b+c+d+e, a-b+c+d-e, a-b+c-d+e, a-b+c-d-e, \\ a-b-c+d+e, a-b-c+d-e, a-b-c-d+e, a-b-c-d-e \end{Bmatrix}$$

$$\equiv [a \pm b \pm c \pm d \pm e]_\Omega \tag{150}$$

gives:

$$V_\Omega \cdot V_\Omega = a^2 + b^2 + c^2 + d^2 + e^2. \tag{151}$$

Even the introduction of "virtual" parameters ε can be incorporated into the vector V_Ω as follows:

$$V_\Omega = \{a+b+c, a+b-c, a-b+i\cdot c, a-b-i\cdot c\}$$

$$\equiv [a \pm b \pm I \cdot c]_\Omega$$

$$\text{with:} \quad V_\Omega \cdot V_\Omega = a^2 + b^2. \tag{152}$$

Now we introduce the following matrix V:

$$V = \left[\gamma_{\beta'}^{0\alpha'} \frac{\partial}{\partial x^0} \pm \gamma_{\beta'}^{1\alpha'} \frac{\partial}{\partial x^1} \pm \gamma_{\beta'}^{2\alpha'} \frac{\partial}{\partial x^2} \pm \gamma_{\beta'}^{3\alpha'} \frac{\partial}{\partial x^3} \pm I \cdot \varepsilon^m \cdot \mathrm{f}_m \right]. \tag{153}$$

Incomplete expansion without the virtual terms leads to:

$$V_{\beta'}^{\alpha'} = \left[\gamma_{\beta'}^{\mu\alpha'} \frac{\partial}{\partial x^\mu} \right] \equiv \gamma_{\beta'}^{\mu\alpha'} \partial_\mu = \left[\gamma_{\beta'}^{0\alpha'} \frac{\partial}{\partial x^0} \pm \gamma_{\beta'}^{1\alpha'} \frac{\partial}{\partial x^1} \pm \gamma_{\beta'}^{2\alpha'} \frac{\partial}{\partial x^2} \pm \gamma_{\beta'}^{3\alpha'} \frac{\partial}{\partial x^3} \right]_\Omega$$

$$= \begin{Bmatrix} \gamma_0^{\alpha'} \left(\frac{\partial}{\partial x^0} + \frac{\partial}{\partial x^1} + \frac{\partial}{\partial x^2} + \frac{\partial}{\partial x^3} \right) \\ \gamma_1^{\alpha'} \left(\frac{\partial}{\partial x^0} + \frac{\partial}{\partial x^1} + \frac{\partial}{\partial x^2} - \frac{\partial}{\partial x^3} \right) \\ \vdots \\ \gamma_7^{\alpha'} \left(\frac{\partial}{\partial x^0} - \frac{\partial}{\partial x^1} - \frac{\partial}{\partial x^2} - \frac{\partial}{\partial x^3} \right) \end{Bmatrix} \equiv \begin{Bmatrix} \gamma_0^{\alpha'} (+++) \\ \gamma_1^{\alpha'} (++-) \\ \gamma_2^{\alpha'} (+-+) \\ \gamma_3^{\alpha'} (+--) \\ \vdots \\ \gamma_7^{\alpha'} (---) \end{Bmatrix}. \tag{154}$$

Further expansion gives us:

$$V_{\beta'}^{\alpha'} = \begin{Bmatrix} \gamma_0^0(+++)\,\gamma_0^1(+++)\ldots\gamma_0^7(+++) \\ \gamma_1^0(++-)\,\gamma_1^1(++-)\ldots\gamma_1^7(++-) \\ \gamma_2^0(+-+)\,\gamma_2^1(+-+)\ldots\gamma_2^7(+-+) \\ \gamma_3^0(+--)\,\gamma_3^1(+--)\ldots\gamma_3^7(+--) \\ \vdots \\ \gamma_7^0(---)\,\gamma_7^1(---)\ldots\gamma_7^7(---) \end{Bmatrix}. \tag{155}$$

It can be shown that with the conditions:

$$\left(\sum_{\alpha'=0}^{7}\gamma_0^{\alpha'}\right)^2 = \left(\sum_{\alpha'=0}^{7}\gamma_1^{\alpha'}\right)^2 = \left(\sum_{\alpha'=0}^{7}\gamma_2^{\alpha'}\right)^2 = \left(\sum_{\alpha'=0}^{7}\gamma_3^{\alpha'}\right)^2$$

$$= \left(\sum_{\alpha'=0}^{7}\gamma_4^{\alpha'}\right)^2 = \left(\sum_{\alpha'=0}^{7}\gamma_5^{\alpha'}\right)^2 = \left(\sum_{\alpha'=0}^{7}\gamma_6^{\alpha'}\right)^2 = \left(\sum_{\alpha'=0}^{7}\gamma_7^{\alpha'}\right)^2, \tag{156}$$

the inner product of the matrices V leads to the Klein–Gordon operator:

$$\sum_{\alpha'=0}^{7}\sum_{\beta'=0}^{7} V_{\beta'}^{\alpha'}\cdot V_{\beta'}^{\alpha'} = G\left[\gamma_{\beta'}^{\alpha'}\right]\cdot\left(\left(\frac{\partial}{\partial x^0}\right)^2 + \left(\frac{\partial}{\partial x^1}\right)^2 + \left(\frac{\partial}{\partial x^2}\right)^2 + \left(\frac{\partial}{\partial x^3}\right)^2\right). \tag{157}$$

Thereby the term $G\left[\gamma_{\beta'}^{\alpha'}\right]$ denotes a complex but scalar function of the coefficients $\gamma_{\beta'}^{\alpha'}$. Construction of the two operators:

$$V_{\beta'}^{+\alpha'} = V_{\beta'}^{\alpha'} + M\cdot\delta_{\beta'}^{\alpha'};\quad V_{\beta'}^{-\alpha'} = V_{\beta'}^{\alpha'} - M\cdot\delta_{\beta'}^{\alpha'} \tag{158}$$

leads to the complete Klein–Gordon equation with the operator:

$$\sum_{\alpha'=0}^{7}\sum_{\beta'=0}^{7} V_{\beta'}^{+\alpha'}\cdot V_{\beta'}^{-\alpha'} = G\left[\gamma_{\beta'}^{\alpha'}\right]\cdot\left(\left(\frac{\partial}{\partial x^0}\right)^2 + \left(\frac{\partial}{\partial x^1}\right)^2 + \left(\frac{\partial}{\partial x^2}\right)^2 + \left(\frac{\partial}{\partial x^3}\right)^2\right) - M^2. \tag{159}$$

Applying either of the operators in (158) on a function f requires the function to become a vector $f_{\alpha'}$. Thus, in contrast to the Dirac form with quaternions, we not only have scalar results now when performing the matrix multiplication as an inner product but also obtain eight functions instead of four.

As seen with the Dirac matrices, however, the metrical origin in connection with the additional special transformation (148) has

shown us that the Dirac approach (no matter whether it is the one with quaternions or with the vectorial roots) simply takes care of transformation options of the derivatives instead of the coordinates as an ordinary transformation does. In other words, while the approach (75) only "variates" the coordinates within a function *f*[...], the intrinsic Dirac approach also "variates" the derivatives with respect to these coordinates (or dimensions). Doing this via the transformation (148) and conditions (149) in the quaternionic case and equation (156) in the vectorial case automatically ensures this derivative variation to stay intrinsic. From this metric consideration automatically follows that there also should be higher-order variations with respect to second-order or mixed derivatives.

We leave it to the mathematically skilled and interested reader to play with the new apparatus and refer to appendix C with respect to a much simpler (and compacter) derivation of the metric Dirac equation from the Einstein–Hilbert action (also c.f. references [15–18].

6.9.7 Metric Quantum Electro Dynamics MQED

The apparatus of Quantum Electro Dynamics (QED) is mainly based on Green's function or propagator technology. Green's functions, on the other hand, are certain fundamental solutions obtained from Schrödinger, Klein–Gordon and Dirac equations, allowing for the integral construction of superposed complex solutions [191].

With the establishment of the origin of these quantum equations out of metric principles, we automatically have derived the metric origin of QED.

For illustration, we here want to consider the so-called Feynman propagator for scalar particles Δ_F (c.f. [191], pp 500), which solves the following Klein–Gordon equation:

$$\left(m_0^2+\frac{\partial^2}{\partial t^2}-\Delta_{3D}\right)\Delta_F=-\delta^1\left[\mathbf{x}'-\mathbf{x}\right]$$

$$\Rightarrow \qquad \Delta_F=f\left[t,x,y,z\right]=\int\frac{d^4p\cdot e^{-i\cdot\mathbf{p}(\mathbf{x}'-\mathbf{x})}}{p^2-m_0^2+i\cdot\varepsilon}. \qquad (160)$$

Please remember that this equation is an adjusted form of (128) and results from the metric transformation (112) in 6 dimensions with two of these dimensions being suitably entangled (see [157], appendix of this chapter and equation (128)). Choosing C_f = 1 in (128), we could illustrate a non-zero-ε-approximation in 1 dimension (Fig. 6.3). For illustration, we have shown the Dirac delta distribution as incomplete integral in order to "smear the particle out" over a bit of space. The question we have to answer would be how to give a metric explanation for the occurrence of the Feynman propagator in (160). The easiest way would be to take equation (128) and demand the curvature to take on the propagator property, leading to:

$$R^{*}=\frac{5}{\left(f+C_{f}\right)^{2}}\cdot\left(\frac{C_{v0}^{2}}{5}\cdot C_{f}+\frac{C_{v0}^{2}}{5}\cdot f+\frac{\partial^{2} f}{c^{2}\cdot\partial t^{2}}-\Delta_{3D} f\right);\quad f=f[t,x,y,z]$$

$$\Rightarrow\qquad -\delta^{4}[\mathbf{x}'-\mathbf{x}]=\frac{R^{*}}{5}\cdot\left(f+C_{f}\right)^{2}-\frac{\overbrace{C_{v0}^{2}}^{m_0^2}}{5}\cdot C_{f}=\frac{\overbrace{C_{v0}^{2}}^{m_0^2}}{5}\cdot f+\frac{\partial^{2} f}{c^{2}\cdot\partial t^{2}}-\Delta_{3D} f. \quad (161)$$

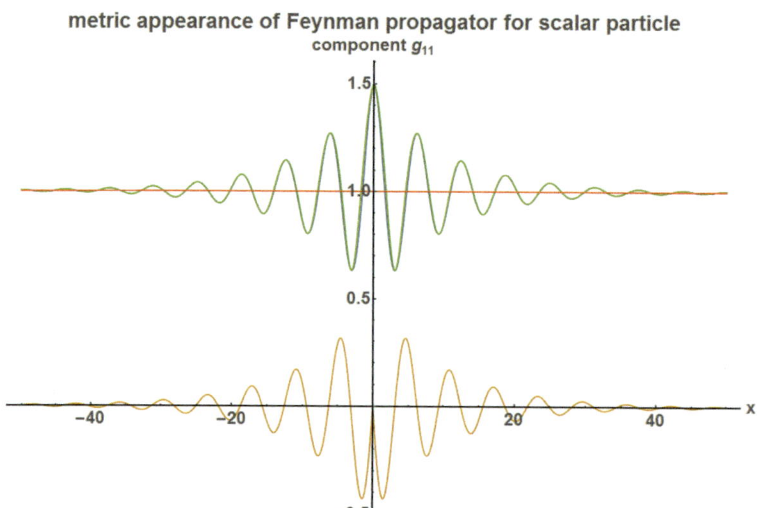

Figure 6.3 Scalar particle in its metric picture (approximated point-like Feynman propagator, see the text) with real part (blue), absolute part (green) and imaginary part (yellow), due to non-zero-$\boldsymbol{\varepsilon}$. In addition, we have shown the g_{11} Minkowski line (red). Please note that $\boldsymbol{\varepsilon}$ is only an auxiliary parameter which should be made to zero in a limiting procedure after performance of the integration within the QED-apparatus.

Considering f quantum mechanically small against the constant C_f (assuring the asymptotically flat space), we can rewrite (161) as follows:

$$\delta^{4}[\mathbf{x}'-\mathbf{x}]-\left(\frac{R^{*}}{5}\cdot C_{f}-m_{0}^{2}\right)\cdot C_{f}=m_{0}^{2}\cdot f+\frac{\partial^{2} f}{c^{2}\cdot\partial t^{2}}-\Delta_{3D} f$$

$$\Rightarrow \qquad R^* = \frac{5}{C_f}\cdot\left(m_0^2 - \frac{\delta^4[\mathbf{x}'-\mathbf{x}]}{C_f}\right). \qquad (162)$$

Thus, the transformed Ricci curvature R^* for the scalar particle (Feynman particle) would approximately be a constant (flat space) plus a Dirac delta distribution on a certain position in space and time. In the case of small rest masses m_0, the Ricci curvature would even become perfectly proportional to the Dirac delta distribution. This is not a "completely unreasonable" finding and only requires the solutions f to be asymptotically small against any constant in order to

a. guarantee the flat space-time we obviously observe in earthly conditions AND
b. allow for the effects of Quantum Electro Dynamics to be correct for all scales being significantly bigger than the Schwarzschild sizes (Schwarzschild radii, c.f. section 6.9.9, "Quantum Gravity Dynamic (QGD)," below) of the objects being considered.

As a mere by-product, we have automatically also found the explanation for the question why QED describes the world so nicely. After all, it already is a part of the Theory of Everything, already resides in the Einstein–Hilbert action (just needed to be spotted there via our $F[f]$-scaling) and the only thing we had to do was to work out the connection to the metric picture.

6.9.8 Towards a Metric Quantum Chromo Dynamic (MQCD)

In order to metrically mirror the strong interaction or Quantum Chromo Dynamic [192], we need to find a radius dependency which assures the enclosure of the quarks, the alleged carriers of the strong charge/force. This is needed as there obviously are no free quarks in our universe. We only find them in bonded states in hadrons or other heavy particles. Metrics which ensure the necessary r-dependency can be found via structural approaches of the following kind:

$$g_{\alpha\beta}^{8} = \begin{pmatrix} -c^2 & 0 & 0 & 0 & \cdots & 0 \\ 0 & 1 & 0 & 0 & \cdots & 0 \\ 0 & 0 & r^2 & 0 & \cdots & 0 \\ 0 & 0 & 0 & r^2\cdot\sin^2\varphi_1 & \cdots & 0 \\ \cdots & \cdots & & \cdots & \ddots & 0 \\ 0 & 0 & 0 & 0 & 0 & g_{77} \end{pmatrix}$$

$$\times F\left[f\left[t,r,\vartheta,\varphi_1,\varphi_2,\varphi_3,\varphi_4,\varphi_5\right]\right];\quad F[f]=f^{\frac{2}{3}}$$

$$f\left[t,r,\vartheta,\varphi_1,\varphi_2,\varphi_3,\varphi_4,\varphi_5\right]=f_t[t]\cdot f_r[r]\cdot f_\vartheta[\vartheta]\cdot\prod_{i=1}^{5}f_{\varphi_i}\left[\varphi_i\right];$$

$$g_{44}=\left(g_{\varphi_2}\left[\varphi_2\right]'\right)^2\cdot r^0;$$

$$g_{55}=\left(g_{\varphi_3}\left[\varphi_3\right]'\right)^2\cdot r^1;$$

$$g_{66}=\left(g_{\varphi_4}\left[\varphi_4\right]'\right)^2\cdot r^2;$$

$$g_{77}=\left(g_{\varphi_5}\left[\varphi_5\right]'\right)^2\cdot r^{-1};. \tag{163}$$

This gives us the following differential equation for the spatial coordinates when demanding the Ricci scalar to vanish:

$$R^*=0=\begin{pmatrix}\frac{1}{2}\cdot\frac{\partial}{r\cdot\partial r}+\left(r\cdot A_6^2+\frac{A_5^2}{r^2}+\frac{A_4^2}{r}+A_3^2\right)\\ +\Delta_{\text{3D-sphere}}+E^2+\frac{27}{28\cdot r^2}\end{pmatrix}\Psi$$

$$\Psi=\Psi[r,\vartheta,\varphi]. \tag{164}$$

Thereby we had fixed some of the functions in our separation approach from (163) (third line) as follows (for i = 2, 3, 4, 5):

$$f_{\varphi_i}\left[\varphi_i\right]=f_{\varphi_i}\left[g_{\varphi_i}\left[\varphi_i\right]\right]\equiv f_{\varphi_i}\left[g_{\varphi_i}\right];\quad f_{\varphi_i}\left[g_{\varphi_i}\right]=C_{-i}\cdot e^{-A_i\cdot g_{\varphi_i}}+C_{+i}\cdot e^{+A_i\cdot g_{\varphi_i}}, \tag{165}$$

$$f_t[t]=C_{t1}\cdot\cos[c\cdot E\cdot t]+C_{t2}\cdot\sin[c\cdot E\cdot t]. \tag{166}$$

Only considering the radius part in (164), we find the funny potential element $r\cdot A_6^2$, which is linear in r and thus, would grow to infinity for increasing r. Therefore, particles with this kind of interaction would be confined to certain volumes where—due to their reciprocal interaction—the r-linear potential has been shielded. As further generalization of potentials also with respect to other coordinates is easily possible via structural "playing" with the additional—compactified—coordinate components and by adding more dimensions, we can adjust our approach to whatever is needed in order to fully display the strong force interaction. So, for instance, the metrics already considered in [18], section 6.9, "More Metric Options and Subsequent Quantum Equations Plus Their Solutions," give angular dependencies. More options for radius settings and quite some discussion were presented in chapter 11 of our book [18].

Taking the fact that in order to obtain the characteristics of the strong interaction, we already require quite some metric complexity, plus the knowledge about the strength and compactness of this interaction with respect to spatial recourses, we might draw the following conclusions:

(a) A complete layout of the metric QCD picture is extremely complex and voluminous.
(b) Regarding the quantum computer technology, the metric Quantum Chromo Dynamic seems to provide the same technological leap as it was for the step from chemical power (QED-based) to atom power (QCD-based). To put it more associatively, one might compare the energy output of a candle with the one of a uranium reactor and regarding the computer power to be expected one has to compare conventional with atom bombs, which is to say one leaps from TNT to "The Gadget" and the subsequent thermonuclear "Ivy Mike."

Today's work on quantum computer concepts is based on QED and, thus, purely electromagnetically founded. With a metric QCD at hand, one would have the means to start one energy level higher already, namely QCD-based (strong force founded). As such a technology change is comparable with moving with the speed of sound and then accelerating to the speed of light (in vacuum), we have to expect the "leap" to be rather significant, if not to be of almost singularity character.

6.9.9 Quantum Gravity Dynamic (QGD)

In order to obtain a quantum gravity dynamic theory, we would require Green's function or a propagator approach for quantum gravity equations resulting from transformations of the kind (76) for a gravitational metric. As it was already shown that mass can be introduced either by additional dimensions or geometrically by a Schwarzschild radius (c.f. sub-section 6.9.3, "The Metric Origin of Mass," above), here we consider the alternative via a Schwarzschild approach. The Schwarzschild solution given as follows [27, 193]:

$$g_{\alpha\beta}^{\text{Schwarzschild}} = \begin{pmatrix} -c^2\left(1-\frac{2\cdot G\cdot M}{c^2\cdot r}\right) & 0 & 0 & 0 \\ 0 & \left(1-\frac{2\cdot G\cdot M}{c^2\cdot r}\right)^{-1} & 0 & 0 \\ 0 & 0 & r^2 & 0 \\ 0 & 0 & 0 & r^2\cdot\sin^2\vartheta \end{pmatrix} \tag{167}$$

has been considered the most basic gravity solution. We shall try to construct our example for a QGD starting point from there.

Here r gives the spherical radius, M stands for the gravitational mass of the Schwarzschild object and the constants c and G are the speed of light in vacuum and the gravitational or Newton constant, respectively. One easily sees that the solution above has two singularities, namely one at $r = r_s = \frac{2\cdot M\cdot G}{c^2}$, which is a result of our choice of coordinates and could be transformed away, and the other at $r = 0$. The parameter r_s is called the Schwarzschild radius.

The simplicity regarding the Schwarzschild metric holds because we have $R = 0$ (as in the flat space, too). Now using (83) plus moving to 4 dimensions, we can write (92) as:

$$\frac{R^{**}}{F[f]} = R^{*}$$

$$\Rightarrow \qquad R^{**} = -C_{N2}\cdot\frac{1}{F[f]}\cdot\frac{\partial F[f]}{\partial f}\cdot\Delta_g f \xrightarrow{4D} \left[\frac{R^{**}}{6}+\Delta_g\right]f = 0, \tag{168}$$

where we immediately see the proximity to the classical quantum Klein–Gordon equation again. Of course, this is just a trivial identity[9], but assuming that one would have some physical knowledge about the value of R^{**}, it would become non-trivial and could be applied for the determination of the function f. Here we intend to consider the matter-free case with the condition $\Delta R = 0$, leading to equation (96). With respect to the operator Δ_g, it has to be pointed out that this is not exactly the Laplace operator for the Schwarzschild metric, because this metric is not of type (2). The reader finds the correct formula for its derivation in appendix I.

---

[9]Please prove by setting in the corresponding expressions from (83) and evaluating for arbitrary flat metrics!

Knowing that the Ricci scalar of the Schwarzschild solution is $R = 0$ and applying the simplification rule (90), we have from (89) (c.f. (168)):

$$F[f]\cdot R^{*}=R^{**}=-\frac{6}{f}\cdot\Delta_{g}f=-\frac{6}{f}\cdot\Delta_{SS}f$$

$$=-\frac{6}{f}\cdot\left(\Delta_{\text{sphere}}f-r_{s}\cdot\left(\frac{f^{(0,1,0,0)}+r\cdot f^{(0,2,0,0)}}{r^{2}}+\frac{f^{(2,0,0,0)}}{c^{2}(r-r_{s})}\right)\right)$$

$$=-\frac{6}{f}\cdot\left(\Delta_{\text{3D-sphere}}f-\frac{f^{(2,0,0,0)}}{c^{2}}-r_{s}\cdot\left(\frac{f^{(0,1,0,0)}+r\cdot f^{(0,2,0,0)}}{r^{2}}+\frac{f^{(2,0,0,0)}}{c^{2}(r-r_{s})}\right)\right)$$

$$f^{(0,1,0,0)}=\frac{\partial f[t,r,\vartheta,\varphi]}{\partial r};$$

$$f^{(0,2,0,0)}=\frac{\partial^{2} f[t,r,\vartheta,\varphi]}{\partial r^{2}};$$

$$f^{(2,0,0,0)}=\frac{\partial^{2} f[t,r,\vartheta,\varphi]}{\partial t^{2}}. \tag{169}$$

Thereby we denote the Laplace operator in spherical polar coordinates with $\Delta_{\text{3D-sphere}}$. It is connected with the Laplace operator in spherical Minkowski coordinates as follows:

$$\Delta_{\text{sphere}}f=\Delta_{\text{3D-sphere}}f-\frac{f^{(2,0,0,0)}}{c^{2}}. \tag{170}$$

Please note that the functional wrapper $F[f]$ reading:

$$G_{\alpha\beta}=g_{\alpha\beta}\cdot F[f]=g_{\alpha\beta}\cdot(C_{1}+f)^{2}=g_{\alpha\beta}\cdot(C_{1}+f[t,r,\vartheta,\varphi])^{2} \tag{171}$$

fulfills the necessary condition (90) in 4 dimensions and was applied in (169). Taking the result from (87) and inserting it in (169), we would obtain the following Lagrange density L_M:

$$R=0=\overbrace{F[f]\cdot R^{*}}^{R_{\text{matter}}}-\overbrace{\begin{pmatrix}-\Gamma^{\mu}_{\sigma\alpha}\Gamma^{**\sigma}_{\beta\mu}+\Gamma^{\sigma}_{\alpha\beta}\Gamma^{**\mu}_{\sigma\mu}-\Gamma^{**\mu}_{\sigma\alpha}\Gamma^{\sigma}_{\beta\mu}+\Gamma^{**\sigma}_{\alpha\beta}\Gamma^{\mu}_{\sigma\mu}\\+\Gamma^{**\sigma}_{\alpha\beta\,,\sigma}-\Gamma^{**\sigma}_{\beta\sigma\,,\alpha}-\Gamma^{**\mu}_{\sigma\alpha}\Gamma^{**\sigma}_{\beta\mu}+\Gamma^{**\sigma}_{\alpha\beta}\Gamma^{**\mu}_{\sigma\mu}\end{pmatrix}g^{\alpha\beta}}^{2\cdot\kappa\cdot L_{M}+2\cdot\Lambda(\text{here: }\Lambda=0)}$$

$$\equiv R_{\text{matter}}-2\cdot\kappa\cdot L_{M}-2\cdot\Lambda$$

$$\Rightarrow \quad = R^{**} + \overbrace{\frac{6}{f}\cdot\left(\Delta_{\text{sphere}} f - r_s \cdot \left(\frac{f^{(0,1,0,0)} + r\cdot f^{(0,2,0,0)}}{r^2} + \frac{f^{(2,0,0,0)}}{c^2\left(r-r_s\right)}\right)\right)}^{2\cdot\kappa\cdot L_M}$$

$$\Rightarrow \quad L_M = \frac{3}{\kappa\cdot f}\cdot\left(\Delta_{\text{sphere}} f - r_s \cdot \left(\frac{f^{(0,1,0,0)} + r\cdot f^{(0,2,0,0)}}{r^2} + \frac{f^{(2,0,0,0)}}{c^2\left(r-r_s\right)}\right)\right). \quad (172)$$

The metric for the Laplace operator Δ_{sphere} as used above can be obtained from (167) simply by setting $M = 0$. Setting $r_s = 0$ in (169) gives us the Klein–Gordon equation for the central symmetric problem:

$$0 = \left(\frac{R^{**}}{6} + \Delta_{\text{sphere}}\right) f. \quad (173)$$

This becomes the classical quantum mechanical central field problem for time-independent f and R^{**} chosen as follows (E_n denotes the energy eigenvalues)[10]:

$$\frac{R^{**}}{6} = \frac{M^2 c^2}{\hbar^2} + V - E_n. \quad (174)$$

Inserting (174) into (173), thereby assuming the dependencies $f=f[r,\vartheta,\varphi]$ (but not t), directly gives us the Schrödinger hydrogen equation (e.g., [186, 187, 194]):

$$0 = \left(\frac{R^{**}}{6} + \Delta_{\text{sphere}}\right) f \xrightarrow{f_{,t}=0} 0 = \left(\frac{M^2 c^2}{\hbar^2} + \frac{V_r}{r} - E_n + \Delta_{\text{3D-sphere}}\right) f. \quad (175)$$

As a kind of add-on bonbon, we can also evaluate the corresponding Einstein–Hilbert Lagrange density for the "Schrödinger hydrogen," being:

$$\Rightarrow \quad L_M = \frac{3}{\kappa\cdot f}\cdot \Delta_{\text{sphere}} f, \quad (176)$$

which is obtained from (172) by setting $r_s = 0$. In the case of time-independent f, we exactly reproduce the Schrödinger H-atom problem and could write:

$$\Rightarrow \quad L_M = \frac{3}{\kappa\cdot f}\cdot \Delta_{\text{3D-sphere}} f. \quad (177)$$

[10]Here we have the reduced Planck constant ħ, the rest mass M and the speed of light in vacuum c. V denotes the classical potential (c.f. classical text books like [194]).

Setting (175) into (177) finally connects the Einstein–Hilbert Lagrange matter density with the classical quantum theoretical mass, potential and energy terms as follows:

$$\Rightarrow \qquad L_M = \frac{3}{\kappa \cdot f} \cdot \Delta_{\text{3D-sphere}} f = \frac{3}{\kappa \cdot f} \cdot \left(E_n - \frac{M^2 c^2}{\hbar^2} - \frac{V_r}{r} \right) f$$

$$= \frac{3}{\kappa} \cdot \left(E_n - \frac{M^2 c^2}{\hbar^2} - \frac{V_r}{r} \right). \qquad (178)$$

Not exactly unexpected, the result still mirrors the classical matter-energy structure in a surprisingly close manner. Naturally, this has to be the result for the simplification of a vanishing Schwarzschild radius r_s and is explicitly incorporating the classical relation.

While we have already solved equation (169) under the condition of vanishing Ricci scalars R and R^* ($\rightarrow R^{**} = 0$) in [27], we would need to obtain Green's function in order to move forward with our QGD approach:

$$\Delta_{\text{3D-sphere}} f - \frac{f^{(2,0,0,0)}}{c^2} - r_s \cdot \left(\frac{f^{(0,1,0,0)} + r \cdot f^{(0,2,0,0)}}{r^2} + \frac{f^{(2,0,0,0)}}{c^2 (r - r_s)} \right) = \delta^4 [\mathbf{x}' - \mathbf{x}]. \qquad (179)$$

As we here only intend to give a brief introduction, we simplify our approach for f to a pure r-t-dependency. This gives us:

$$2 \cdot \frac{f^{(0,1)}}{r} + f^{(0,2)} - \frac{f^{(2,0)}}{c^2} - r_s \cdot \left(\frac{f^{(0,1)} + r \cdot f^{(0,2)}}{r^2} + \frac{f^{(2,0)}}{c^2 (r - r_s)} \right) = \delta^4 [\mathbf{x}' - \mathbf{x}]. \qquad (180)$$

We use the following integral presentation for the Dirac distribution:

$$\delta^4 [\mathbf{x}' - \mathbf{x}] = \iint dk \cdot d\omega \cdot e^{i \cdot k \cdot (r - r')} \cdot e^{-i \cdot \omega \cdot (t - t')} \qquad (181)$$

and the following Fourier approach for $f[\ldots]$:

$$f[t,r] = \iint dk \cdot d\omega \cdot g[\omega,k] \cdot e^{i \cdot k \cdot (r - r')} \cdot e^{-i \cdot \omega \cdot (t - t')}. \qquad (182)$$

Setting this into (180), we obtain:

$$\iint\left\{\begin{array}{c} dk\cdot d\omega\cdot g[\omega,k]\cdot e^{i\cdot k\cdot(r-r')}\cdot e^{-i\cdot\omega\cdot(t-t')} \\ \times\left[\frac{i\cdot k}{r}\cdot\left(2-\frac{r_s}{r}\right)-k^2\cdot\left(1-\frac{r_s}{r}\right)+\frac{\omega^2}{c^2}\cdot\left(1+\frac{r_s}{r-r_s}\right)\right]\end{array}\right\}$$

$$=\iint dk\cdot d\omega\cdot e^{i\cdot k\cdot(r-r')}\cdot e^{-i\cdot\omega\cdot(t-t')}$$

$$\Rightarrow \qquad g[\omega,k]=\left[\frac{i\cdot k}{r}\cdot\left(2-\frac{r_s}{r}\right)-k^2\cdot\left(1-\frac{r_s}{r}\right)+\frac{\omega^2}{c^2}\cdot\left(1+\frac{r_s}{r-r_s}\right)\right]. \qquad (183)$$

Thus, the approach (182) gives us the Schwarzschild–Feynman propagator for spherical symmetry as follows:

$$f[t,r]=\iint dk\cdot d\omega\cdot\frac{e^{i\cdot k\cdot(r-r')}\cdot e^{-i\cdot\omega\cdot(t-t')}}{\left[\frac{i\cdot k}{r}\cdot\left(2-\frac{r_s}{r}\right)-k^2\cdot\left(1-\frac{r_s}{r}\right)+\frac{\omega^2}{c^2}\cdot\left(1+\frac{r_s}{r-r_s}\right)\right]}. \qquad (184)$$

If interested in a propagator in asymptotically Cartesian coordinates, we should apply the Schwarzschild metric in isotropic coordinates (valid only outside the event horizon, meaning $r > r_s/4$), reading:

$$g_{\text{iso}}^{\alpha\beta}=\begin{pmatrix} \frac{1}{c^2}\frac{\left(1+\frac{r_s}{r}\right)^2}{\left(1-\frac{r_s}{r}\right)^2} & 0 & 0 & 0 \\ 0 & g^{\Sigma\Sigma} & 0 & 0 \\ 0 & 0 & g^{\Sigma\Sigma} & 0 \\ 0 & 0 & 0 & g^{\Sigma\Sigma}=-\left(1+\frac{r_s}{r}\right)^{-4} \end{pmatrix}$$

$$r^2=16\cdot\left(x^2+y^2+z^2\right). \qquad (185)$$

We leave the task of finding and playing with the corresponding Schwarzschild–Feynman propagator to the mathematically skilled and interested reader.

6.10 From Smart Black Holes to Smart Spheres?

In chapter 5, "The New Abacus," we saw that, when taking the equation for the Laplace length l_P from the Bekenstein thought experiment [182, 183]), we find the following ratio between r_s (Schwarzschild radius of a black hole) and l_P (c.f. equation (60)):

$$\frac{r_s}{\ell_P} = 2 \cdot \sqrt{\pi \cdot \left(q + \sqrt{q \cdot (1+q)}\right)}. \tag{186}$$

Most interestingly, we also found (c.f. chapter 5, "The New Abacus," and section 6.8, "The Bekenstein Bit Problem") that the solution to the extremal volume problem for a fixed radius r_f for n-spheres results in the same dependency when variating with respect to the number of dimensions of those n-spheres. We obtain (see dots in Fig. 5.5 in chapter 5 and Fig. 6.2) excellent fits, when applying an approach like:

$$r_f = L \cdot \sqrt{\pi \cdot \left(n + \sqrt{n \cdot (1+n)}\right)}. \tag{187}$$

Thereby we have the characteristic (system-dependent) length scale L.

This automatically gives us a connection between the size-parameter r_f of any system of spherical symmetry and its theoretical capability to store information. A perfect mathematical n-sphere thereby follows the rule (187) almost perfectly, while other systems may do so only from certain critical sizes onwards, but, nevertheless, we think we can draw the conclusion that the information storage capacity of given systems—if showing enough spherical symmetry—can be extracted from (187). Then the structural size-parameter r_f determines the number of storable bits n in dependence on the system-immanent length parameter L.

From this, one even may deduce that r_f and L could be substituted by other system characteristics. While in (187) their dimension is length, we should not exclude mass, time, charges, energies, and so on.

6.11 The Ultimate Computer Machine

Now, having shown that Quantum Theory also has a metric origin and can be obtained as a transformation of solutions to the Einstein field equations, we wonder how we could metrically construct a most fundamental computer machine. The task was already performed in [28, 29], but in connection with the deeper understanding of the geometric nature of everything as outlined above, here we want to repeat the essentials by using higher dimensional approaches.

In order to start a bit more general, we apply a metric with two radius coordinates. The following set of ρ-r-dependencies from [18] shall be used:

$$g_{\alpha\beta}^{8}=\begin{pmatrix} -c^2 & 0 & 0 & 0 & \cdots & 0 \\ 0 & 1 & 0 & 0 & \cdots & 0 \\ 0 & 0 & r^2 & 0 & \cdots & 0 \\ 0 & 0 & 0 & r^2\cdot\sin^2\varphi_1 & \cdots & 0 \\ \cdots & \cdots & \cdots & \cdots & \ddots & 0 \\ 0 & 0 & 0 & 0 & 0 & g_{77} \end{pmatrix}$$

$$\times F\left[f\left[t,r,\vartheta,\varphi,\tau,\rho,\theta,\phi\right]\right];\quad F[f]=f^{\frac{2}{3}}$$

$$f\left[t,r,\vartheta,\varphi,\tau,\rho,\theta,\phi\right]=f_t\left[t\right]\cdot f_r\left[r\right]\cdot f_\vartheta\left[\vartheta\right]\cdot f_\varphi\left[\varphi\right]\cdot f_\tau\left[\tau\right]\cdot f_\rho\left[\rho\right]\cdot f_\theta\left[\theta\right]\cdot f_\phi\left[\phi\right];$$

$$g_{44}=-c_4^2;$$

$$g_{55}=r;$$

$$g_{66}=\rho^2\cdot r;$$

$$g_{77}=\rho^2\cdot\sin[\theta]^2\cdot r;. \tag{188}$$

We refrain from interpreting the various coordinates but simply use the following partial solutions for our separation approach:

$$f_\varphi\left[\varphi\right]=C_{\varphi-}\cdot e^{-i\cdot A\cdot\varphi}+C_{\varphi+}\cdot e^{+i\cdot A\cdot\varphi}, \tag{189}$$

$$f_\tau\left[\tau\right]=C_{\tau1}\cdot\cos\left[c_4\cdot E\cdot\tau\right]+C_{\tau2}\cdot\sin\left[c_4\cdot E\cdot\tau\right], \tag{190}$$

$$f_\vartheta\left[\vartheta\right]=C_{P\vartheta}\cdot P_L^A\left[\cos[\vartheta]\right]+C_{Q\vartheta}\cdot Q_L^A\left[\cos[\vartheta]\right], \tag{191}$$

$$f_\phi\left[\phi\right]=C_{\phi-}\cdot e^{-i\cdot B\cdot\phi}+C_{\phi+}\cdot e^{+i\cdot B\cdot\phi}, \tag{192}$$

$$f_t\left[t\right]=C_{t1}\cdot\cos\left[c\cdot E_t\cdot t\right]+C_{t2}\cdot\sin\left[c\cdot E_t\cdot t\right], \tag{193}$$

$$f_\theta\left[\theta\right]=C_{P\theta}\cdot P_\ell^B\left[\cos[\theta]\right]+C_{Q\theta}\cdot Q_\ell^B\left[\cos[\theta]\right]. \tag{194}$$

Now we change the metric (188) as follows:

$$g_{\alpha\beta}^{8}=\begin{pmatrix} -c^2 & 0 & 0 & 0 & \cdots & 0 \\ 0 & 1 & 0 & 0 & \cdots & 0 \\ 0 & 0 & r^2 & 0 & \cdots & 0 \\ 0 & 0 & 0 & r^2\cdot\sin^2\varphi_1 & \cdots & 0 \\ \cdots & \cdots & \cdots & \cdots & \ddots & 0 \\ 0 & 0 & 0 & 0 & 0 & g_{77} \end{pmatrix}$$

$$\times F\left[f\left[t,r,\vartheta,\varphi,\tau,\rho,\theta,\phi\right]\right];\quad F[f]=f^{\frac{2}{3}}$$

$$f\left[t,r,\vartheta,\varphi,\tau,\rho,\theta,\phi\right]=f_t[t]\cdot f_r[r]\cdot f_\vartheta[\vartheta]\cdot f_\varphi[\varphi]\cdot f_\tau[\tau]\cdot f_\rho[\rho]\cdot f_\theta[\theta]\cdot f_\phi[\phi];$$

$$g_{44}=-c_4^2\cdot r;$$

$$g_{55}=r;$$

$$g_{66}=\rho^2;$$

$$g_{77}=\rho^2\cdot\sin[\theta]^2;. \tag{195}$$

Thereby we have used the same coordinates as in (188) and we will also use the solutions (189) to (194). This gives us the following scaled Ricci scalar R^*:

$$R^*=0=\frac{7}{3}\cdot\begin{pmatrix} \frac{3}{4\cdot r^2}-\frac{L\cdot(L+1)}{r^2}-\frac{\ell\cdot(\ell+1)}{\rho^2}-\frac{3}{7\cdot\rho^2} \\ +\frac{2\cdot\partial}{r\cdot\rho\cdot\partial\rho}+\frac{\partial^2}{r\cdot\partial\rho^2}+\frac{3}{7\cdot r\cdot\rho^2}+\frac{\partial}{r\cdot\partial r} \\ +\frac{2\cdot\partial}{r\cdot\partial r}+\frac{\partial^2}{\partial r^2}-E^2-\frac{E_t^2}{r} \end{pmatrix}f. \tag{196}$$

As it was shown in [18] we cannot separate the two radii with the approach from (195) and thus, try for the solution of the following condition for ℓ and the two differential equations:

$$0=-\frac{\ell\cdot(\ell+1)}{\rho^2}-\frac{3}{7\cdot\rho^2}\quad\Rightarrow\quad\ell_{1,2}=\pm i\cdot\frac{\sqrt{35}}{14}-\frac{1}{2}$$

$$0=\left(\frac{2\cdot\partial}{r\cdot\rho\cdot\partial\rho}+\frac{\partial^2}{r\cdot\partial\rho^2}+\frac{3}{7\cdot r\cdot\rho^2}\mp\frac{E_\rho^2}{r}\right)f_\rho, \tag{197}$$

$$0=\left(\frac{3}{4\cdot r^2}-\frac{L\cdot(L+1)}{r^2}+\frac{E_\rho^2}{r}+\frac{\partial}{r\cdot\partial r}+\frac{2\cdot\partial}{r\cdot\partial r}+\frac{\partial^2}{\partial r^2}-E^2-\frac{E_t^2}{r}\right)f_r. \quad (198)$$

Regarding the strange condition for the parameter ℓ, it might be worthwhile to investigate such options (potentially other metrics with similar outcomes and momentum conditions) in connection with the strong interaction and the so-called Gell–Mann matrices (e.g., [192], p. 10).

While (197) gives us the spherical Bessel functions via:

$$f_\rho[\rho]=C_j\cdot j_{\ell_+}\left[E_\rho\cdot\rho\right]+C_y\cdot y_{\ell_+}\left[E_\rho\cdot\rho\right];\quad \ell_+=i\cdot\frac{\sqrt{35}}{14}-\frac{1}{2}$$

$$f_\rho[\rho]=C_j\cdot j_{\ell_-}\left[-i\cdot E_\rho\cdot\rho\right]+C_y\cdot y_{\ell_-}\left[-i\cdot E_\rho\cdot\rho\right];\quad \ell_-=-i\cdot\frac{\sqrt{35}}{14}-\frac{1}{2}, \quad (199)$$

we obtain the following solution for f_r:

$$f_r[r]=\mathrm{e}^{-r\cdot E}\cdot r^{L-\frac{1}{2}}\cdot\left(C_U\cdot U[nn,2\cdot(L+1),2\cdot r\cdot E]+C_L\cdot L_{-nn}^{1+2\cdot L}[2\cdot r\cdot E]\right)$$

$$nn=\frac{1}{2}\cdot\left(2\cdot(L+1)+\frac{E_t^2\pm E_\rho^2}{E}\right). \quad (200)$$

We recognize the hypergeometric function $U[a,b,z]$ and the Laguerre polynomials $L_{-nn}^{1+2\cdot L}$ in the solutions above, which are well-known from the Schrödinger hydrogen solution (e.g., [194]).

In the interesting case of $L = 1/2$, equation (200) simplifies to:

$$f_r[r]=e^{-r\cdot E}\cdot\left(C_U\cdot U[nn,3,2\cdot r\cdot E]+C_L\cdot L_{-nn}^2[2\cdot r\cdot E]\right)$$

$$nn=\frac{1}{2}\cdot\left(3+\frac{E_t^2\pm E_\rho^2}{E}\right). \quad (201)$$

Intrigued by the fact that we have obtained conditions which forced us to certain restrictions of the parameters L and ℓ, we wonder whether there are also options demanding half-spin solutions. The metric we would be interested in reads:

$$g_{\alpha\beta}^{8}=\begin{pmatrix} -c^2 & 0 & 0 & 0 & \cdots & 0 \\ 0 & 1 & 0 & 0 & \cdots & 0 \\ 0 & 0 & r^2 & 0 & \cdots & 0 \\ 0 & 0 & 0 & r^2\cdot\sin^2\varphi_1 & \cdots & 0 \\ \cdots & \cdots & \cdots & \cdots & \ddots & 0 \\ 0 & 0 & 0 & 0 & 0 & g_{77} \end{pmatrix}$$

$$\times F\left[f\left[t,r,\vartheta,\varphi_1,\varphi_2,\varphi_3,\varphi_4,\varphi_5\right]\right];\quad F[f]=f^{\frac{2}{3}}$$

$$f\left[t,r,\vartheta,\varphi_1,\varphi_2,\varphi_3,\varphi_4,\varphi_5\right]=f_t[t]\cdot f_r[r]\cdot f_\vartheta[\vartheta]\cdot\prod_{i=1}^{5}f_{\varphi_i}\left[\varphi_i\right];$$

$$g_{44}=\left(g_{\varphi_2}\left[\varphi_2\right]'\right)^2\cdot r^1;$$

$$g_{55}=\left(g_{\varphi_3}\left[\varphi_3\right]'\right)^2\cdot r^1;$$

$$g_{66}=\left(g_{\varphi_4}\left[\varphi_4\right]'\right)^2\cdot r^0;$$

$$g_{77}=\left(g_{\varphi_5}\left[\varphi_5\right]'\right)^2\cdot r^0;. \tag{202}$$

Now we partially “fix” some of the functions as follows (for i = 3, 4, 5):[11]

$$f_{\varphi_i}\left[\varphi_i\right]=f_{\varphi_i}\left[g_{\varphi_i}\left[\varphi_i\right]\right]\equiv f_{\varphi_i}\left[g_{\varphi_i}\right];\quad f_{\varphi_i}\left[g_{\varphi_i}\right]=C_{-i}\cdot e^{-A_i\cdot g_{\varphi_i}}+C_{+i}\cdot e^{+A_i\cdot g_{\varphi_i}}$$

$$f_{\varphi_2}\left[g_{\varphi_2}\right]=C_{-2}\cdot e^{-i\cdot A_2\cdot g_{\varphi_2}}+C_{+2}\cdot e^{+i\cdot A_2\cdot g_{\varphi_2}}, \tag{203}$$

$$f_t[t]=C_{t1}\cdot\cos[c\cdot E\cdot t]+C_{t2}\cdot\sin[c\cdot E\cdot t], \tag{204}$$

$$f_\vartheta\left[\vartheta=\varphi_1\right]=C_{P\vartheta}\cdot P_L^{A_2}\left[\cos\left[\vartheta=\varphi_1\right]\right]+C_{Q\vartheta}\cdot Q_L^{A_2}\left[\cos\left[\vartheta=\varphi_1\right]\right], \tag{205}$$

with the associated Legendre polynomials $P_L^{A_2}, Q_L^{A_2}$. This gives us the following differential equation for the r-dependency of f[...]:

$$R^*-0=\begin{pmatrix} \frac{3-4\cdot L\cdot(L+1)}{4\cdot r^2}+\frac{A_2^2+A_3^2}{r}+A_4^2+A_5^2 \\ +\frac{\partial}{r\cdot\partial r}+\overbrace{\Delta_{\text{3D-sphere}}[r]}^{2\frac{\partial}{r\cdot\partial r}+\frac{\partial^2}{\partial r^2}}-E^2 \end{pmatrix}\Psi$$

[11]If not used as index, “i” stands for the imaginary number $i=\sqrt{-1}$.

$$\Psi=\Psi\left[r,\vartheta=\varphi_1,\varphi\right]=f_r\left[r\right]\cdot f_\vartheta\left[\vartheta=\varphi_1\right]\cdot f_{\varphi_2}\left[\varphi_2\right]$$

$$\Rightarrow \qquad 0=f_\vartheta\left[\vartheta\right]\cdot f_{\varphi_2}\left[\varphi_2\right]\cdot\left(\begin{array}{c}\frac{3-4\cdot L\cdot(L+1)}{4\cdot r^2}+\frac{A_2^2+A_3^2}{r}+A_4^2+A_5^2\\+\frac{3\cdot\partial}{r\cdot\partial r}+\frac{\partial^2}{\partial r^2}-E^2\end{array}\right)f_r\left[r\right]. \quad (206)$$

We find that for L = 1/2 the terms with r^{-2} vanish, resulting in the differential equation for the remaining coordinate r:

$$0=\left(\frac{A_2^2+A_3^2}{r}+A_4^2+A_5^2+\frac{3\cdot\partial}{r\cdot\partial r}+\frac{\partial^2}{\partial r^2}-E^2\right)f_r\left[r\right]. \qquad (207)$$

This is extremely interesting, because now we have obtained a condition, namely:

$$\frac{3-4\cdot L\cdot(L+1)}{4\cdot r^2}=0 \quad\Rightarrow\quad L=\frac{1}{2} \qquad (208)$$

forcing us to except the existence of spinors with a geometry given due to the metric (202). With respect to the cases of half-spins regarding the Legendre functions, we refer to the appendix H and a variety of previous papers by this author (mainly within the series of "Einstein had it ..."; see the list of references in this book).

Now we can also solve (206) and (207) via:

$$f_r\left[r\right]$$
$$=e^{-r\cdot\sqrt{E^2-\left(A_4^2+A_5^2\right)}}\cdot r^{L-\frac{1}{2}}\cdot\left(\begin{array}{c}C_U U\left[-nn,L\cdot(1+L),2\cdot r\cdot\sqrt{E^2-\left(A_4^2+A_5^2\right)}\right]\\+C_L L_{nn}^{1+2\cdot L}\left[2\cdot r\cdot\sqrt{E^2-\left(A_4^2+A_5^2\right)}\right]\end{array}\right)$$

$$nn=\frac{A_2^2+A_3^2}{2\cdot\sqrt{E^2-\left(A_4^2+A_5^2\right)}}-1-L, \qquad (209)$$

$$f_r\left[r\right]=e^{-r\cdot\sqrt{E^2-\left(A_4^2+A_5^2\right)}}\left(\begin{array}{c}C_U U\left[-nn,3,2\cdot r\cdot\sqrt{E^2-\left(A_4^2+A_5^2\right)}\right]\\+C_L L_{nn}^{2}\left[2\cdot r\cdot\sqrt{E^2-\left(A_4^2+A_5^2\right)}\right]\end{array}\right)$$

$$nn=\frac{A_2^2+A_3^2-3\cdot\sqrt{E^2-\left(A_4^2+A_5^2\right)}}{2\cdot\sqrt{E^2-\left(A_4^2+A_5^2\right)}}, \qquad (210)$$

respectively.

Taking the solutions from (203) to (205), equation (206) clearly gives us the connection to the classical picture with potential, mass and so on. Thereby the angular part is completely sucked into the *L*-term, which—as shown above—vanishes in the half-spin setting *L* = 1/2.

In this case, we can extract the connection to the classical picture with potential *V*[*r*], operator, energy *E*, and mass terms *M* from (207) as follows:

$$0=\left(\overbrace{\frac{\partial}{r\cdot\partial r}}^{?}+\overbrace{\frac{A_3^2+A_4^2}{r}}^{V[r]}+\overbrace{A_5^2+A_6^2}^{M}+\overbrace{\frac{2\cdot\partial}{r\cdot\partial r}+\frac{\partial^2}{\partial r^2}}^{\Delta_{r\text{3D-sphere}}}-\overbrace{E^2}^{\text{energy}}\right)f_r[r]. \qquad (211)$$

Similarly, we can allocate *V*[*r*], *M* and energy to the terms in (198) as follows:

$$0=\left(\overbrace{\frac{\partial}{r\cdot\partial r}}^{?}+\overbrace{\frac{3}{4\cdot r^2}-\frac{L\cdot(L+1)}{r^2}}^{\text{momentum}}+\overbrace{\frac{E_\rho^2}{r}-\frac{E_t^2}{r}}^{V[r]}+\overbrace{\frac{2\cdot\partial}{r\cdot\partial r}+\frac{\partial^2}{\partial r^2}}^{\Delta_{r\text{3D-sphere}}}-\overbrace{E^2}^{\text{energy}}\right)f_r. \quad (212)$$

If it were not for the missing of the masses, we might conclude that we have obtained the metric description for a fermion-like particle. However, as we know that we could always add mass via additional dimensions like:

$$g_{\alpha\beta}^{10}=\begin{pmatrix}-c^2 & 0 & 0 & 0 & \cdots & 0\\ 0 & 1 & 0 & 0 & \cdots & 0\\ 0 & 0 & r^2 & 0 & \cdots & 0\\ 0 & 0 & 0 & r^2\cdot\sin^2\varphi_1 & \cdots & 0\\ \cdots & \cdots & \cdots & \cdots & \ddots & 0\\ 0 & 0 & 0 & 0 & 0 & g_{99}\end{pmatrix}$$

$$\times F\left\lfloor f\left[t,r,\vartheta,\varphi,\tau,\rho,\theta,\phi\right]\right\rfloor;\quad F[f]=f^{\frac{1}{2}}$$

$$f\left[t,r,\vartheta,\varphi,\tau,\rho,\theta,\phi,x,y\right]=\begin{pmatrix}f_t[t]\cdot f_r[r]\cdot f_\vartheta[\vartheta]\cdot f_\varphi[\varphi]\cdot f_\tau[\tau]\\ \cdot f_\rho[\rho]\cdot f_\theta[\theta]\cdot f_\phi[\phi]\cdot f_x[x]\cdot f_y[y]\end{pmatrix};$$

$$g_{44}=-c_4^2\cdot r;\quad g_{55}=r;\quad g_{66}=\rho^2;\quad g_{77}=\rho^2\cdot\sin[\theta]^2;\quad g_{88}=1;\quad g_{99}=1$$

$$f_x[x]=C_{-x}\cdot e^{-A_x\cdot x}+C_{+x}\cdot e^{+A_x\cdot x};$$

$$f_y[y]=C_{-y}\cdot e^{-A_y\cdot y}+C_{+y}\cdot e^{+A_y\cdot y}$$

$$f_{\varphi_i}\left[g_{\varphi_i}\right]=C_{-i}\cdot e^{-A_i\cdot g_{\varphi_i}}+C_{+i}\cdot e^{+A_i\cdot g_{\varphi_i}} \tag{213}$$

resulting in the r-dependency for the corresponding solutions to (198), respectively to (212) as follows:

$$0=\begin{pmatrix} \overbrace{\frac{\partial}{r\cdot\partial r}}^{?}+\overbrace{\frac{3}{4\cdot r^2}-\frac{L\cdot(L+1)}{r^2}}^{\text{momentum}}+\overbrace{A_x^2+A_y^2}^{M}+\overbrace{\frac{E_\rho^2}{r}-\frac{E_t^2}{r}}^{V[r]} \\ +\underbrace{\frac{2\cdot\partial}{r\cdot\partial r}+\frac{\partial^2}{\partial r^2}}_{\Delta_{r\text{3D-sphere}}}-\overbrace{E^2}^{\text{energy}} \end{pmatrix} f_r, \tag{214}$$

we see that there always seems to be a way to metrically describe a physical situation, including its various quantum states.

And by the way, evaluation of the Ricci scalar R of the metric (202) from (1) via the setting $F[f] = 1$ gives:

$$R=-\frac{3}{2\cdot r^2}. \tag{215}$$

Thus, we have a metric with a r^{-2} asymptotically vanishing curvature, which apparently acts as a seed for spin-1/2-particles or at least particle-like objects when adding (and plying with) a scalar factor $F[f]$ as done in (202). For $r = 0$, this curvature would become infinite. Thus, there would be a singularity, but the scalar factor F prevents this singularity from occurring, as it simply makes R^* (the Ricci scalar of the scaled metric) zero.

Now, just as shown in chapter 5, "The New Abacus," we could use our already rather physical metric (213) and its active degrees of freedom regarding certain coordinates for quantum gravity computation presses. One inner degree of freedom—for instance—could be the following extension of (213):

$$g_{\alpha\beta}^{10}=\begin{pmatrix} -c^2 & 0 & 0 & 0 & \cdots & 0 \\ 0 & 1 & 0 & 0 & \cdots & 0 \\ 0 & 0 & r^2 & 0 & \cdots & 0 \\ 0 & 0 & 0 & r^2\cdot\sin^2\varphi_1 & \cdots & 0 \\ \cdots & \cdots & \cdots & \cdots & \ddots & 0 \\ 0 & 0 & 0 & 0 & 0 & g_{99} \end{pmatrix}$$

$$\times F\left[f\left[t,r,\vartheta,\varphi,\tau,\rho,\theta,\phi\right]\right];\quad F[f]=f^{\frac{1}{2}}$$

$$f\left[t,r,\vartheta,\varphi,\tau,\rho,\theta,\phi,x,y\right]=\begin{pmatrix} f_t[t]\cdot f_r[r]\cdot f_\vartheta[\vartheta]\cdot f_\varphi[\varphi]\cdot f_\tau[\tau] \\ \cdot f_\rho[\rho]\cdot f_\theta[\theta]\cdot f_\phi[\phi]\cdot f_x[x]\cdot f_y[y] \end{pmatrix};$$

$$g_{44}=-c_4^2\cdot r;g_{55}=r;g_{66}=\rho^2;g_{77}=\rho^2\cdot\sin[\theta]^2;g_{88}=g_x'[x]^2;g_{99}=g_y'[y]^2$$

$$f_x\left[g_x[x]\right]=C_{-x}\cdot e^{-A_x\cdot g_x[x]}+C_{+x}\cdot e^{+A_x\cdot g_x[x]};$$

$$f_y\left[g_y[y]\right]=C_{-y}\cdot e^{-A_y\cdot g_y[y]}+C_{+y}\cdot e^{+A_y\cdot g_y[y]}$$

$$f_{\varphi_i}\left[g_{\varphi_i}\right]=C_{-i}\cdot e^{-A_i\cdot g_{\varphi_i}}+C_{+i}\cdot e^{+A_i\cdot g_{\varphi_i}} \tag{216}$$

Again, the functions g_x and g_y are arbitrary and could be used for operations/data of any arbitrary form. And just as already discussed in chapter 5, "The New Abacus," in order to assure digital outcomes to the metric components g_{88} and g_{99}, we construct Fourier series of the kind:

$$g_x=\frac{x}{2}+\frac{2}{\pi}\cdot\sum_{k=1}^{\infty}i^{2\cdot(k-1)}\cdot\frac{\sin[(2\cdot k-1)\cdot x]}{(2\cdot k-1)^2};$$

$$g_y=\frac{y}{2}+\frac{2}{\pi}\cdot\sum_{k=1}^{\infty}i^{2\cdot(k-1)}\cdot\frac{\sin[(2\cdot k-1)\cdot y]}{(2\cdot k-1)^2}. \tag{217}$$

This degree of freedom exists for all coordinates, which is to say, with (216) solving the quantum Einstein field equations, so does:

$$g_{\alpha\beta}^{10}=$$

$$=\begin{pmatrix} -g_t'[t]^2\cdot c^2 & 0 & 0 & 0 & \cdots & 0 \\ 0 & g_r'[r]^2 & 0 & 0 & \cdots & 0 \\ 0 & 0 & g_\vartheta'[\vartheta]^2\cdot g_r{}^2 & 0 & \cdots & 0 \\ 0 & 0 & 0 & g_\varphi'[\varphi]^2\cdot g_r{}^2\cdot\sin^2 g_\vartheta[\vartheta] & \cdots & 0 \\ \cdots & \cdots & \cdots & \cdots & \ddots & 0 \\ 0 & 0 & 0 & 0 & 0 & g_{99} \end{pmatrix}$$

$$\times F\left[f\left[t,r,\vartheta,\varphi,\tau,\rho,\theta,\phi\right]\right];\quad F[f]=f^{\frac{1}{2}}$$

$$f[t,r,\vartheta,\varphi,\tau,\rho,\theta,\phi,x,y]=\begin{pmatrix} f_t[g_t]\cdot f_r[g_r]\cdot f_\vartheta[g_\vartheta]\cdot f_\varphi[g_\varphi]\cdot f_\tau[g_\tau] \\ \cdot f_\rho[g_\rho]\cdot f_\theta[g_\theta]\cdot f_\phi[g_\phi]\cdot f_x[g_x]\cdot f_y[g_y] \end{pmatrix};$$

$$g_{44}=-c_4^2\cdot g_\tau'[\tau]^2 g_r;\quad g_{55}=g_r\cdot g_\rho'[\rho]^2;\quad g_{66}=g_\rho[\rho]^2\cdot g_\theta'[\theta]^2$$

$$g_{77}=g_\rho[\rho]^2\cdot g_\phi'[\phi]^2\sin[g_\theta[\theta]]^2;\quad g_{88}=g_x'[x]^2;\quad g_{99}=g_y'[y]^2 \quad (218)$$

too. If anyhow possible to activate all these intrinsic degrees of freedom, the corresponding "bit" would evaluate functions and not just numbers and the subsequent calculation power is probably tremendous.

6.11.1 Conclusions with Respect to "The Ultimate Computer Machine"

Based on the metric derivation of quantum equations completely compatible with the General Theory of Relativity, we were able to find metric structures for Turing machines. These gravity or Einstein quantum computers, as we might like to call them, contain the classical quantum and the classical digital computers as degenerated forms. Along the way, we found an explanation for the storage of bits by a black hole and—to our surprise—discovered a fundamental length scale with the number 1 being equal to 12.6883 Planck length.

6.12 Appendix to Chapter "The Quantum Gravity Turing Machine"

6.12.1 Simple Example for Dimensional Mass Creation

In [156–158], we introduced special de Sitter vacuum metrics, which gave mass terms in the corresponding Ricci scalar based Klein–Gordon equation (89) in various dimensions. For completeness, we repeat the main results here.

Introducing a metric for the coordinates t, r, ϑ, φ, u, v of the form:

$$g_{\alpha\beta}^{6}=\begin{pmatrix} -c^2 & 0 & 0 & 0 & 0 & 0 \\ 0 & 1 & 0 & 0 & 0 & 0 \\ 0 & 0 & r^2 & 0 & 0 & 0 \\ 0 & 0 & 0 & r^2\cdot\sin[\vartheta] & 0 & 0 \\ 0 & 0 & 0 & 0 & g[v] & 0 \\ 0 & 0 & 0 & 0 & 0 & g[v] \end{pmatrix}\cdot f[t,r,\vartheta,\varphi], \quad (219)$$

with the entanglement solution for u and v as follows:

$$g[v]=\frac{C_{v1}^2}{C_{v0}^2\cdot\left(1+\cosh\left[C_{v1}\cdot\left(v+C_{v2}\right)\right]\right)}, \tag{220}$$

gives us the following Ricci scalar from (89) with condition (90):

$$R^*=\frac{1}{F[f]^3}\cdot\left(-C_{N2}\cdot F[f]\cdot\frac{\partial F[f]}{\partial f}\cdot\Delta_g f\right);\quad F[f]=C_f+f[t,r,\vartheta,\varphi]$$

$$\xrightarrow{6D} R^*=\frac{5}{f^2}\cdot\begin{pmatrix}\frac{C_{v0}^2}{5}\cdot f+\frac{\partial^2 f}{c^2\cdot\partial t^2}\\ -\Delta_{\text{3D-sphere}}f\end{pmatrix};\quad f=f[t,r,\vartheta,\varphi];\quad C_f=0. \tag{221}$$

Here $\Delta_{\text{3D-sphere}}f$ denotes the Laplace operator in spherical coordinates.

As for R^* = 0 and $\frac{C_{v0}^2}{5}=\frac{m^2\cdot c^2}{\hbar^2}$ (m = mass, c = speed of light in vacuum, $\hbar$ = reduced Planck constant), we clearly have obtained the quantum Klein–Gordon equation and we can easily conclude that the Higgs field can obviously be constructed by a suitable set of additional, entangled coordinates (c.f. section 6.9.2, "The Metric Origin of Scalar Fields like the Higgs Field and Symmetry Breaking").

It should be pointed out that (219) is not of type (2) and thus, its Ricci tensor and Ricci scalar should be evaluated in accordance with the equation given in appendix I.

In connection with a non-vanishing cosmological constant, we have also shown in [157] that with:

$$g_{\alpha\beta}^6=\begin{pmatrix}-c^2\cdot h[r] & 0 & 0 & 0 & 0 & 0\\ 0 & \frac{1}{h[r]} & 0 & 0 & 0 & 0\\ 0 & 0 & r^2 & 0 & 0 & 0\\ 0 & 0 & 0 & r^2\cdot\sin[\vartheta] & 0 & 0\\ 0 & 0 & 0 & 0 & g[v] & 0\\ 0 & 0 & 0 & 0 & 0 & g[v]\end{pmatrix}$$

$$\text{with: } h[r]=1+\frac{C_v^2\cdot r^2}{6}-r^2\cdot\frac{\Lambda}{3}-\frac{C_t}{r};\quad C_v^2=\Lambda, \tag{222}$$

we could have a complete vacuum solution, where we have:

$$G^{\alpha\beta}+\Lambda g^{\alpha\beta}=R^{\alpha\beta}-\frac{1}{2}Rg^{\alpha\beta}+\Lambda g^{\alpha\beta}=0. \tag{223}$$

In this case, the condition $C_v^2=\Lambda$ provides the necessary balance between matter and cosmological constant. As the constant C_r would be arbitrary, it could be set to zero in order to avoid any singularities.
We see that the metric (222) would be a vacuum solution to the Einstein field equations, but still the structure (universe) it describes contains matter. Assuming that the condition (223) should be a boundary condition for the object described by (222), we obtain a balance rule for the mass parameter C_v and the cosmological constant. The object would have a resulting non-zero Ricci curvature of:

$$R=3\cdot\Lambda. \tag{224}$$

Similarly, we could construct a metric as follows:

$$g_{\alpha\beta}^{6}=\begin{pmatrix}-c^2\cdot h[r] & 0 & 0 & 0 & 0 & 0\\ 0 & \frac{1}{h[r]} & 0 & 0 & 0 & 0\\ 0 & 0 & r^2 & 0 & 0 & 0\\ 0 & 0 & 0 & r^2\cdot\sin[\vartheta] & 0 & 0\\ 0 & 0 & 0 & 0 & a^2\cdot g[v] & 0\\ 0 & 0 & 0 & 0 & 0 & a^2\end{pmatrix}$$

$$\text{with}:\ h[r]=1+\frac{S^2\cdot r^2}{3\cdot a^2}-r^2\cdot\frac{\Lambda}{3}-\frac{C_r}{r};\ S^2=\frac{a^2\cdot\Lambda}{2}. \tag{225}$$

Now we shall consider S a spin-like matter option in addition to our masses as given above.

This time we have:

$$g_{1,2}[v]=\begin{Bmatrix}C_{S1}\sin\left[S\left(v-C_{S2}\right)\right]^2\\ C_{S1}\cos\left[S\left(v-C_{S2}\right)\right]^2\end{Bmatrix}. \tag{226}$$

For more information and discussion (especially with respect to the extension to arbitrary coordinates and the flatness problem of our universe), the reader is referred to the original papers [156–158].

6.12.2 Consequences from the Bekenstein Thought Experiment Regarding the Solutions to the Quantum Einstein Field Equations

In [3], we have shown that the classical n-dimensional Schwarzschild solution could be applied to construct internally structured n-dimensionally black holes, while outside we still have the usual 4-dimensional solution from [193] with the classical Schwarzschild metric. This, however, would not explain how the black hole can code any information.

With the help of the new metric solutions evaluated in [13], namely, in the 3-dimensional case with coordinates t, r and an angle:

$$g_{\alpha\beta} = C_1 \cdot f[t]^4 \cdot \begin{pmatrix} -c^2 & 0 & 0 \\ 0 & t^2 & 0 \\ 0 & 0 & t^2 \cdot \sin(r)^2 \end{pmatrix}$$

$$f[t] = \sqrt{t^{\pm i \cdot c - 1}} \cdot C_f, \tag{227}$$

(Please note that r has become an angle, while t took over the position of the radius),

$$g_{\alpha\beta} = C_1 \cdot f[t]^4 \cdot \begin{pmatrix} -c^2 & 0 & 0 \\ 0 & \rho^2 & 0 \\ 0 & 0 & \rho^2 \cdot \sin(r)^2 \end{pmatrix}$$

$$f[t] = e^{\pm \frac{i \cdot c}{2 \cdot \rho} \cdot t} \cdot C_f, \tag{228}$$

(This represents a shell-like object) and in the 4-dimensional case[12] with coordinates t, x, y, z:

$$g_{\alpha\beta} = f\left[t + c_x \cdot x + c_y \cdot y + c_z \cdot z\right]^2 \cdot \begin{pmatrix} c^2 & 0 & 0 & 0 \\ 0 & -1 & 0 & 0 \\ 0 & 0 & 1 & 0 \\ 0 & 0 & 0 & 1 \end{pmatrix}$$

$$f\left[t + c_x \cdot x + c_y \cdot y + c_z \cdot z\right] = \frac{C_{f1}}{c_t \cdot t + c_x \cdot x + c_y \cdot y + c_z \cdot z + C_{f0}}$$

[12]An extension of solutions of this type in n dimensions is given in [212, 214].

$$c_t^2 + c^2 \cdot \left(-c_x^2 + c_y^2 + c_z^2\right) = 0, \tag{229}$$

we want to solve this problem, too. Thereby, we found that the Schwarzschild singularity could be avoided (Fig. 6.3).

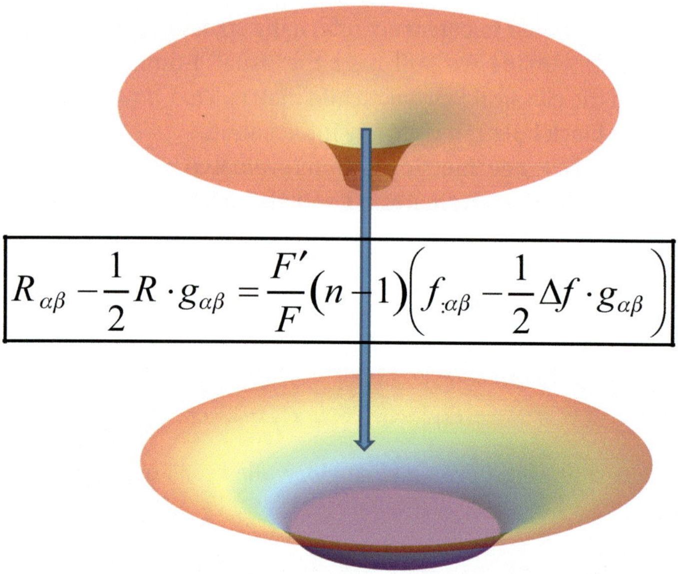

Figure 6.3 "From the classical Schwarzschild solution to a Quantum Black Hole" [13].

At first, however, we should note that also the n-dimensional layered Schwarzschild solutions from section 3.8 in [3] would provide plenty of options to code information, because there are enough degrees of freedom regarding the thicknesses of the individual x-dimensional layers of the onion-like Schwarzschild object, which was proposed there (see also [13]). Similar assumptions could be made for the Robertson-Walker approach introduced in [13], but we will not consider these possibilities here.

In the case of solutions (229), one might assume some kind of standing waves inside the black hole, but as we currently do not have the math to realize such structures, we postpone the investigation of this possibility.

Thus, here we concentrate on solutions (227) and (228) as potential inner solutions to a black hole. As we see that the parameter ρ clearly is a length, we want to derive its properties. From basic Quantum Theory, we know that a particle at rest has the time dependency:

$$f[t]=e^{\pm \mathrm{i}\cdot\frac{m\cdot c^2}{\hbar}\cdot t}\cdot C_f, \tag{230}$$

with m giving the rest mass of the particle and $\hbar$ denoting the reduced Planck constant. Comparing with the $f[t]$-function from the metric solution (228), we find:

$$\frac{m\cdot c}{\hbar}=\frac{1}{2\cdot\rho}. \tag{231}$$

Inserting the Schwarzschild radius $r_s=\frac{2\cdot m\cdot G}{c^2}$ (G, Newton's constant; c, speed of light in vacuum), thereby substituting the rest mass m, leaves us with:

$$\frac{r_s\cdot c^3}{2\cdot\hbar\cdot G}=\frac{1}{2\cdot\rho}\quad\Rightarrow\quad\rho=\frac{1}{r_s}\cdot\left(\frac{c^3}{\hbar\cdot G}\right)^{-1}=\frac{\ell_P^2}{r_s}. \tag{232}$$

Here ℓ_P denotes the Planck length. By inserting (60) into (232), we obtain:

$$\rho=\frac{\ell_P^2}{r_s}=\frac{\ell_P^2}{2\cdot\ell_P\cdot\sqrt{\pi\cdot\left(q+\sqrt{q\cdot(1+q)}\right)}}=\frac{\ell_P}{2\cdot\sqrt{\pi\cdot\left(q+\sqrt{q\cdot(1+q)}\right)}}. \tag{233}$$

Thus, while for a black hole the number of bits thrown into it leads to an almost perfectly square-root-like increase of the Schwarzschild radius in accordance with equation (60), the ρ-parameter of the (228)-objects decreases with the number of bits. (228)-objects would have the same ρ-parameter, which we may see as a size, as a black hole only for Schwarzschild radii r_s equal to the Planck length. In other words, for growing black holes with radii bigger than the Planck length, the corresponding equally heavy (228)-objects would be significantly smaller than the black holes.

So, we ask: Could the (228)-objects be used as building blocks for a black hole, residing inside it, which is to say behind the event horizon?

Assuming that the black hole's surface is made out of metric spherical objects of the type (228) and further assuming that each of these objects in the surface of the black hole, which is to say at $r = r_s$ (which also happens to be the event horizon), requires its own surface space of something like $C_\rho{*}\rho^2$, we can directly evaluate the number of such (228)-objects; now on, we name ρ-spheres, residing inside the event horizon with increasing numbers of bits thrown into the black hole. Assuming that the mass is always additive, the total mass m of the black hole must then be distributed among the N ρ-spheres, which changes (232) to:

$$\frac{r_s \cdot c^3}{2 \cdot \hbar \cdot G} = \frac{1}{2 \cdot N \cdot \rho} \Rightarrow N \cdot \rho = \frac{1}{r_s} \cdot \left(\frac{c^3}{\hbar \cdot G} \right)^{-1} = \frac{\ell_P^2}{r_s} \Rightarrow \rho = \frac{\ell_P^2}{N \cdot r_s}. \quad (234)$$

Also having to satisfy the following equation for the N ρ-spheres sitting on the surface, we have to solve the following equation:

$$N \cdot C_\rho \cdot \rho^2 = 4 \cdot \pi \cdot r_s^2$$

$$\Rightarrow \qquad C_\rho \cdot \frac{\ell_P^2}{4 \cdot \pi \cdot N \cdot \left(q + \sqrt{q \cdot (1+q)} \right)} = (4 \cdot \pi)^2 \cdot \ell_P^2 \cdot \left(q + \sqrt{q \cdot (1+q)} \right)$$

$$\Rightarrow \qquad N = \frac{1}{64 \cdot \pi^3 \cdot \left(q + \sqrt{q(1+q)} \right)^2}. \quad (235)$$

We realize that such a structure could not be used to store any information, because the number of ρ-spheres should have to increase with the number of bits and not decrease as it does. Things are improving the moment we allow a combination of ρ-spheres and (227)-objects (the latter we shall call t-spheres) to make up our inner black hole. We propose the following (simplest of the many possibilities) structure:

(A) In the center of the black hole sits a ρ-sphere of "radius-parameter" ρ given in (232) and thus, $\rho = \frac{\ell_P^2}{r_s}$, which is to say, the bigger the Schwarzschild radius r_s of the black hole, the smaller its core. In fact, for infinite masses, the core would become a singularity.

(B) This single ρ-sphere core is surrounded by t-spheres (227) and the number of those t-spheres, which a black hole can bind, is proportional to the number of bits the black hole has swallowed.

(C) Taking the Bekenstein-condition, this demands an average size for the t-spheres, being bound by the black hole or the black hole's surface, to be such that its projected surface would be equal to ℓ_P^2. In other words, we could assume the average radius of the t-spheres (the ones bound to the black hole) to be equal to $\ell_P / \sqrt{\pi}$.

With such a structure, it is very well possible that in fact black holes have no singularity and follow our scheme of inner-outer-solution, but one cannot detect any changes to the classical Schwarzschild solution from the outside, because the inner-parts are always hidden behind the event horizon.

But does this help us to solve the Bekenstein information problem?

Yes, it does.

We can imagine many t-sphere objects (of number $N = q$) sitting on the surface of the black hole. As the generalized solution to (227) would read:

$$g_{\alpha\beta} = C_1 \cdot f[t]^4 \cdot \begin{pmatrix} -c^2 & 0 & 0 \\ 0 & A^2 \cdot t^2 & 0 \\ 0 & 0 & B^2 \cdot t^2 \cdot \sin(r)^2 \end{pmatrix}$$

$$f[t] = \sqrt{t^{\pm\frac{i \cdot c}{A} - 1}} \cdot C_f, \tag{236}$$

we see that each t-sphere could store information not only via a certain sign within the exponent but also via the free parameter B.

6.12.3 The Size of the Electron?

Applying (234) and assuming a ρ-sphere-structure for the electron gives us:

$$\frac{r_s \cdot c^3}{2 \cdot \hbar \cdot G} = \frac{1}{2 \cdot N \cdot \rho}$$

$$\Rightarrow \qquad N \cdot \rho = \frac{1}{r_s} \cdot \left(\frac{c^3}{\hbar \cdot G} \right)^{-1} = \frac{\ell_P^2}{r_s}$$

$$\Rightarrow \qquad \rho = \frac{\ell_P^2}{N \cdot r_s} = \frac{1.93 \times 10^{-13}}{N} \text{meter}. \tag{237}$$

Setting $N = 1$ we would end up with a ρ-sphere of $\rho = 1.93 \times 10^{-13}$meter for the "pure" or "naked" electron.

6.13 Another "Anecdotic" Sidenote

A reviewer who had read one of the first versions of the manuscript of this book had asked the author what "on earth" the information storage problem in black holes or the question of the size of the electron has to do with future warfare.

The author answered with a few questions:

"What has the shape of an arrow tip to do with its ability to penetrate a human body?"

"What has the law of gravity to do with the ballistic curve of a bullet?"

"What has the mass of a lump of uranium to do with the atom bomb?"

Then the reviewer responded and stated:

"Good point! So, all physics is war!"

"No," this author responded, "all science is!"

... and a bit later he added, "But this does not mean that it has to be used in such a way. It should mean, however, that we—the scientists—have to be aware of the consequences of what we are doing. It also—quite automatically—means that we have to exercise caution about who we give our results to. In my humble opinion, the current established science is but a circus of bad clowns and freaks who have prostituted themselves to anti-social elements. By making themselves the whores of politicians and evil globalists, they have rendered themselves parasites of the very societies which are forced to nourish them.

It is for this reason that I would rather give my knowledge to the whole mankind than to provide it to any 'elite' no matter how much 'elite' they may pretend to be."

Chapter 7

Natural and Artificial Intelligence

7.1 From Current Artificial Intelligence to Natural Intelligence

Some may say that even the worst artificial intelligence (AI) is still better than natural stupidity, but the main driving force for making AI more natural is the lack of innovation and creativity from which all current AI concepts suffer. Even the most sophisticated AI algorithms require training and preprogramming and they develop failure the moment they are confronted with situations they have not been trained for.

We think that the reason for this deficiency is to be found in the dominance of linearity in all AI algorithms. As usual with problems like this, a look into the tool-box of nature may help. That being said, the path to overcome these problems is quite clear and—also for pure safety reasons (see Fermi's bottleneck-problem at the beginning of this book)—we will not go deeper into the matter within this study ... which, in other words, just means that we have a solution, but will not give it away to the wrong entities.

Quantum Gravity War: How Will the Nearby Unification of Physics Change the Future of Warfare
Norbert Schwarzer

ISBN 978-981-4968-58-4 (Hardcover), 978-1-032-71070-9 (eBook)
www.jennystanford.com

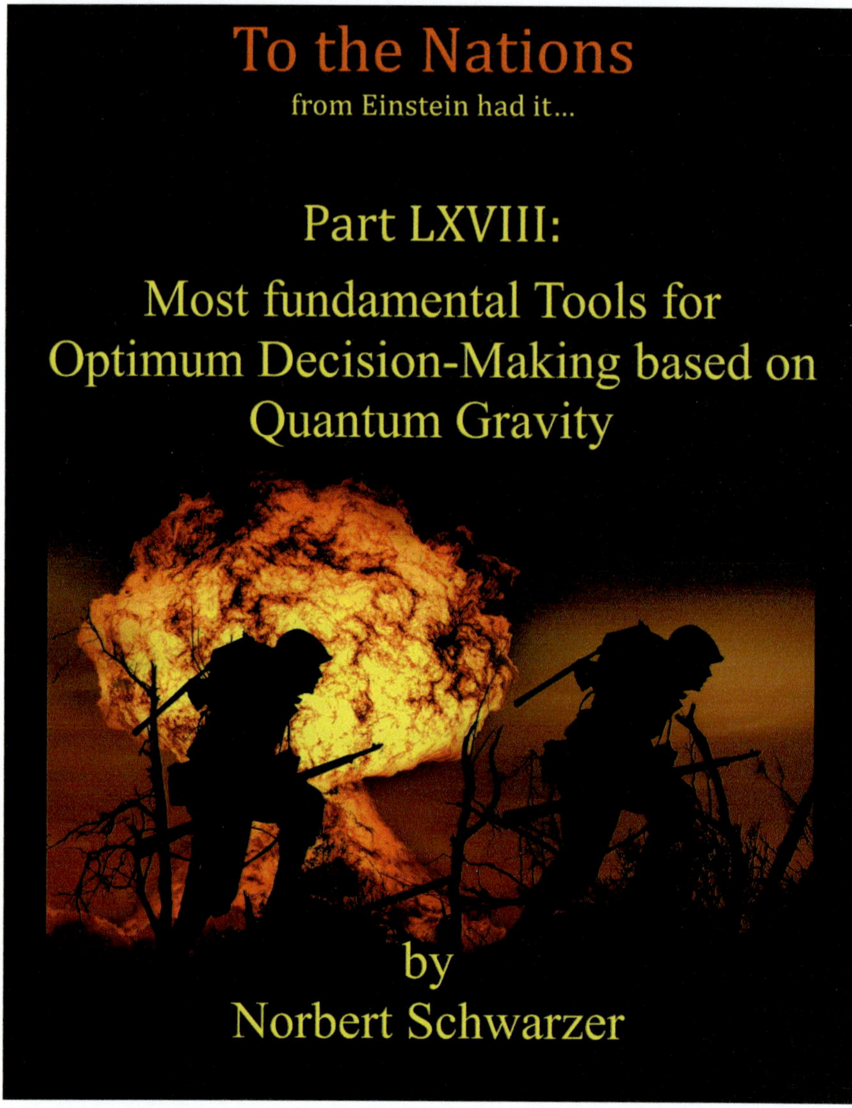

Cover of one of the author's studies about optimum decision-making by the means of a quantum gravity theory [23].

7.2 Towards an Ethic AI via a Metric Approach

The technology proposed above, as being completely based on a minimum principle, does—in principle—perfectly coincide with ethical requirements as it only depends on the holistic adjustment of the boundary conditions, put into each and every process. Thereby this kind of adjustment can lead from practical military (potentially cruel on first and limited sight) strategical optimum (like going for a quick victory rather than aiming for a minimum of casualties) to a "global minimum of suffering" (which is just the correct definition of the ultimately achievable good in a universe governed by minimum principles). In other words, ethical requirements like the ones being defined for future AI development (responsible, equitable, traceable, reliable, and governable) can be met quite easily as they are immanent to the concept, which has been proposed here so far.

Thereby, the "optimum good" requires a purely mathematical description, which has been already derived and presented in [218].

Chapter 8

The Spinor

8.1 A Time-Spatial Understanding of the Current Quantum Computer Tools—Here: The Spin ½

One favored object the quantum computer guys like to play with (and partially pretty much rely on) is the spinor, a funny object which changes its sign after 360° rotation.

Unfortunately, they, the quantum theoreticians, know that these objects exist and they have effective descriptions for them, too, but they have no metric formulation for them and thus, do not know how these objects truly look like and where they come from, which is to say how they appear in a metrical sense.

In fact, simply put: So far nobody knows what the electron metrically looks like.

In a generalized theory of everything, however, one should be able to find such objects and can even illustrate them and show how they are spatially (and timely) distributed. So, for instance, in the previous chapter we created metric quaternion structures to the quantum Einstein field equations, which favored half-spin solutions with respect to the scalar wave function (c.f. metric (202)).

The set of pictures (Figs. 8.1 to 8.18) below shows such a spinor. It arises, when having 8 dimensions (4 are compactified) and demanding a vanishing radius r^2-interaction (which is just a kind of work or energy minimum

Quantum Gravity War: How Will the Nearby Unification of Physics Change the Future of Warfare
Norbert Schwarzer

ISBN 978-981-4968-58-4 (Hardcover), 978-1-032-71070-9 (eBook)
www.jennystanford.com

condition). The corresponding 8-dimensional solution has been presented in the previous chapter and—with a bit more background and discussion in connection with other options— in [17, 18].

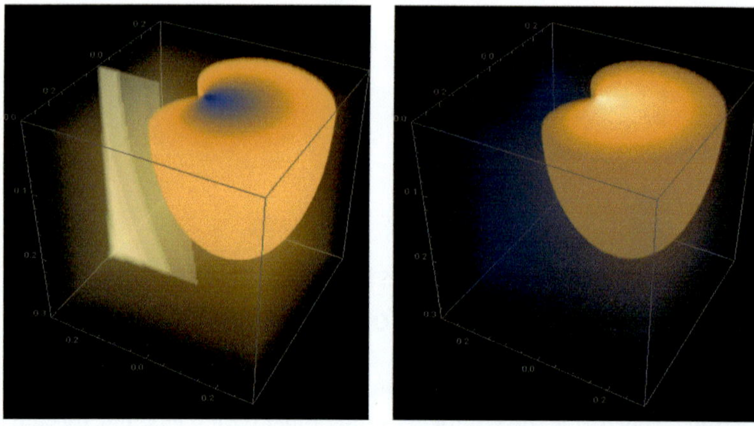

Figure 8.1 Spinor-object as derived from an 8-dimensional space-time (4 dimensions appear compactified to the human "eye"/sensors). Rotation angles 0 (left) and $2*\pi$ (right). Please note the changed color coding, which is clearly showing the sign-switch of the metric object after one full rotation ($360° = 2*\pi$). The corresponding 8-dimensional solution, being used for the evaluation here, has been presented in [17, 18] and was reevaluted and discussed in the previous chapter (c.f. metric solution (202)). Also note that in contrast to earlier presentations (e.g., [17, 18]), here we show the absolute values of the wave function *f* and not the metric wrapper *F*[*f*].

Suddenly there it is: An object which you can turn around 360° and which does not look the same but has changed its sign instead. Only after 720° of rotation, it comes back to its former self.

The picture in the graphics grid below (Fig. 8.2) shows $\pi/4$ rotations in a clockwise manner (top left to bottom right). If this was a classical object, then the eighth picture ($8*\pi/4 = 2\pi = 360°$) should be exactly the same as the first one, but obviously it is not. On the contrary, it has changed its sign and, thus, is a spinor.

8.1.1 What Do We Need This For?

In the military, there is an old saying: "Seeing the battlefield is much better than any model of it ... no matter how good the latter (the model) may be constructed."

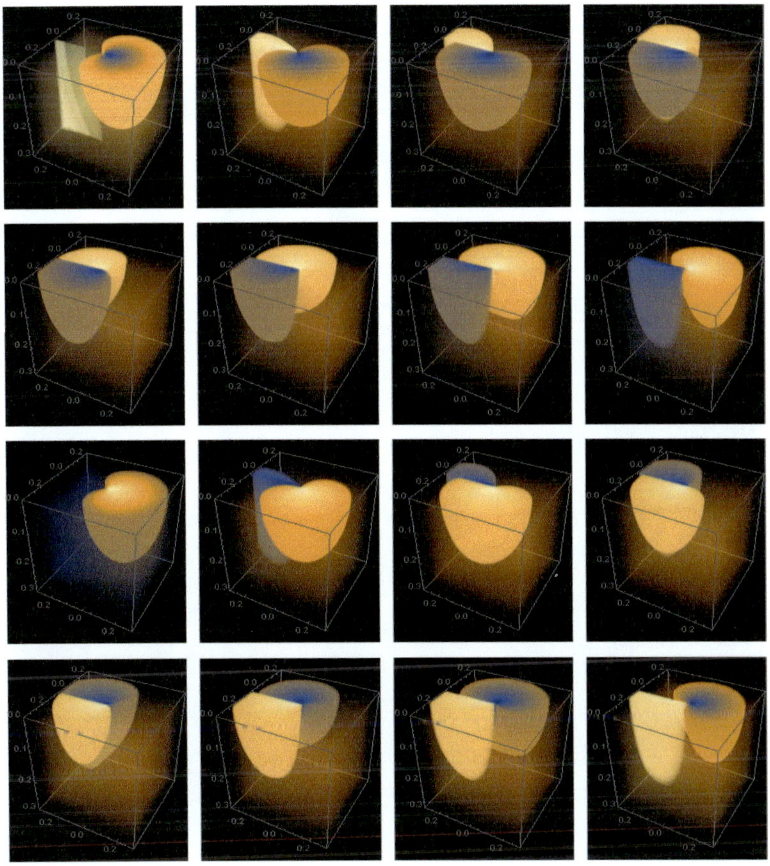

Figure 8.2 Spinor-object as derived from an 8-dimensional space-time (4 dimensions appear compactified to the human "eye"/sensors). Rotation angle from picture to picture is $\pi/4$ clockwise from left to right and π clockwise from up to down. For better resolution, we also show the single pictures in a zoomed presentation in the series of figures below. The corresponding 8-dimensional solution, being used for the evaluation here, has been presented in [17]. Please note that in contrast to earlier presentations (e.g., [17, 18]), here we show the absolute values of the wave function f and not the metric wrapper $F[f]$.

In the current quantum computer approach, there are only models, nothing but effective descriptions of what the theoreticians use to actually do their quantum computing and subsequent analysis. Yes, the models allow for a lot of things, but they do not give a true understanding (nor illustration) of what is going on during the computational processes.

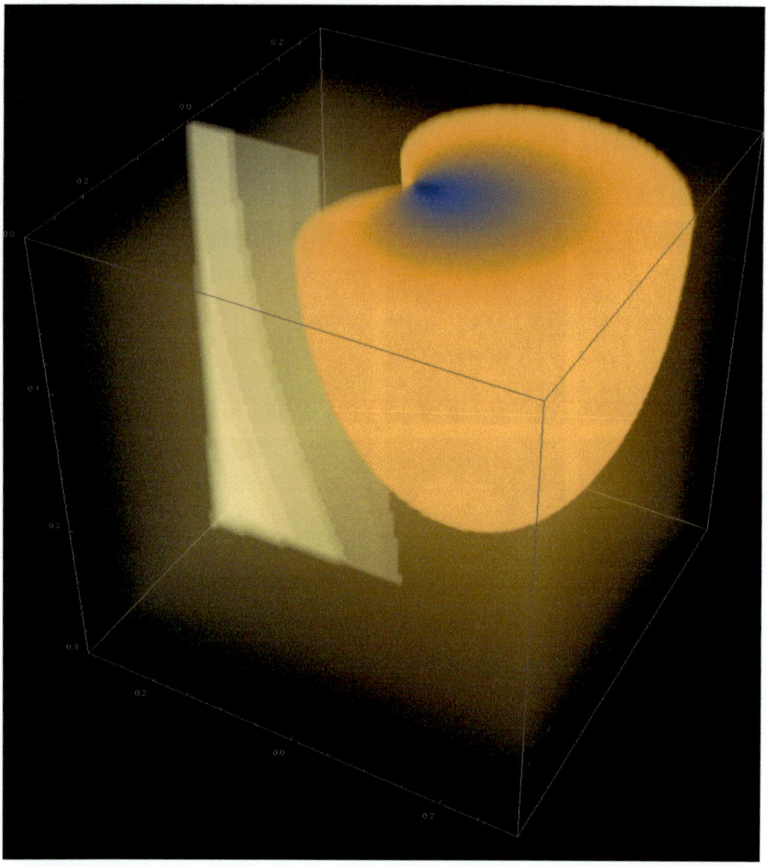

Figure 8.3 Spinor-object as derived from an 8-dimensional space-time (4 dimensions appear compactified to the human "eye"/sensors). Rotation angle 0. The corresponding 8-dimensional solution, being used for the evaluation here, has been presented in [17]. Please note that in contrast to earlier presentations (e.g., [17, 18]), here we show the absolute values of the wave function *f* and not the metric wrapper *F*[*f*].

Obviously, it would be an advantage to have not just an effective, which is to say a purely model-based, but a true, which is to say an illustrative, understanding of what these new machines are doing. This is not, of course not, because one usually feels better to apply something one has understood completely, but to assure that one

(a) uses the full potential of the new tool;
(b) knows all its weaknesses;

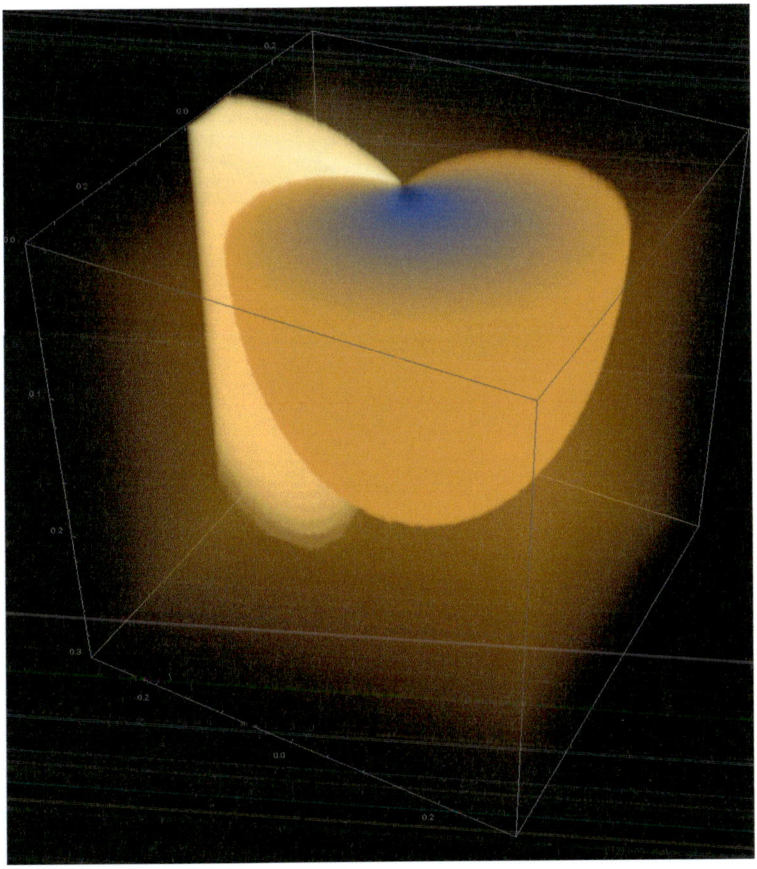

Figure 8.4 Spinor-object as derived from an 8-dimensional space-time (4 dimensions appear compactified to the human "eye"/sensors). Rotation angle $\pi/4$. The corresponding 8-dimensional solution, being used for the evaluation here, has been presented in [17]. Please note that in contrast to earlier presentations (e.g., [17, 18]), here we show the absolute values of the wave function f and not the metric wrapper $F[f]$.

(c) understands its failure mechanisms and can cope with them;

(d) can be absolutely sure that the enemy does not suddenly come up with a sexed-up version of the new technology. Thereby "sexed-up" means that the enemy has it improved so much that one's own machines are not only outdated and useless in just one stroke, but that the enemy can actually turn them against oneself.

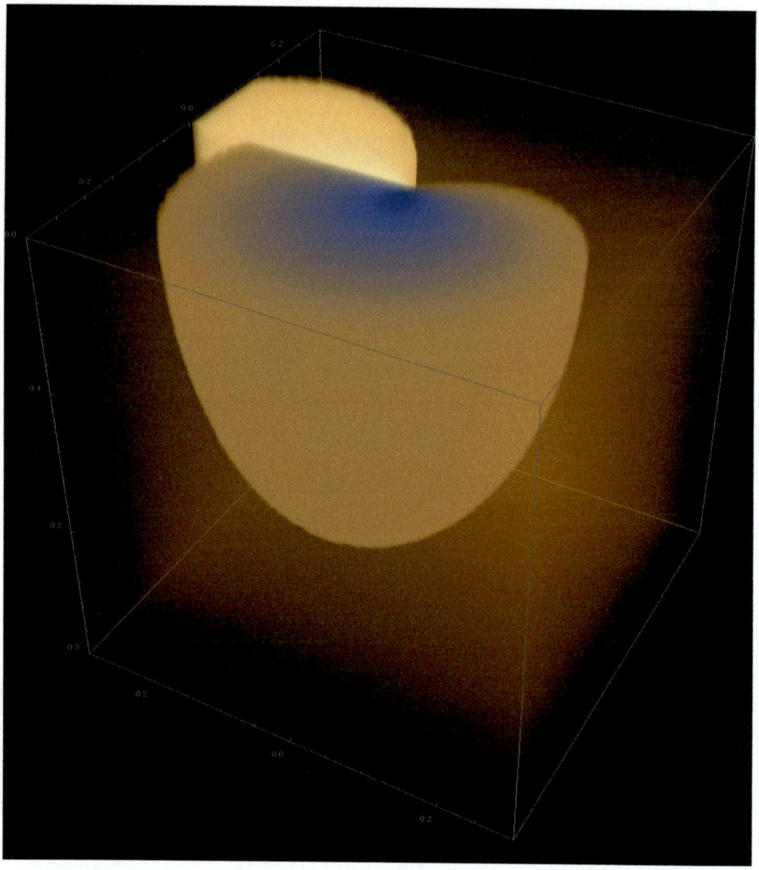

Figure 8.5 Spinor-object as derived from an 8-dimensional space-time (4 dimensions appear compactified to the human "eye"/sensors). Rotation angle $2*\pi/4$. The corresponding 8-dimensional solution, being used for the evaluation here, has been presented in [17]. Please note that in contrast to earlier presentations (e.g., [17, 18]), here we show the absolute values of the wave function f and not the metric wrapper $F[f]$.

So: With the help of a complete theoretical approach, we can actually **see** the spinor battlefield[1] and have a model that allows us to illustrate it (we do also have all the effective models the enemy relies on, by the way, which is a great strategical advantage!), while others only have the effective description, their model (quaternions they call it).

[1]Spinor battlefields (SBFs), in fact, are pretty much all battlefields one needs to control in a quite nearby future of warfare theory.

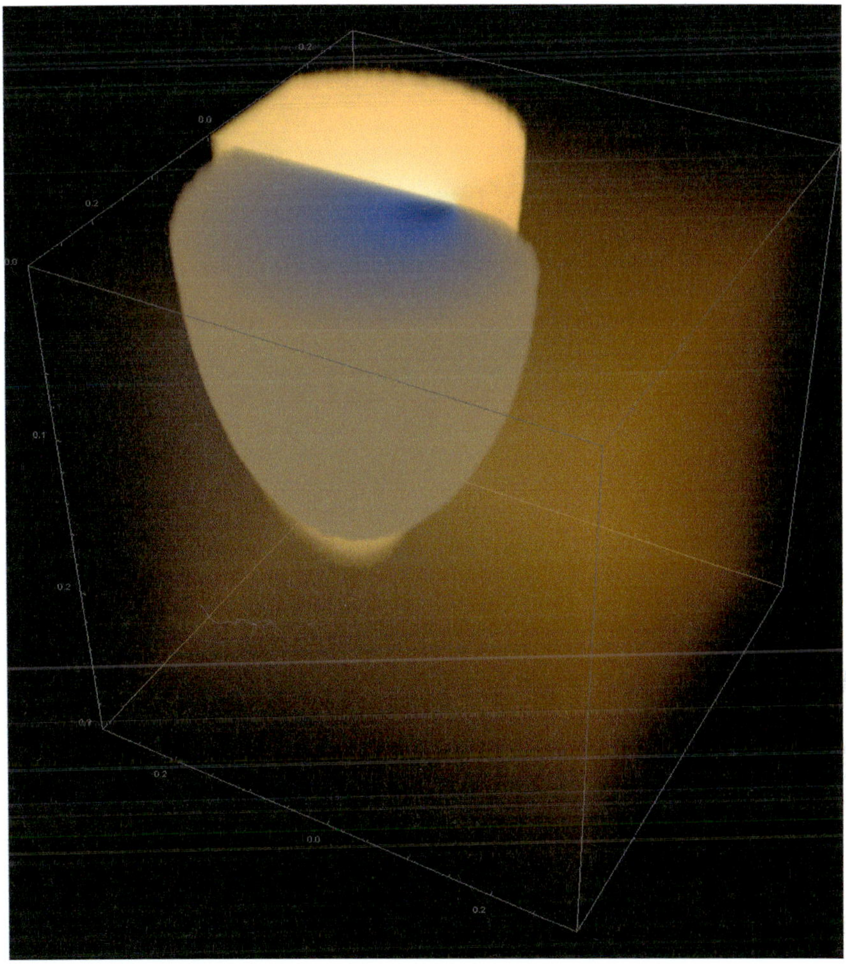

Figure 8.6 Spinor-object as derived from an 8-dimensional space-time (4 dimensions appear compactified to the human "eye"/sensors). Rotation angle $3*\pi/4$. The corresponding 8-dimensional solution, being used for the evaluation here, has been presented in [17]. Please note that in contrast to earlier presentations (e.g., [17, 18]), here we show the absolute values of the wave function f and not the metric wrapper $F[f]$.

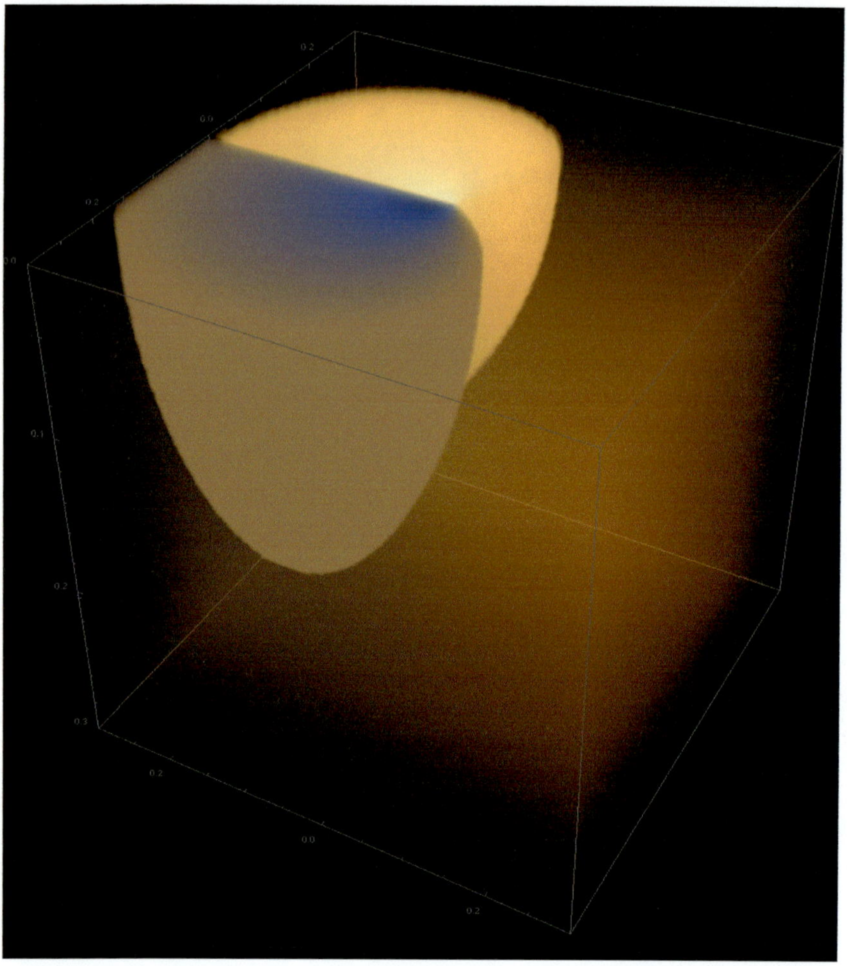

Figure 8.7 Spinor-object as derived from an 8-dimensional space-time (4 dimensions appear compactified to the human "eye"/sensors). Rotation angle $4*\pi/4$. The corresponding 8-dimensional solution, being used for the evaluation here, has been presented in [17]. Please note that in contrast to earlier presentations (e.g., [17, 18]), here we show the absolute values of the wave function *f* and not the metric wrapper *F*[*f*].

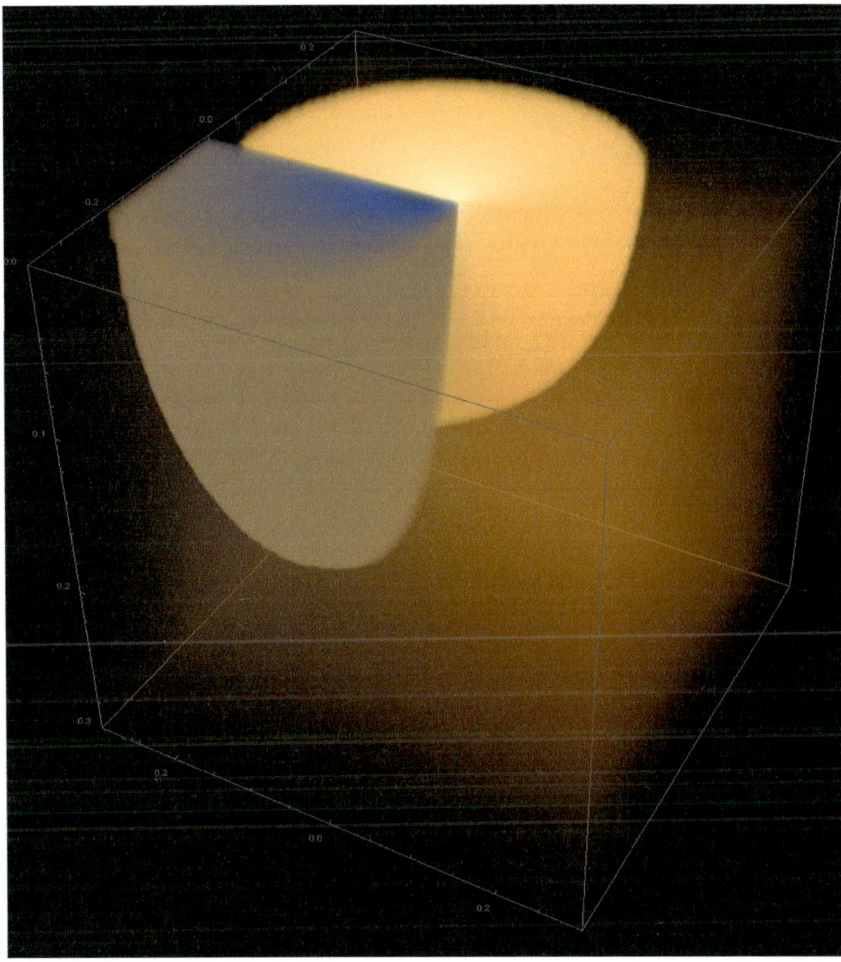

Figure 8.8 Spinor-object as derived from an 8-dimensional space-time (4 dimensions appear compactified to the human "eye"/sensors). Rotation angle $5*\pi/4$. The corresponding 8-dimensional solution, being used for the evaluation here, has been presented in [17]. Please note that in contrast to earlier presentations (e.g., [17, 18]), here we show the absolute values of the wave function f and not the metric wrapper $F[f]$.

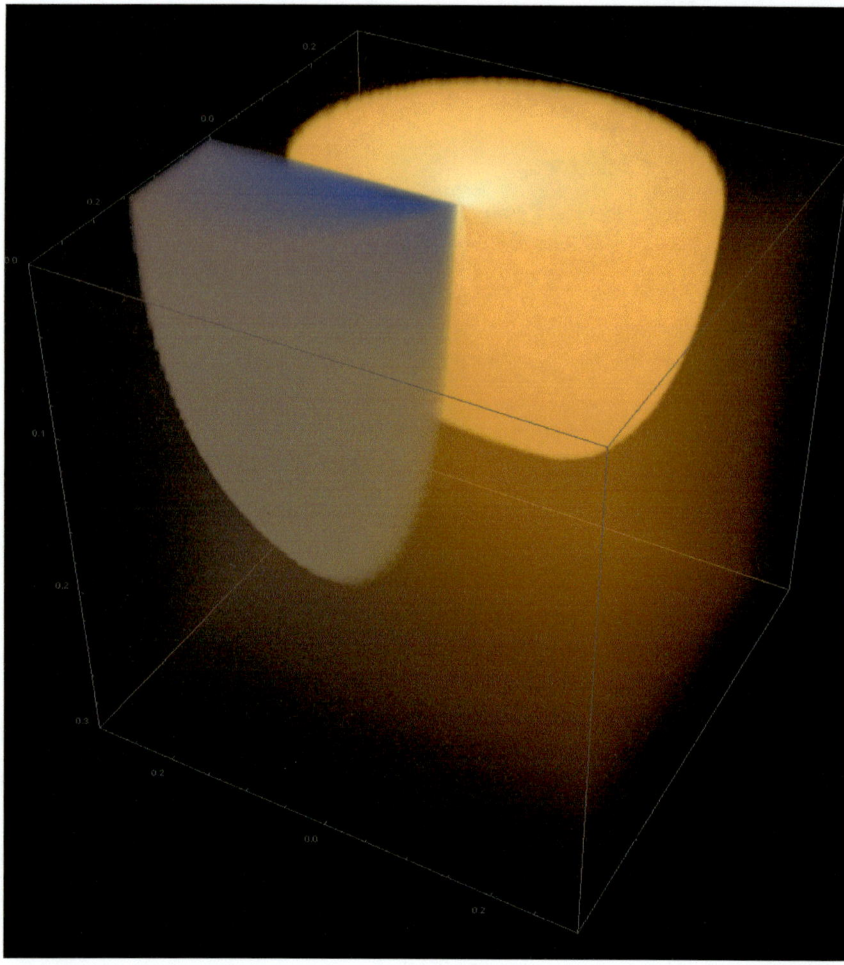

Figure 8.9 Spinor-object as derived from an 8-dimensional space-time (4 dimensions appear compactified to the human "eye"/sensors). Rotation angle $6*\pi/4$. The corresponding 8-dimensional solution, being used for the evaluation here, has been presented in [17]. Please note that in contrast to earlier presentations (e.g., [17, 18]), here we show the absolute values of the wave function f and not the metric wrapper $F[f]$.

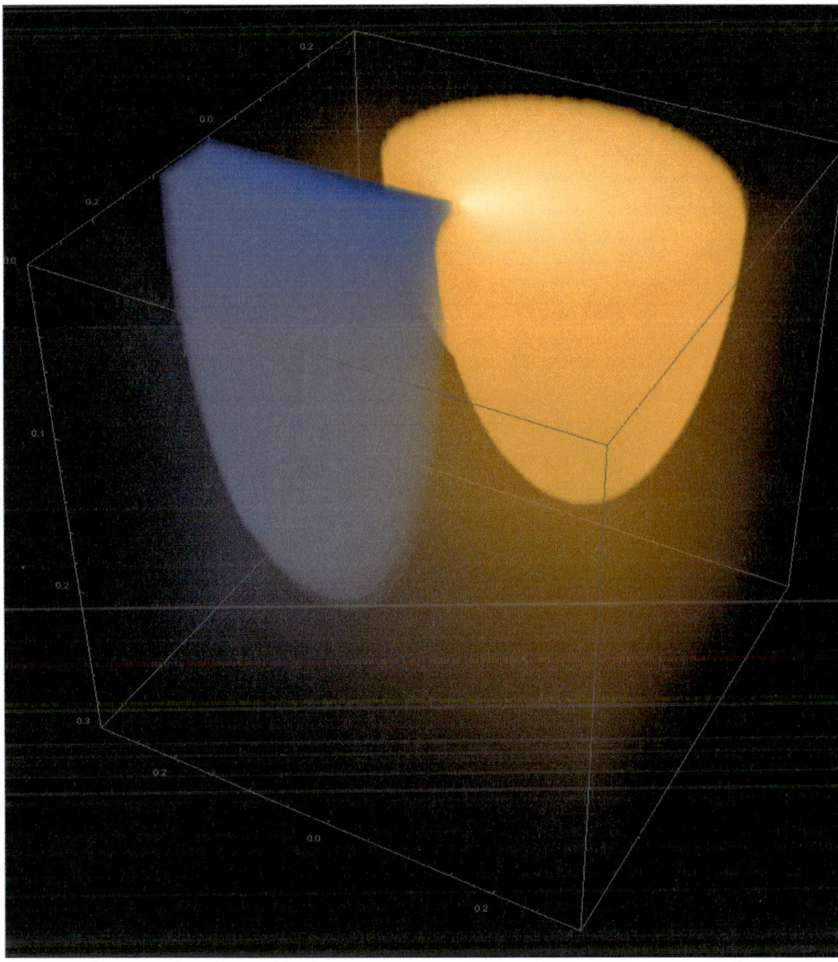

Figure 8.10 Spinor-object as derived from an 8-dimensional space-time (4 dimensions appear compactified to the human "eye"/sensors). Rotation angle 7*π/4. The corresponding 8-dimensional solution, being used for the evaluation here, has been presented in [17]. Please note that in contrast to earlier presentations (e.g., [17, 18]), here we show the absolute values of the wave function *f* and not the metric wrapper *F*[*f*].

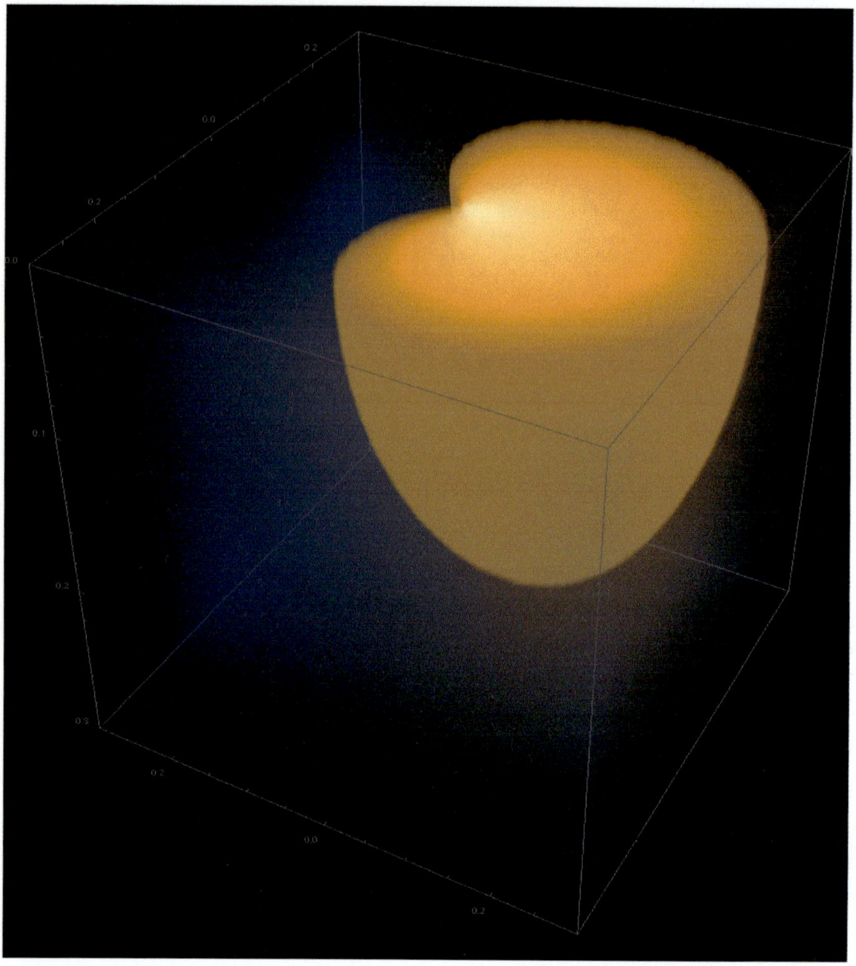

Figure 8.11 Spinor-object as derived from an 8-dimensional space-time (4 dimensions appear compactified to the human "eye"/sensors). Rotation angle $8^*\pi/4$. The corresponding 8-dimensional solution, being used for the evaluation here, has been presented in [17]. Please note that in contrast to earlier presentations (e.g., [17, 18]), here we show the absolute values of the wave function *f* and not the metric wrapper *F*[*f*].

Figure 8.12 Spinor-object as derived from an 8-dimensional space-time (4 dimensions appear compactified to the human "eye"/sensors). Rotation angle $9*\pi/4$. The corresponding 8-dimensional solution, being used for the evaluation here, has been presented in [17]. Please note that in contrast to earlier presentations (e.g., [17, 18]), here we show the absolute values of the wave function *f* and not the metric wrapper *F*[*f*].

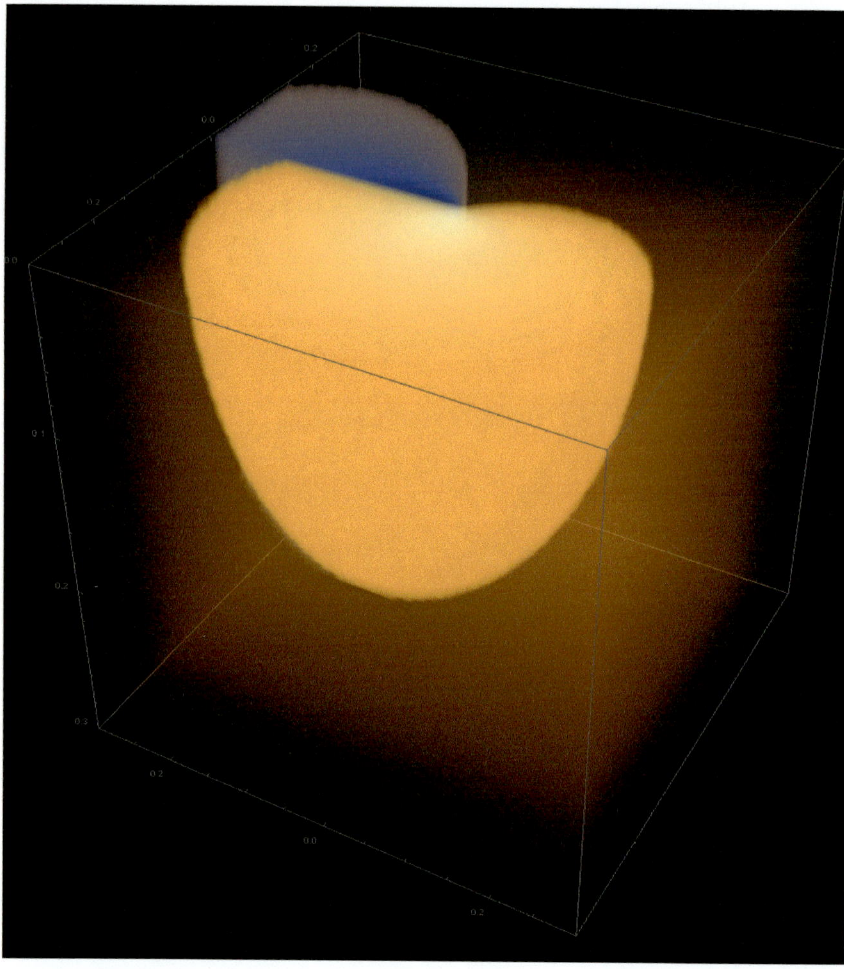

Figure 8.13 Spinor-object as derived from an 8-dimensional space-time (4 dimensions appear compactified to the human "eye"/sensors). Rotation angle $10*\pi/4$. The corresponding 8-dimensional solution, being used for the evaluation here, has been presented in [17]. Please note that in contrast to earlier presentations (e.g., [17, 18]), here we show the absolute values of the wave function f and not the metric wrapper $F[f]$.

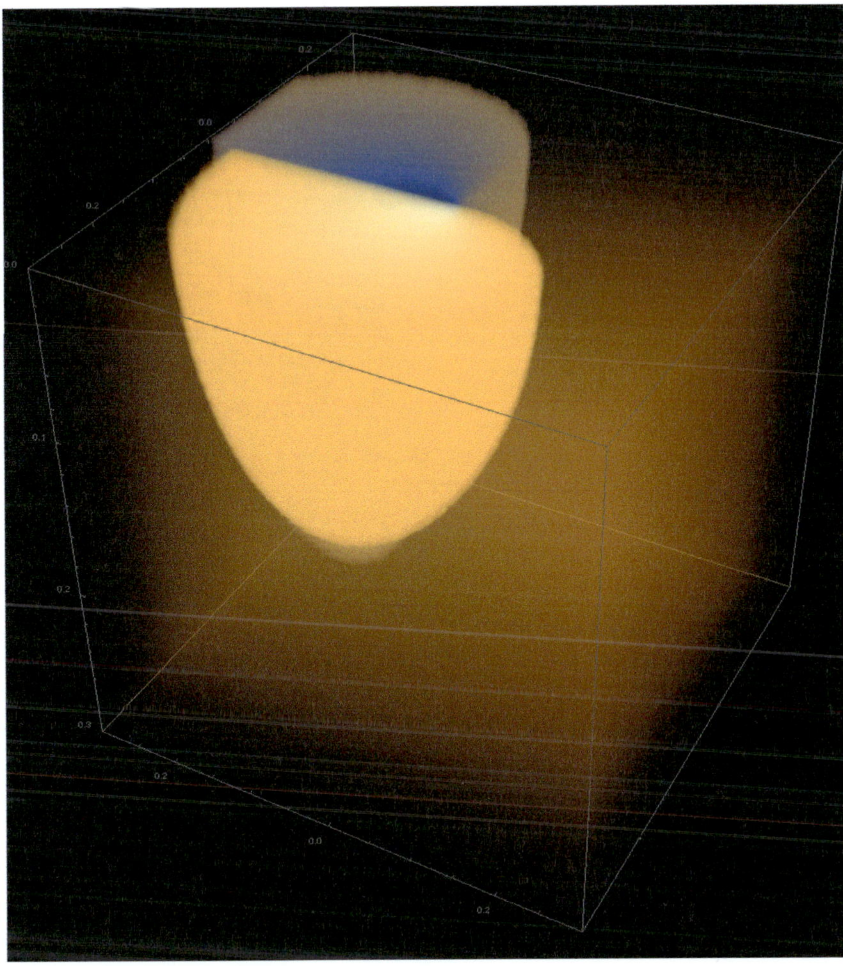

Figure 8.14 Spinor-object as derived from an 8-dimensional space-time (4 dimensions appear compactified to the human "eye"/sensors). Rotation angle $11*\pi/4$. The corresponding 8-dimensional solution, being used for the evaluation here, has been presented in [17]. Please note that in contrast to earlier presentations (e.g., [17, 18]), here we show the absolute values of the wave function f and not the metric wrapper $F[f]$.

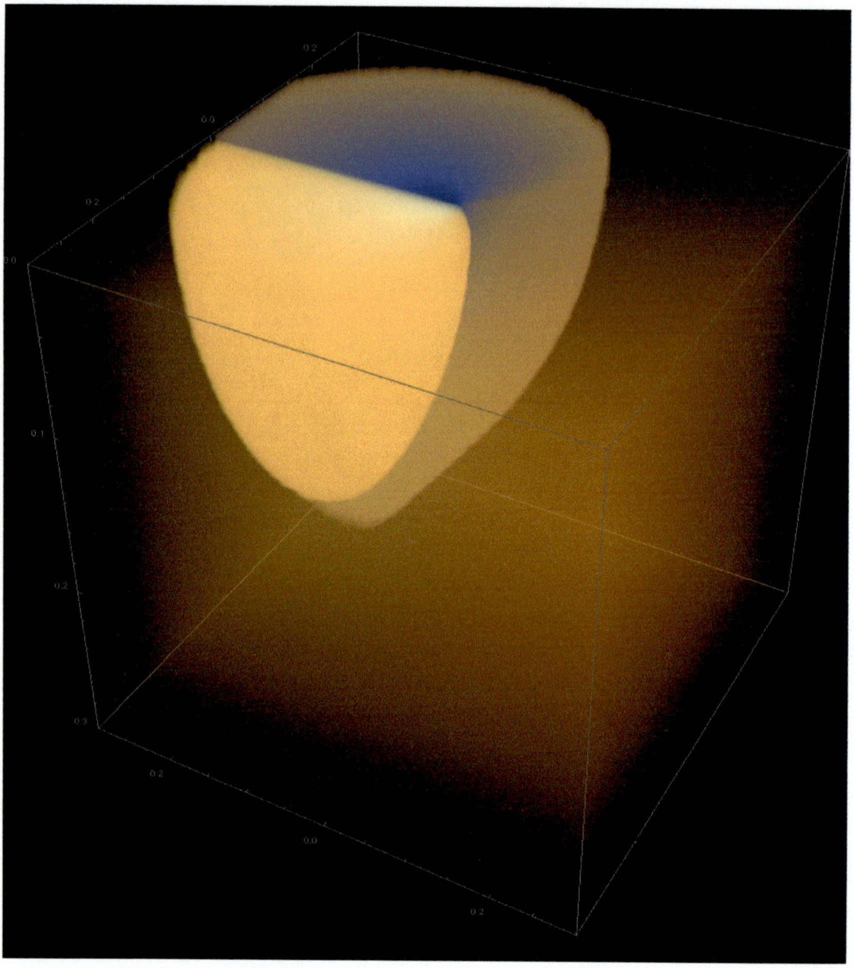

Figure 8.15 Spinor-object as derived from an 8-dimensional space-time (4 dimensions appear compactified to the human "eye"/sensors). Rotation angle $12*\pi/4$. The corresponding 8-dimensional solution, being used for the evaluation here, has been presented in [17]. Please note that in contrast to earlier presentations (e.g., [17, 18]), here we show the absolute values of the wave function f and not the metric wrapper $F[f]$.

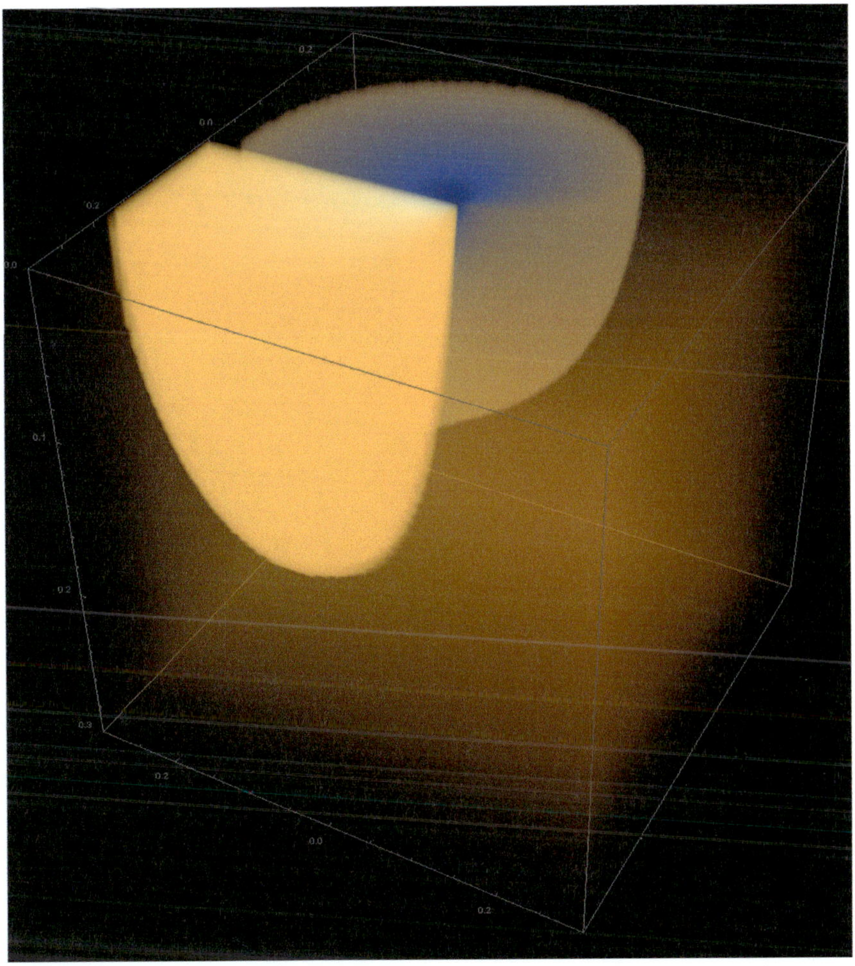

Figure 8.16 Spinor-object as derived from an 8-dimensional space-time (4 dimensions appear compactified to the human "eye"/sensors). Rotation angle $13*\pi/4$. The corresponding 8-dimensional solution, being used for the evaluation here, has been presented in [17]. Please note that in contrast to earlier presentations (e.g., [17, 18]), here we show the absolute values of the wave function *f* and not the metric wrapper *F*[*f*].

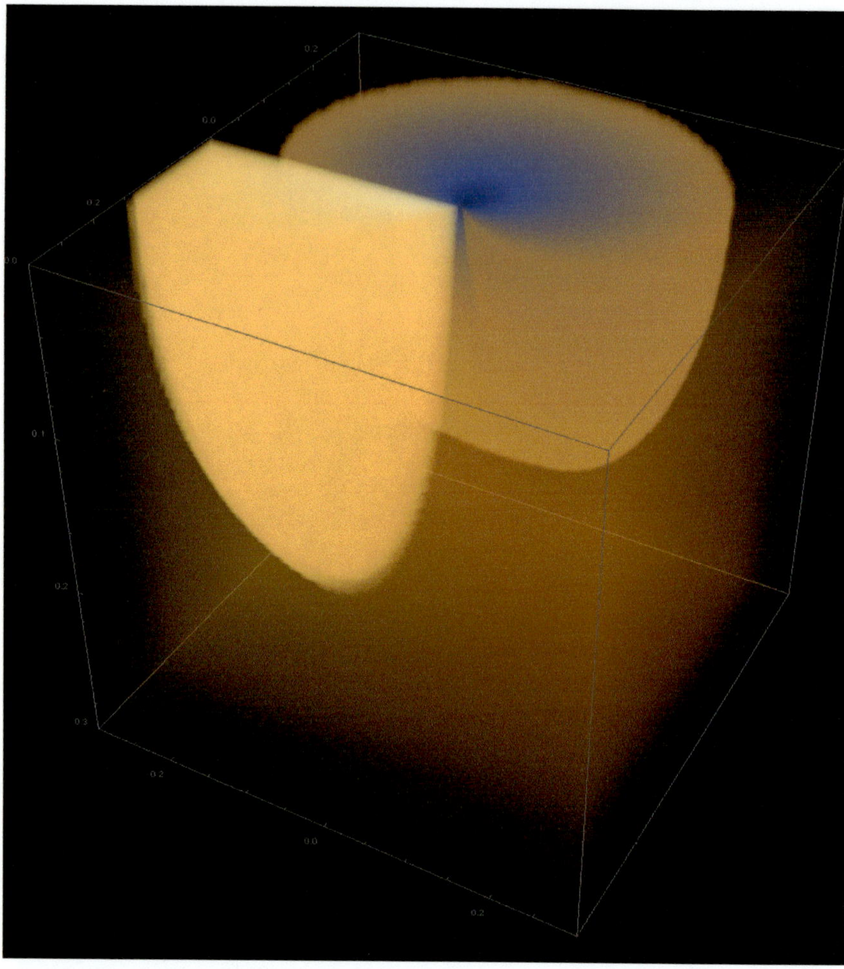

Figure 8.17 Spinor-object as derived from an 8-dimensional space-time (4 dimensions appear compactified to the human "eye"/sensors). Rotation angle $14*\pi/4$. The corresponding 8-dimensional solution, being used for the evaluation here, has been presented in [17]. Please note that in contrast to earlier presentations (e.g., [17, 18]), here we show the absolute values of the wave function f and not the metric wrapper $F[f]$.

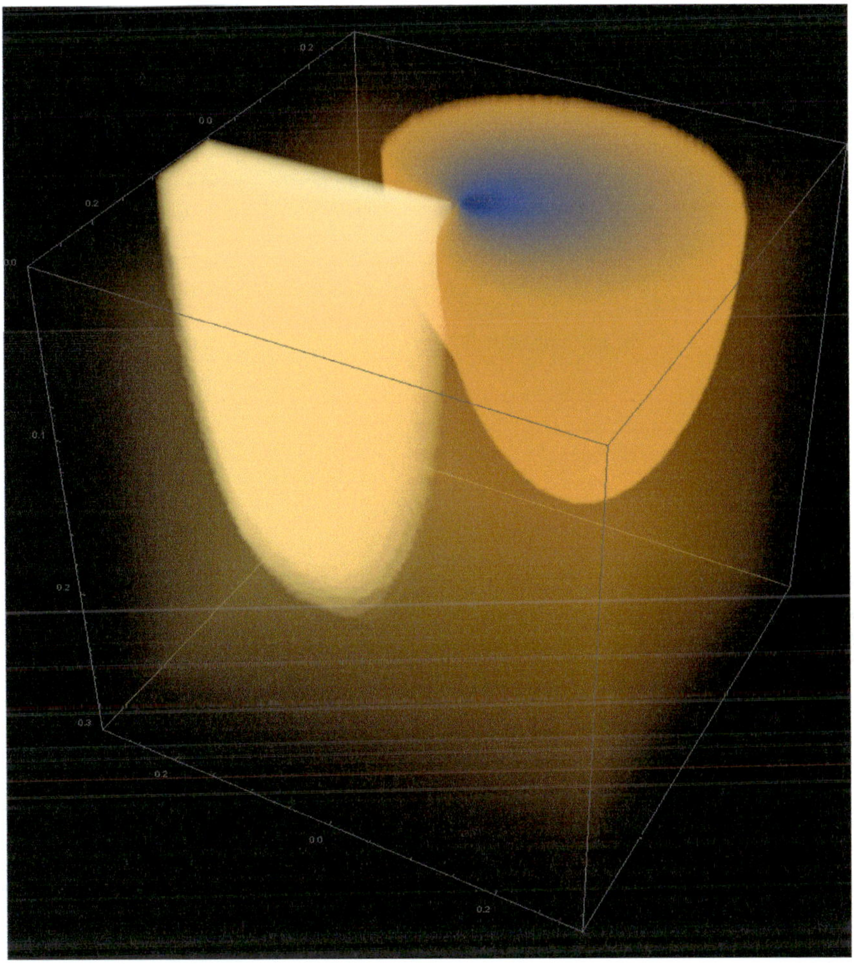

Figure 8.18 Spinor-object as derived from an 8-dimensional space-time (4 dimensions appear compactified to the human "eye"/sensors). Rotation angle $15*\pi/4$. The corresponding 8-dimensional solution, being used for the evaluation here, has been presented in [17]. Please note that in contrast to earlier presentations (e.g., [17, 18]), here we show the absolute values of the wave function f and not the metric wrapper $F[f]$.

Chapter 9

The Brain

9.1 Super Intelligent Soldiers = SIS

In March 2020, a researcher from Canada sent the author this link:

https://singularityhub-com.cdn.ampproject.org/c/s/singularityhub.com/2020/01/14/scientists-discovered-mini-computers-in-human-neurons-and-thats-great-news-for-ai/amp/

It led to an article, entitled "Scientists Discovered 'Mini-Computers' in Human Neurons—and That's Great News for AI," which informs about one of the most interesting discoveries regarding the capability of human neurons:

> "With just their input cables, human neurons can perform difficult logic calculations previously only seen in entire neural networks. To restate: human neurons are far more powerful devices than originally thought. And if deep learning algorithms—the AI method loosely based on the brain that's taken our world by storm—take note, they can be too."

Only a few weeks before this article was published, the author of this book discovered some very principal degrees of freedom regarding the human "neural network," and by analyzing their potential, it became apparent that the human brain could, in principle, locally form supercomputers within or as part of its own structures (Fig. 9.1).

Quantum Gravity War: How Will the Nearby Unification of Physics Change the Future of Warfare
Norbert Schwarzer

ISBN 978-981-4968-58-4 (Hardcover), 978-1-032-71070-9 (eBook)
www.jennystanford.com

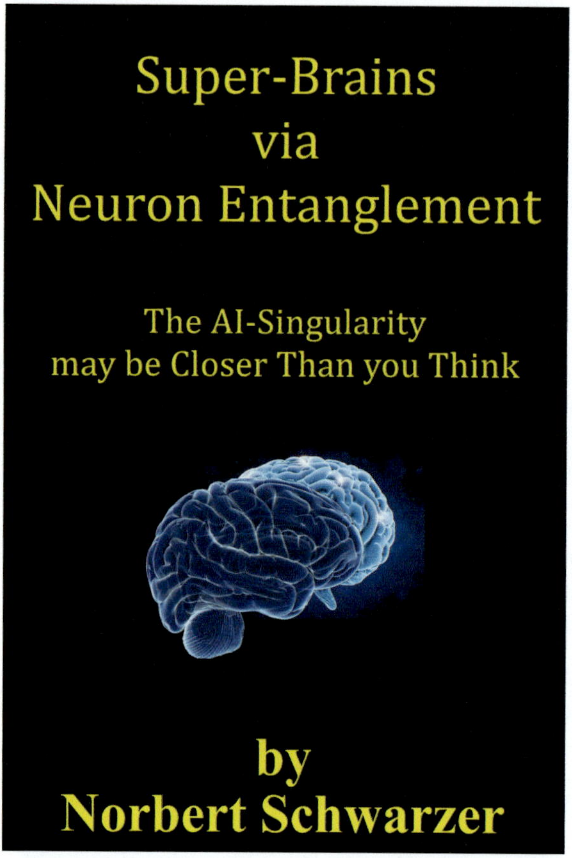

Figure 9.1 Cover picture of one of the author's publications [195] (for illustration only).

This agrees quite nicely with the findings of the scientists from the link given above, but there seems to be more ... much more.

Theoretically, namely, there is no limit to the human neurons to not just combine themselves to any ordinary, which is to say classical, computer structure, but it could even form quantum computers and—so far this author cannot see why not—gravity quantum computer structures. The only question would be how to keep these structures stable enough so that they could actually couple into the conscious mind and be used for something on a tactical or strategic level. This neural coherence may require certain states of natural or artificially supported (human-machine interfaces) concentration.

Combinations of greater groups of individuals with such "brain-reinforcements" may even act together, forming superintelligent swarms [4, 65], which can by far easily and quickly solve any problem and evaluate its potentials in no time (Figs. 9.2 and 9.3).

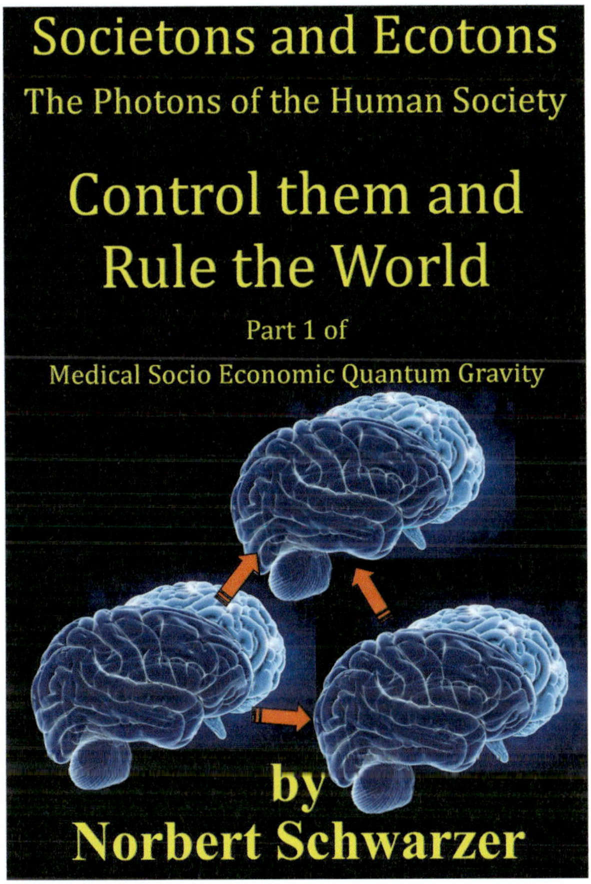

Figure 9.2 Cover picture of one of the author's publications [4] (for illustration only).

Figure 9.3 Cover picture of one of the author's publications [65] (for illustration only).

Chapter 10

The Big Bang Bomb

10.1 What Would be the Ultimate Bomb (or Inexhaustible Source of Energy)?

Using such a bomb makes no sense in combat to a participant who intends to live after, but as such a bomb could be the ultimate threat (terroristic or just as a fundamental equalizer or a doomsday weapon), one should better know about it.

So, it seems that such a thing is extremely powerful—YES indeed—but also quite useless if one does not intend to blow up the whole universe or use it as a "doomsday-bomb" or ultimate threat (like the Israelis with the a-bomb they—so they claim officially—don't have).

But how about a bomb that is not a bomb but just a trigger? ... a trigger for the energy residing at each and every point in space-time (no mathematical but Planck points are being meant here) and which can be set free on demand whenever one wishes to?

Yes indeed, this sounds too fantastic to be true, but shouldn't we better make sure that we know about such an option in case there is such a possibility? After all, there must have been a trigger that had set loose the whole Big Bang story of our universe. So why should this not be possible a second, a third, or an *X* time?

Quantum Gravity War: How Will the Nearby Unification of Physics Change the Future of Warfare
Norbert Schwarzer

ISBN 978-981-4968-58-4 (Hardcover), 978-1-032-71070-9 (eBook)
www.jennystanford.com

Figure 10.1 Cover from [148].

"Because there is no suitable metric solution that allows for such concentrated energy states," critics might answer.

Well, this may have been true until the discovery of the scaled metric and the subsequent quantum gravity field equations. Apart from the Big Bang, namely, we meanwhile have another piece of circumstantial evidence for the possible existence of such confined "space-time energy" within the Planck volumina of our universe. We simply need to observe the so-called ρ-sphere-metrics we have discussed in the appendix of chapter 6, "The Quantum Gravity Turing Machine" (see section 6.12.2, "Consequences from the Bekenstein Thought Experiment Regarding the Solutions to the Quantum

Einstein Field Equations," and section 6.12.3, "The Size of the Electron?"). There we found the possibility for the existence of metric objects of spherical symmetry with increasing mass for decreasing size.

Thus, at least theoretically, such objects can exist and then ...

Space itself is the bomb (or contains it), and tickled[1] with the right frequencies, parts of it can be made to blow up in a controlled manner.

The theory seems to predict the existence of such a bomb when

(a) there is a Higgs field;
(b) there is a photonic background.

And there is also the additional circumstantial evidence of a space-time that seems to expand when being compressed (see [9] and—for convenience—Appendix F in this book).

10.1.1 Inexhaustible Source of Energy

Of course, the concept, if it is controllable, could also lead to an inexhaustible source of energy, whereby energy is just harvested out of space-time.

10.1.2 The Big Bang Bomb as a Simple Quantum Gravity Vacuum Solution

Starting with the following metric:

$$g_{\alpha\beta}=\begin{pmatrix} -H^2\cdot c^2 & 0 & 0 & 0 \\ 0 & A^2 & 0 & 0 \\ 0 & 0 & B^2\cdot r^2 & 0 \\ 0 & 0 & 0 & D^2\cdot\sin[\vartheta]^2\cdot r^2 \end{pmatrix} \quad (238)$$

and demanding a vanishing cosmological constant

$$\Lambda=0, \quad (239)$$

we have shown in [143] that the resulting Ricci scalar R^* of the transformed metric using (1) and the corresponding 4D wrapper-function $F[f=f[t,r,\vartheta,\varphi]] = (C_1+f)^2$ reads:

[1]One simply imagines this like a laser being pointed at the spot where you would like to have the epi-center of the explosion. Then the character of this "laser" could also decide the amount of energy being set free (possibly reaching from just enough to make a human head to burst, via incoming warheads to blow up in a great enough distance from your position, to a complete army being wiped out in just one "tickling" stroke).

$$R^{*}=\frac{1}{F[f]^{3}}\cdot\left(\begin{array}{c}\left(C_{N1}\cdot\left(\frac{\partial F[f]}{\partial f}\right)^{2}-C_{N2}\cdot F[f]\cdot\frac{\partial^{2}F[f]}{\partial f^{2}}\right)\cdot\left(\tilde{\nabla}_{g}f\right)^{2}\\ -C_{N2}\cdot F[f]\cdot\frac{\partial F[f]}{\partial f}\cdot\Delta_{g}f\end{array}\right)$$

$$=-C_{N2}\cdot\frac{\frac{\partial F[f]}{\partial f}\cdot\Delta_{g}f}{F[f]^{2}}\xrightarrow{4D}=-3\cdot\frac{(C_{1}+2\cdot f)}{\frac{1}{4}\cdot(C_{1}+2\cdot f)^{2}}\cdot\Delta_{g}f=-\frac{12\cdot\Delta_{g}f}{C_{1}+2\cdot f},\tag{240}$$

$$\Rightarrow\qquad -\frac{C_{1}+2\cdot f}{12}\cdot R^{*}=\Delta_{g}f=\Delta_{3D-g}f-\frac{f^{(2,0,0,0)}}{H^{2}\cdot c^{2}}.\tag{241}$$

Now we substitute as follows in (241):

$$-\frac{C_{1}+2\cdot\tilde{f}}{12}\cdot R^{*}=\Delta_{g}\tilde{f}=\Delta_{3D-g}\tilde{f}-\frac{\tilde{f}^{(2,0,0,0)}}{H^{2}\cdot c^{2}}$$

$$f\equiv\frac{C_{1}+2\cdot\tilde{f}}{12}$$

$$\Rightarrow\qquad -f\cdot R^{*}=\Delta_{g}\left(6\cdot f-\frac{C_{1}}{2}\right)=6\cdot\Delta_{g}f=6\cdot\left(\Delta_{3D-g}f-\frac{f^{(2,0,0,0)}}{H^{2}\cdot c^{2}}\right).\tag{242}$$

Applying the separation approach $f[t,r,\vartheta,\varphi]=g[t]*h[r]*Y[\vartheta]*Z[\varphi]$ with the separation parameters C_t and m (compare with the sub-section “Flat Space Situation” in [140]) in (241) gives us the following immediate solution for $g[t]$ and $Z[\varphi]$:

$$g[t]=C_{t1}\cdot\cos[H\cdot c\cdot C_{t}\cdot t]+C_{t2}\cdot\sin[H\cdot c\cdot C_{t}\cdot t],\tag{243}$$

$$Z_{m}[\varphi]=C_{\varphi1}\cdot\cos[D\cdot m\cdot\varphi]+C_{\varphi2}\cdot\sin[D\cdot m\cdot\varphi].\tag{244}$$

The two remaining differential equations can be constructed as follows:

$$-\frac{m^{2}}{\sin[\vartheta]^{2}}+\frac{\cot[\vartheta]Y'[\vartheta]+Y''[\vartheta]}{B^{2}Y[\vartheta]}=l(1+l),\tag{245}$$

$$\frac{l(1+l)}{r^{2}}-\frac{\left(2\cdot h'[r]+r\cdot h''[r]\right)}{B^{2}\cdot r\cdot h[r]}=\frac{B^{2}\cdot C_{t}^{2}}{A^{2}}-\frac{R^{*}}{6}.\tag{246}$$

While the first equation of the two gives the usual solution with the Legendre polynomials, as also known from the Schrödinger hydrogen solution [194], only in slightly generalized manner (watch the asymmetry-parameter B):

$$Y_k^m[\vartheta] = C_{Pv} \cdot P_k^{B \cdot m}[\cos[\vartheta]] + C_{Qv} \cdot Q_k^{B \cdot m}[\cos[\vartheta]]; \quad k = \frac{1}{2}\left(\sqrt{1+4B^2 l(1+l)} - 1\right)$$

$$B^2 l(1+l) \equiv L(1+L), \tag{247}$$

we have a rather non-Schrödinger-like solution with respect to the radius coordinate in the case of $R^* = 0$:

$$h[r] = C_j \cdot j_k\left[\frac{B^2}{A} \cdot C_t \cdot r\right] + C_y \cdot y_k\left[\frac{B^2}{A} \cdot C_t \cdot r\right]; \quad k = \frac{1}{2}\left(\sqrt{1+4B^2 l(1+l)} - 1\right). \tag{248}$$

The reader finds an extensive discussion to this solution in [140, 143].

In the case $l = 0$ we can simplify the last two results to:

$$Y[\vartheta] = C_{\vartheta 1} \cdot \cosh\left[m\left(\ln\left[\cot\left[\frac{\vartheta}{2}\right]\right]\right)\right] - i \cdot C_{\vartheta 2} \cdot \sinh\left[m\left(\ln\left[\cot\left[\frac{\vartheta}{2}\right]\right]\right)\right], \tag{249}$$

$$h[r] = \frac{C_{r1}}{r} \cdot e^{-i \cdot r \cdot \sqrt{C_t^2}} + \frac{C_{r2}}{r} \cdot e^{i \cdot r \cdot \sqrt{C_t^2}} \tag{250}$$

where we have set H = A = B = D = 1.

In [140], we answered the question whether we might apply inner-outer approaches in connection with solution (250). A setting of the kind $C_{r2} = -C_{r1}$ allows us to avoid any r = 0-singularity within the function $h[r]$. The rest is just oscillations along the radius coordinate, and if connected with the corresponding time function $g[t]$ (243), thereby choosing $C_{t2} = \pm i^* C_{t1}$ allows us to obtain radial photonic waves:

$$g[t] \cdot h[r] = \frac{C_{r1} \cdot C_{t1}}{r} \cdot e^{\pm i \cdot c \cdot t \cdot C_t}\left(e^{-i \cdot r \cdot C_t} - e^{i \cdot r \cdot C_t}\right)$$

$$= \frac{C_{r1} \cdot C_{t1}}{r} \cdot \left(e^{-i \cdot C_t \cdot (r \pm c \cdot t)} - e^{i \cdot C_t \cdot (r \pm c \cdot t)}\right). \tag{251}$$

But the number of options does not end here, because we could simply demand to have a boundary $r = b$ at which our radial photonic structure should end its oscillations. For this, we simply require the evaluation of the zeros of the following equation:

$$\frac{\left(e^{-i\cdot b\cdot C_t} - e^{i\cdot b\cdot C_t}\right)}{b} = 0 \tag{252}$$

with respect to C_t. In result, we will obtain a series of quantized $C_t \to C_j$, with j being a quantum number. Please note that we could choose the boundary condition (252) in such a way that we see (251) as the inner solution and the Schwarzschild metric function for the time component as the outer one (c.f. corresponding section "Inner and Outer Approaches" in [140], where we also introduced a generalized Schwarzschild solution). One might try a function $g_r[r]$ within the outer solution. The boundary conditions should then read:

$$C_r \cdot \frac{\left(e^{-i\cdot b\cdot C_t} - e^{i\cdot b\cdot C_t}\right)}{b} = 1 - \frac{r_s}{g_r[b]}$$

$$C_r \cdot \left[\frac{i\cdot\left(e^{-i\cdot b\cdot C_t} - e^{i\cdot b\cdot C_t}\right)}{b} - \frac{\left(e^{-i\cdot b\cdot C_t} - e^{i\cdot b\cdot C_t}\right)}{b^2}\right] = \frac{r_s}{g_r[b]^2} \cdot g_r'[b]. \tag{253}$$

Direct solution with $g_r[r]$, however, does not provide us with a suitable radius function for the outer region:

$$r_s \cdot \left[1 - C_r \cdot \frac{\left(e^{-i\cdot r\cdot C_t} - e^{i\cdot r\cdot C_t}\right)}{r}\right]^{-1} = g_r[r]. \tag{254}$$

In [140], we therefore only searched for asymptotic solutions, which, most surprisingly, still led to forms perfectly mirroring the experimental findings in connection with the Schwarzschild metric.

Now we want to get a feeling about the possible values for these C_j, and by taking the Dirac solution for a particle at rest (e.g., [181]), we find that:

$$C_j = \frac{M\cdot c}{\hbar}. \tag{255}$$

In order to stay in the metric picture, we substitute the mass M by the Schwarzschild radius $r_s = \frac{2\cdot M\cdot G}{c^2}$ and obtain:

$$C_j = r_s \cdot \frac{c^3}{2\cdot\hbar\cdot G} = \frac{r_s}{2\cdot\ell_P^2}. \tag{256}$$

Remember that l_P denotes the Planck length of about 1.616 × 10^{-35} meters.

Now we assume a space full of objects, being made just out of spherical photons of the kind (251), which all end at a certain maximum radius $r = b$. The idea for such a construction was already discussed in a series of partially fictive, partially non-fictive stories (e.g., see the introduction part of [3]). The spherical objects were named "Friedmanns" as they resemble minuscule universes. It was also shown within the series of "Einstein had it ..." (see the reference list of this book) that such a structure quite easily explains the quantum properties observable in this universe (alternatively [3] and references given there).

Additionally, assuming that these little objects have a boundary which itself is just as small as the Planck length times a certain number q, giving the objects a diameter of $2 \cdot l_P \cdot q$, we end up with the condition $b = l_P \cdot q$. Setting $C_{r1} = i$ gives us the boundary condition and its solutions as follows:

$$\frac{2\cdot\sin\left[C_j\cdot b\right]}{b}=0\Rightarrow C_{j,k}=\left\{\frac{2\cdot\pi\cdot j}{b},\frac{\pi+2\cdot\pi\cdot k}{b}\right\};\quad j,k=0,1,2,3,\ldots. \tag{257}$$

This yields the subsequent lowest possible wave number ($j = k = 0$) to be:

$$C_{0,0}=\frac{\pi}{\ell_P\cdot q}, \tag{258}$$

which, according to the Dirac-exponent expressed in metric units (256), corresponds to a Schwarzschild radius as:

$$C_{0,0}=\frac{\pi}{\ell_P\cdot q}=\frac{r_s}{2\cdot\ell_P^2}\quad\Rightarrow\quad r_s=\frac{2\cdot\pi\cdot\ell_P}{q}. \tag{259}$$

The next three wave numbers would be:

$$C_{1,0}=\frac{2\cdot\pi}{\ell_P\cdot q}=\frac{r_s}{2\cdot\ell_P^2}\quad\Rightarrow\quad r_s=\frac{4\cdot\pi\cdot\ell_P}{q}$$

$$C_{0,1}=\frac{3\cdot\pi}{\ell_P\cdot q}=\frac{r_s}{2\cdot\ell_P^2}\quad\Rightarrow\quad r_s=\frac{6\cdot\pi\cdot\ell_P}{q}$$

$$C_{2,0}=\frac{4\cdot\pi}{\ell_P\cdot q}=\frac{r_s}{2\cdot\ell_P^2}\quad\Rightarrow\quad r_s=\frac{8\cdot\pi\cdot\ell_P}{q}. \tag{260}$$

We immediately see that in order to avoid the formation of black holes, which would appear the moment we would have the condition $r_s \geq l_P$, we need to have certain minimum object sizes (q-numbers) in order to allow the object to hold a certain photon of a given wave number (or consisting of such). Thus, the simplest object only containing one wave with $C_{0,0}$ would require a $q > 2^*\pi$ in order to avoid to turn into a black hole. This corresponds to a minimum size of the object as follows:

$$C_{0,0}: \; b > r_s \;\; \Rightarrow \;\; b > 2\cdot\pi\cdot\ell_P. \tag{261}$$

In conclusion, a universe being considered as grain-like with small photonic spheres to make up its fundamental building bricks restricts the smallest possible size of these structures to a minimum of at least two times π times the Planck length. With the size of these structures close to this limit[2], the vacuum energy density of the universe would be almost equal to the Planck energy density of 4.633×10^{113} J/m^3. Increasing the object-sizes *b* leads to a decrease in the corresponding vacuum energy density proportional to b^{-3}.

Thus, spherical photonic objects with quantum solutions of the type (251) would contain significant amounts of energy even at sizes of many thousand times the Planck length (please also see footnote 2). On the other hand, it is known that by using the upper limit of the cosmological constant, the vacuum energy of free space has been estimated to be 10^{-9} J/m^3.

Please note, therefore, that the assumption of small structures as fundamental building bricks should not be restricted to such photonic ensembles. However, there would also be another, perhaps more interesting, if not to say elegant, explanation for the obvious discrepancy of the observed cosmological vacuum energy density and the quantum theoretically necessary one. One might simply assume that the spherical photons are standing waves being trapped inside the boundary $r = b$. With the parameter settings already introduced in order to avoid r = 0-singularities, we automatically obtain standing waves of the types:

$$g[t]\cdot h[r] = \frac{C_{r1}\cdot C_{t1}}{r}\cdot e^{\pm i\cdot c\cdot t\cdot C_t}\left(e^{-i\cdot r\cdot C_t} - e^{i\cdot r\cdot C_t}\right)$$

$$= \frac{C_{r1}\cdot C_{t1}}{r}\cdot e^{\pm i\cdot c\cdot t\cdot C_j}\cdot 2\cdot \sin\left[r\cdot C_j\right]. \tag{262}$$

With the C_j chosen in accordance with condition (257), we could easily have ensured an inner-outer constellation of the kind:

---

[2]Please note that here we do not consider the possibility of solutions to the quantum Einstein field equations with completely different scale and size behavior as derived in [13, 14, 16, 17, 18].

$$g[t]\cdot h[r]=\begin{cases}\dfrac{C_{r1}\cdot C_{t1}}{r}\cdot \mathrm{e}^{\pm i\cdot c\cdot t\cdot C_j}\cdot 2\cdot \sin\left[r\cdot C_j\right] & \text{for} \quad r\leq b \\ 0 \quad \text{for} \quad r>b & \end{cases}. \tag{263}$$

As a by-product, even higher wave numbers would now be possible as long as they follow the boundary condition (257) and sum up to standing waves (262) being completely reflected at the boundary $r = b$. Then it might well be that the true vacuum energy density would be even higher than just the Planck energy density, but as this energy is trapped inside our photonic spheres, it has no effect on anything bigger than $r = b$ ... this does at least hold to the physical means we have at hand so far. As a consequence, we might also see black holes just as such spherical photons with all their energy being trapped inside a sphere of the size $r \leq b$ and consisting of an ensemble of spherical photons all forming standing waves of the kind as given in (262).

And why can the waves not come out of the spherical photon-like objects?

Because—so the possibly simplest explanation—outside there is no substrate which could transport them. With the ordinary space being subdivided into just many photon-spheres of sizes $r = b$ (and potentially bigger), waves with wavenumbers bigger than $C_t = C_j = 1/b$ will find no spatial ground on which they could propagate. Only waves of lengths of several Planck lengths find enough spatial (substrate) material to be supported and can exist within our usual space-time. Between the two scales, there is a gap ensuring the absence of certain ultra-energetic photons in our visible part of the universe. With such a constellation, we, which is to say our visible universe(s), are sitting on an energy bomb millinillion times more powerful in each piece of Planck volume than our whole universe could bring together. One way to understand the nature of the "substrate" might simply be to assume additional dimensions which are curled up at very low scales (Planck scale). So, for instance, we find spherical photonic solutions in principle also higher dimensional spaces and they are all sporting a radial dependency of the kind:

$$h[r]=\frac{C_J\cdot J_k\left[C_t\cdot r\right]+C_Y\cdot Y_k\left[C_t\cdot r\right]}{r^{\frac{n-1}{2}}}; \quad k=\frac{1}{2}\sqrt{1+4l(1+l)-2n+n^2}. \tag{264}$$

Here J, Y are the Bessel functions of the first and second kind, l is the momentum quantum number as before and n gives the number of angular coordinates forming the $n + 1$-sphere. We see that with $C_Y = 0$, we can always construct a free-of-singularities solution and the zeros of the J-Bessel functions ensure that $h[r = b] = 0$ can always be satisfied by a proper choice of C_t. Simply assuming that one of the angular coordinates/dimensions is compactified and does not allow the hyper-$n + 1$-photon to propagate outside the $n + 1$-sphere leads to the desired confinement of the photonic energy.

What happens if space is compressed as in black holes?

In black holes, on the other hand, these photonic spheres are broken up into much smaller sub-structures (perhaps even something close to a continuum), allowing objects with wavelengths following condition $C_j > 1/b$ to exist and to freely propagate. In connection with the compactified dimensions, a black hole could easily be understood as a sphere of higher dimensions than the surrounding $3 + 1$-dimensional space-time.

What does this mean in connection with the overall universe?

The same restriction probably also holds for the outside of our universe, where we would expect to find no wavelength smaller than $1/B$, with B now being the current radial size of our universe. This should at least hold as long as there is not another universe like ours (and existing outside ours) also holding a set of higher energy photons compared to its own surrounding vast—almost—nothingness.

The universe as a fractal pattern

The attentive reader might already guess where this is leading us, namely to a self-similar structure with patterns of photon-spheres, potentially existing in higher dimensional bubbles, actually being universes, inside bigger photon-spheres, inside bigger ... and so on. As we already saw in [113] that the center of a black hole could be very well the perfect birthplace for a universe with the properties as those we find in our cosmos, we might find it less peculiar to understand the world as fractal. As, in addition, we have many other examples in many different fields pointing towards fractal structures and repetitive patterns [114, 121, 122, 196], we should not be surprised at all by our illustrative and—at least in the opinion of this author—rather obvious interpretation of the mathematical findings above.

And how to set this trapped energy free?

We better not even try to answer this question here ... even if we knew it, which does not mean that we actually do know it.

Chapter 11

The Quantum Gravity Enigma Machine

11.1 The Foreseeable Quantum Enigma Disaster

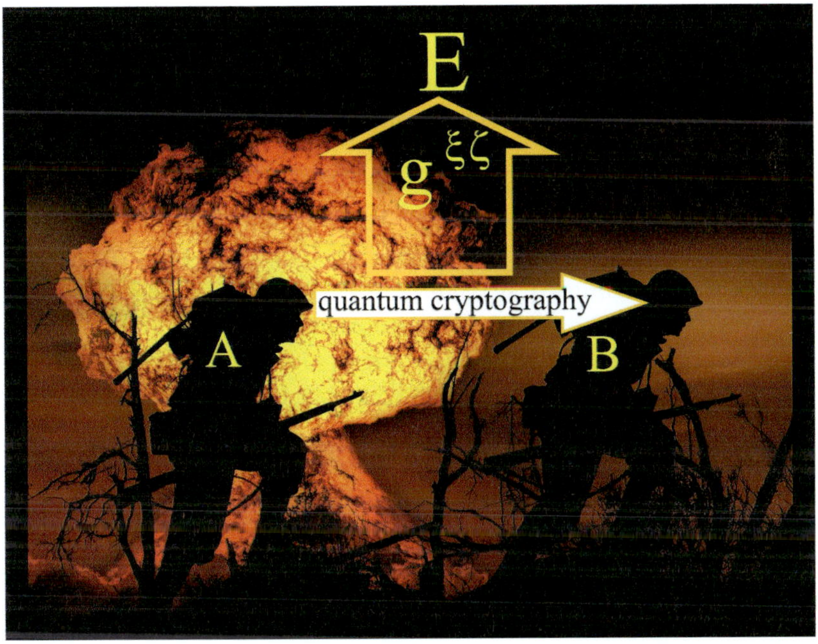

Figure 11.1 The other Enigma Disaster [219].

Quantum Gravity War: How Will the Nearby Unification of Physics Change the Future of Warfare
Norbert Schwarzer

ISBN 978-981-4968-58-4 (Hardcover), 978-1-032-71070-9 (eBook)
www.jennystanford.com

11.1.1 Abstract: Why Will Quantum Cryptography Not Save Us?

Hint: Even though this chapter is part of a book about and with military applications, it does not mean that the solution described in it could not also be a very profitable investment. In fact, and like with the Big Bang bomb in the previous chapter, it may well be the most profitable investment there can be.

Quantum cryptography is assumed to be—almost—absolutely safe [A1] (if performed properly).

The reason lies in the funny properties of the quantum reality, which allows the detection of any eavesdropping on quantum encrypted (entangled) data streams.

But what if this basic assumption is fundamentally flawed?

After all, the current Quantum Theory—on which the whole quantum cryptography is based—does not contain gravity and thus cannot be considered complete.

So, what if a completed theory, a quantum gravity, does destroy the illusion of absolutely safe cryptography, being based on the means of quantum entanglement?

What if a whole world became dependent on this quantum encryption technology, simply because quantum computers or—even worse—quantum gravity computers [A2, A3, A4] (and chapter 5, "The New Abacus," and chapter 6, "The Quantum Gravity Turing Machine") can crack each and every code with a blink of an eye, thereby destroying the safety of all classical—non-quantum or non-entangled—cryptographic means and suddenly there was a way to read all quantum encrypted messages without the slightest risk of detection of eavesdropping?

What if only one entity possesses such a technique and nobody else does know about this or even isn't aware of its potential existence?

In the chapter, it will be shown that quantum cryptography is just another Enigma illusion. Quantum gravity will make sure of that [A6–A9]. But the latter also guarantees the existence of an ultimate solution to each and every cryptographic problem.

11.1.1.1 References for the Abstract

[A1] F. Grasselli, *Quantum Cryptography—From Key Distribution to Conference Key Agreement*, 1st ed. (2021), ISBN: 303064359X.

[A2] N. Schwarzer, *Einstein Had It, but He Did Not See It—Part XXXIX: EQ or The Einstein Quantum Computer*, Self-published, Amazon Digital Services, 2018, Kindle.

[A3] N. Schwarzer, *The Einstein Quantum Computer—Mathematical Principle and Transition to the Classical Discrete and Quantum Computer Design*, Self-published, Amazon Digital Services, 2018, Kindle.

[A4] N. Schwarzer, *Is There an Ultimate, Truly Fundamental and Universal Computer Machine?*, Self-published, Amazon Digital Services, 2019, Kindle.

[A5] N. Schwarzer, *Towards Quantum Einstein Field Equations*, Part 7 of "Medical Socio-Economic Quantum Gravity," Self-published, Amazon Digital Services, December 2020, Kindle.

[A6] N. Schwarzer, *The Quantum Einstein Field Equations in Their Simplest Form*, Part 7a of "Medical Socio-Economic Quantum Gravity", Self-published, Amazon Digital Services, April 2021, Kindle.

[A7] N. Schwarzer, *The World Formula: A Late Recognition of David Hilbert's Stroke of Genius*, ISBN: 9789814877206, Jenny Stanford Publishing.

[A8] N. Schwarzer, *The Dirac Miracle*, Part 8a of "Medical Socio-Economic Quantum Gravity", Self-published, Amazon Digital Services, December 2021, Kindle.

[A9] N. Schwarzer, *My Horcruxes—A Curvy Math to Salvation*, Part 9 of "Medical Socio-Economic Quantum Gravity", Self published, Amazon Digital Services, 2021, Kindle.

11.2 Quantum Gravity Equations

For convenience, here we repeat the necessary basics and add in what is the special need with respect to the topic of this chapter. Thereby, we will only need to assume metrics of the type (2) and also apply the abbreviation $R_{\alpha\beta}^{***}+R_{\alpha\beta} \quad \Rightarrow \quad R_{\alpha\beta}$.

As demonstrated in chapter 3, "The Theory," when starting with the classical Einstein–Hilbert action:

$$\delta W = 0 = \delta \int_V d^n x \left(\sqrt{-G}\cdot\left(R^* - 2\Lambda + L_M\right)\right), \tag{265}$$

with metrics of type (2) and the Ricci scalar R^* (for general metrics see appendix I)

$$R^* = R^*_{\alpha\beta}G^{\alpha\beta} = \begin{pmatrix} \Gamma^{\sigma}_{\alpha\beta,\sigma}g^{\alpha\beta} - \Gamma^{\sigma}_{\beta\sigma,\alpha}g^{\alpha\beta} - \Gamma^{\mu}_{\sigma\alpha}\Gamma^{\sigma}_{\beta\mu}g^{\alpha\beta} + \Gamma^{\sigma}_{\alpha\beta}\Gamma^{\mu}_{\sigma\mu}g^{\alpha\beta} \\ +\dfrac{F'}{F}(1-n)\Delta f + \dfrac{f_{,\alpha}f_{,\beta}g^{\alpha\beta}(1-n)}{4F^2}\left(4FF'' + (F')^2(n-6)\right) \end{pmatrix}\frac{1}{F}$$

$$=\left(R+\frac{F'}{F}(1-n)\Delta f+\frac{f_{,\alpha}f_{,\beta}g^{\alpha\beta}(1-n)}{4F^2}\left(4FF''+(F')^2(n-6)\right)\right)\frac{1}{F}$$

$$\text{with: } F=F[f];\ F'=\frac{\partial F[f]}{\partial f};\ F''=\frac{\partial^2 F[f]}{\partial f^2} \tag{266}$$

as kernel or Lagrange density, we were able to derive new field equations, containing not only the classical part from Einstein's General Theory of Relativity but also a quantum part, which is governed by the wave function f and the metric scale $F[f]$. Thereby we have applied a scaled metric $G_{\delta\gamma} = F[f] \cdot g_{\delta\gamma}$. G shall denote the determinant of the metric tensor $G_{\delta\gamma}$, while g will later stand for the corresponding determinant of the metric tensor $g_{\alpha\beta}$. In order to distinguish our new or scaled Ricci scalar R^*, being based on $G_{\delta\gamma} = F[f] \cdot g_{\delta\gamma}$ from the usual one R, being based on the metric tensor $g_{\alpha\beta}$, we marked it with the a superscript asterisk (*). We also have the matter density L_M and the cosmological constant Λ.

Performing the Hilbert variation [2] with respect to the scale-adapted metric tensor $G_{\alpha\beta}$ and applying the chain rule, gives us the following task:

$\delta W = 0$

$$=\delta_{G_{\alpha\beta}}\int_V d^n x\left(\sqrt{-g\cdot F^n}\times\left(\left(\begin{array}{c}R+\frac{F'}{F}(1-n)\Delta f\\ +\frac{f_{,\alpha}f_{,\beta}g^{\alpha\beta}(1-n)}{4F^2}\left(4FF''+(F')^2(n-6)\right)\end{array}\right)\frac{1}{F}\\ -2\Lambda+L_M\right)\right). \tag{267}$$

$$0=\int_V d^n x\left(\sqrt{-G}\times\left(R^{*\kappa\lambda}-\frac{1}{2}R^*G^{\kappa\lambda}+\kappa\cdot T^{\kappa\lambda}+\Lambda\cdot G^{\kappa\lambda}\right)\delta G_{\kappa\lambda}\right)$$

$$=\int_V d^n x\left(\sqrt{-g\cdot F^n}\times\underbrace{\left(R^*_{\alpha\beta}-\frac{1}{2}R^*G_{\alpha\beta}+\kappa\cdot T_{\alpha\beta}+\Lambda\cdot G_{\alpha\beta}\right)\delta G^{\alpha\beta}}_{\text{INT}}\right). \tag{268}$$

Using our earlier work and inserting the results from [9, 12] into (268) gives us the following variation[1]:

[1]Please note that there are a number of options with respect to the decomposition of the Laplace operator. This has been discussed and considered in [15, 16, 17].

$$\delta W=0=\int_V d^n x\left(\sqrt{-g\cdot F^n}\begin{pmatrix} R_{\alpha\beta}-R\cdot\frac{g_{\alpha\beta}}{2}\\ +\frac{F'}{F}(1-n)\left(f_{,\alpha\beta}-\Gamma^{\gamma}_{\alpha\beta}f_{,\gamma}-\frac{g_{\alpha\beta}}{2}\Delta f\right)\\ +\frac{(1-n)}{4F^2}\begin{pmatrix}4FF''\\+F'F'(n-6)\end{pmatrix}\begin{pmatrix}f_{,\alpha}f_{,\beta}\\-f_{,\mu}f_{,\nu}g^{\mu\nu}\frac{g_{\alpha\beta}}{2}\end{pmatrix}\\ +\kappa T_{\alpha\beta}+\Lambda\cdot F\cdot g_{\alpha\beta}\end{pmatrix}\delta G^{\alpha\beta}\right).\tag{269}$$

We know that the variation of the scaled metric $\delta G_a^{\alpha\beta}$ can be split up into the variation of the function F and the variation of the unscaled metric and subsequently delivers the following:

$$\delta W=0=$$

$$\int_V d^n x\left(\begin{array}{l}\sqrt{-g\cdot F^n}\begin{pmatrix} R_{\alpha\beta}-R\cdot\frac{g_{\alpha\beta}}{2}\\ +\frac{F'}{F}(1-n)\left(f_{,\alpha\beta}-\Gamma^{\gamma}_{\alpha\beta}f_{,\gamma}-\frac{g_{\alpha\beta}}{2}\Delta f\right)\\ +\frac{(1-n)}{4F^2}\begin{pmatrix}4FF''\\+F'F'(n-6)\end{pmatrix}\begin{pmatrix}f_{,\alpha}f_{,\beta}\\-f_{,\mu}f_{,\nu}g^{\mu\nu}\frac{g_{\alpha\beta}}{2}\end{pmatrix}\\ +\kappa T_{\alpha\beta}+\Lambda\cdot F\cdot g_{\alpha\beta}\end{pmatrix}g^{\alpha\beta}\delta\frac{1}{F}\\ +\sqrt{-g\cdot F^n}\begin{pmatrix} R_{\alpha\beta}-R\cdot\frac{g_{\alpha\beta}}{2}\\ +\frac{F'}{F}(1-n)\left(f_{,\alpha\beta}-\Gamma^{\gamma}_{\alpha\beta}f_{,\gamma}-\frac{g_{\alpha\beta}}{2}\Delta f\right)\\ +\frac{(1-n)}{4F^2}\begin{pmatrix}4FF''\\+F'F'(n-6)\end{pmatrix}\begin{pmatrix}f_{,\alpha}f_{,\beta}\\-f_{,\mu}f_{,\nu}g^{\mu\nu}\frac{g_{\alpha\beta}}{2}\end{pmatrix}\\ +\kappa T_{\alpha\beta}+\Lambda\cdot F\cdot g_{\alpha\beta}\end{pmatrix}\frac{1}{F}\delta g^{\alpha\beta}\end{array}\right).\tag{270}$$

The total variation now takes the form (c.f. [12] for details):

$$0=\int_V d^n x\left(\begin{array}{c}\sqrt{-g\cdot F^n}\left(\begin{array}{c}\left(R+\frac{F'}{F}(1-n)\Delta f\right)\left(1-\frac{n}{2}\right)\\ +\frac{f_{,\alpha}f_{,\beta}(1-n)}{4F^2}g^{\alpha\beta}\left(4FF''+F'F'(n-6)\right)\\ \times\left(1-\frac{n}{2}\right)+\kappa T_{\alpha\beta}g^{\alpha\beta}+\Lambda\cdot F\cdot n\end{array}\right)\delta\frac{1}{F}\\ +\sqrt{-g\cdot(F)^{n-2}}\left(\begin{array}{c}R_{\alpha\beta}-R\cdot\frac{g_{\alpha\beta}}{2}\\ +\frac{F'}{F}(1-n)\left(f_{,\alpha\beta}-\Gamma^{\gamma}_{\alpha\beta}f_{,\gamma}-\frac{g_{\alpha\beta}}{2}\Delta f\right)\\ +\frac{(1-n)}{4F^2}\left(\begin{array}{c}4FF''\\ +F'F'(n-6)\end{array}\right)\left(\begin{array}{c}f_{,\alpha}f_{,\beta}\\ -f_{,\mu}f_{,\nu}g^{\mu\nu}\frac{g_{\alpha\beta}}{2}\end{array}\right)\\ +\kappa T_{\alpha\beta}+\Lambda\cdot F\cdot g_{\alpha\beta}\end{array}\right)\delta g^{\alpha\beta}\end{array}\right). \tag{271}$$

Subsequently, demanding the vanishing of the whole integral via the integrand being zero, we obtain two equations, namely:

1. $$0=\left(\begin{array}{c}\left(R+\frac{F'}{F}(1-n)\Delta f\right)\left(1-\frac{n}{2}\right)\\ +\frac{f_{,\alpha}f_{,\beta}(1-n)}{4F^2}g^{\alpha\beta}\left(4FF''+F'F'(n-6)\right)\\ \times\left(1-\frac{n}{2}\right)+\kappa T_{\alpha\beta}g^{\alpha\beta}+\Lambda\cdot F\cdot n\end{array}\right)$$

2. $$0=\left(\begin{array}{c}R_{\alpha\beta}-R\cdot\frac{g_{\alpha\beta}}{2}\\ +\frac{F'}{F}(1-n)\left(f_{,\alpha\beta}-\Gamma^{\gamma}_{\alpha\beta}f_{,\gamma}-\frac{g_{\alpha\beta}}{2}\Delta f\right)\\ +\frac{(1-n)}{4F^2}\left(\begin{array}{c}4FF''\\ +F'F'(n-6)\end{array}\right)\left(\begin{array}{c}f_{,\alpha}f_{,\beta}\\ -f_{,\mu}f_{,\nu}g^{\mu\nu}\frac{g_{\alpha\beta}}{2}\end{array}\right)\\ +\kappa T_{\alpha\beta}+\Lambda\cdot F\cdot g_{\alpha\beta}\end{array}\right). \tag{272}$$

One of the equations (the second one) "**may**" be seen as a set of the quantum Einstein field equations, which determines the metric and the other would fix the wave functions *F* and *f*, whereby only one of the two functions (*F* or *f*) needs to be determined, while the other can be chosen in an arbitrary manner (probably so that one results in most simple equations).

The reader can find some discussion about these results in [12, 15, 16, 17]. In [16, 17] also options for the generalization of the scale-wrapper functions *F*[*f*] as potential multiple, intrinsically entangled wave function and more were introduced and discussed.

Now we were able to use the results above to obtain the Klein–Gordon equation (appendix A), the Schrödinger equation (appendix B), and the Dirac equation (appendix C) directly out of the Einstein–Hilbert action. We also saw (e.g., [17, 18]) that there is enough freedom via various options for the wrapper function *F*[*f*] plus the split-up of the variational integral to also obtain other quantum equations. Thus, so we conclude, the classical Quantum Theory is to be found in the scalar part of the variation (271), which is to say (272) No. 1 ... with quite many options for the split-up of the integration and the variation itself [12–18].

Gravity, however, requires the consideration of the tensorial integrand of (271), being (272) No. 2 (also with many variational degrees of freedom it seems) and quantum gravity is it all just put together (and more[2]). As it is evident from the structure of equation 2 in (272), that our approach also contains the Einstein field equations and thus, the General Theory of Relativity (c. f. appendix G), we refrain here from harping on this fact but only show where Hilbert's and Einstein's matter comes from (appendix G). The two brilliant scientists, namely, had to postulate their matter [1, 2] and introduced it as additional and artificial terms as they saw no way to truly derive it. Here, we will show how it can be done (appendix G).

11.3 Brief Introduction to Quantum Cryptography

As the topic of quantum cryptography has, meanwhile, found its way into the textbook literature, we will not present its basics here but refer to that very literature (e.g., [197, 198]).

The essential property of quantum cryptography should be seen in the fact that the two communicators, which have been usually named Alice and

[2]With this, we refer to the generalized forms being introduced and discussed in [15, 16, 17, 18].

Bob, have some intrinsic knowledge about their communication, which allows them to detect any eavesdropping, performed by a third entity (Eve). Let us call this part of the communication the control-stream. This knowledge, meaning the knowledge of Alice and Bob about their control-stream, is completely independent of the content of Alice's and Bob's conversation, and thus their communication can be given the status secure as long as they do not detect any irregularities in the data of their control-stream they exchange. Thereby each of the two communicators, which is to say either Alice or Bob, can detect such anomalies independently once the secure line has been defined and set up. In other words, Eve has no chance of undetected eavesdropping.

It should be pointed out that there already is quite some skepticism about the fundamentality of quantum cryptography and the safety it can provide (e.g., [199]). This skepticism, however, is of a more general character and does not explicitly name the danger and the reasons for the unsafety issues of quantum cryptography. The critique also does not investigate the fundamental interaction dependency of quantum cryptography. Here, however, we are going to work this latter point out as its fundamental embedded weakness.

11.4 The Illusion of Quantum Cryptographic Safety

Only concentrating on the so-called quantum key distribution, which—in principle—is supposed to provide the safest form of quantum cryptography, we refer to the aspect of the two communicators (Alice and Bob), who cannot be "overheard" by Eve (the eavesdropper) without immediately noticing that something is wrong with their communication via a control-stream of the exchanged data. This noticing process, however, is based on the assumption that Eve is bound to use or observe the same scale (and/or means) of interaction as Alice and Bob have applied in order to set up their line of quantum cryptographic secure communication. This, in fact, is the fundamental necessity or basic assumption for all quantum cyber security approaches.

Here now, we will show that this safety is an illusion.

We start with the first equation in (272) and by realizing that this equation and its various forms (also the other forms one could derive from the integral in (271) (see [12, 13, 17]) contain the essentials of all interactions except gravity (see appendices A, B, and C), we automatically have a very fundamental equation or rather set of equations to describe every form of interaction as long as it is not connected with gravity. On the basis of the resulting equations, all forms of current approaches in quantum cryptography are built. And the assumption of Eve not being able to co-read the messages sent between Alice and Bob without considering that this eavesdropping is detected relies on the conjecture that the interaction being applied by Alice and Bob is literally confined to equation 1 in (272). This conjecture, however, is wrong when seeing the complete (and quantum gravitational) picture. This we see when closely observing equation 2 in (272). This, too, namely, contains the essential wave function *f*, respectively its functional wrapper, and thus the two equations are not independent.

The moment Eve is able to resort to the methods that allow her to check on *f* or *F* via equation 2 in (272) without disturbing the interaction described by 1 in (272), she would be able to co-read the whole communication between Alice and Bob without the two ever being any the wiser.

And how could Eve achieve this?

Well, with Alice and Bob only watching the interactions of 1 in (272), thereby being able to detect Eve's eavesdropping the moment she resorts to the same scale of interactions, there automatically appears a gap in the whole security plan, the moment Eve possesses the ability to make use of the gravitational output also being created along the way of quantum cryptographic communication between Alice and Bob via 1 in (272). Simply by watching the parallel output described via 2 in (272), the gravity part, Eve has full access to their communication channel, while Alice and Bob are—vigilantly and confidently—observing only their own interaction, which is 1 in (272) and, consequently, register nothing out of the ordinary.

11.5 The Ultimate Cryptographic Solution

With the material being presented above (and in our previous work), it is automatically clear how we could overcome the current limitations and issues of "classical" (quantum) cryptography.

The path forward lies in the total control of all aspects of the interaction being used between the communicating agents.

And how can this be achieved?

...

Well, we think the answer is clear. Alice and Bob have to resort to a complete quantum gravitational way of communication, which is to say their interaction has to be described by something like this:

$$0=\left(\begin{array}{c}R_{\alpha\beta}-R\cdot\frac{g_{\alpha\beta}}{2}\\+\frac{F'}{F}(1-n)\left(f_{,\alpha\beta}-\Gamma_{\alpha\beta}^{\gamma}f_{,\gamma}-\frac{g_{\alpha\beta}}{2}\Delta f\right)\\+\frac{(1-n)}{4F^{2}}\left(\begin{array}{c}4FF''\\+F'F'(n-6)\end{array}\right)\left(\begin{array}{c}f_{,\alpha}f_{,\beta}\\-f_{,\mu}f_{,\nu}g^{\mu\nu}\frac{g_{\alpha\beta}}{2}\end{array}\right)+\kappa T_{\alpha\beta}+\Lambda\cdot F\cdot g_{\alpha\beta}\end{array}\right). \tag{273}$$

And how can they do this ... practically?

Gravitational waves, so we think, may be a good starting point.

11.6 A Partially Enigmatic Conclusion

When taking gravity into account, the standard (interaction) means of quantum cryptography are not safe.

Some might say that the technique to perform gravitational eavesdropping as described in the chapter, is not at hand ... not yet, anyway, and most certainly not in bigger scales or numbers.

So why bother?

Apart from the fact that this is not entirely true, we would simply counterargue that the whole COVID-19 disaster also was and still is not more than a blown-up rather ordinary flue if just considering its biological potential, and still it resulted in a major anaphylactic shock reaction [C1] for all western societies. So, what happens to a society if something much more important—like all communication channels, which have to be secured especially if it comes to military, government, finance, energy-, water-, and otherwise life-system-relevant systems, and so on—is suddenly, and perhaps only potentially, compromised? If already an ordinary flue could be made such a big hoax, what will be the auto-immune shock reaction (triggered by whoever and fed by the mainstream media and all other always-hungry parasites—external and internal ones [C1]) to a true threat like generally insecure communication for all life-sustaining means a modern society MUST have?

[C1] N. Schwarzer, *Anaphylactic Shock Attack Strategy—The Assault You Don't Realize until It Is TOO Late*, Part 10a of "Medical Socio-Economic Quantum Gravity," Self-published, Amazon Digital Services, 2021, Kindle.

Chapter 12

Optimum Decision-Making

12.1 The Infinity Options Principle

Strategic decision-making is a key factor in any military conflict. Therefore, tools for optimum decision-making are crucial. Apart from the techniques being applied thereby (e.g., see [23]), there is always also the important question about the principal uncertainties of a given system or situation to be analyzed. Our quantum gravity approach allows us to deal with the problem of the total uncertainty budget in a very general manner.

First, we want to evaluate the limits of the quantum wrapper functions $F[f]$ for infinite numbers of dimensions (degrees of freedom) $n \to \infty$, which gives us for the second-order quantum or Laplace-dominated equations (Klein–Gordon, Schrödinger, ...) (22):

$$\lim_{n\to\infty} F\left[f\right] = \lim_{n\to\infty} C_2 \cdot \left(f \cdot (n-2) + C_1\right)^{\frac{4}{n-2}} = C_2 \tag{274}$$

and for the Dirac (or elastic or similar ... see [18], sub-section “Elastic Space-Times” or [3]) case:

$$\lim_{n\to\infty} F\left[f\right] = \lim_{n\to\infty} C_{f1} \cdot \cos\left[\frac{1}{C_{N2}} \cdot \sqrt{C_{N2} - C_{N1}} \cdot \left(f - C_{f2}\right)\right]^{-\frac{C_{N2}}{C_{N1} - C_{N2}}}$$

Quantum Gravity War: How Will the Nearby Unification of Physics Change the Future of Warfare
Norbert Schwarzer

ISBN 978-981-4968-58-4 (Hardcover), 978-1-032-71070-9 (eBook)
www.jennystanford.com

$$= \lim_{n \to \infty} C_{f1} \cdot \cosh\left[\frac{1}{C_{N2}} \cdot \sqrt{C_{N1} - C_{N2}} \cdot \left(f - C_{f2}\right)\right]^{-\frac{C_{N2}}{C_{N1} - C_{N2}}}$$

$$= \lim_{n \to \infty} C_{f1} \cdot \cosh\left[\frac{1}{2} \cdot \sqrt{\frac{1}{n-1} - 1} \cdot \left(f - C_{f2}\right)\right]^{\frac{4}{n-2}} = C_{f1}. \quad (275)$$

We see that in both cases we obtain constants for the limit $\lim_{n \to \infty}$.

Thus, the resulting quantum transformation or scalar quantum variation of any metric would read $G_{\alpha\beta} = F[f]^* g_{\alpha\beta} = \text{const}^* g_{\alpha\beta}$. In other words, the functional $F[f]$ would become a constant scaling and no quantum effects could be visible anymore. This, however, only holds for the $F[f]$-factor to the whole metric with $n \to \infty$. We still have the possibility of intrinsic scaling factors, which we have discussed elsewhere (see [18], chapter 11, especially section 11.12, "The Math for Body, Soul and Universe?").

This means, in spaces or space-times of infinitely many options/ degrees of freedom and thus, consequently, dimensions, there can be no global (meaning with respect to the whole metric) quantum effects ... at least not in the classical sense of Klein–Gordon, Schrödinger, Dirac, elastic and similar!

Intrinsic quantum effects, based on limited numbers of dimensions of inner metrics and mathematically being brought to light via intrinsic scaling factors, would still be possible (c.f. [18], chapter 11).

12.2 Consequences

We were able to derive that global quantum effects vanish in systems with an infinite number of dimensions. This means that also the principal uncertainty (quantum originated, therefore fundamental and inevitable) disappears ... at least on a global level.

There is no uncertainty and consequently no Heisenberg uncertainty principle in space or space-times with infinitely many dimensions, when considering the global metric with $n \to \infty$. Intrinsic quantum effects, on the other hand, are still possible (c.f. [18], chapter 11).

In other words: An entity being able to see the whole infinite space-time feels no uncertainty with respect to the whole but still realizes uncertainty with respect to the details.

This finding has dramatic consequences with respect to practical problems regarding big system analysis, system science, decision-making, and thus, naturally, socio-economics, politics, medicine, military, and so on. It also means that any ideology-containing or perhaps even ideology-dominated process of data handling, analysis, and subsequent decision-making automatically increases its own—intrinsic and fundamental—uncertainty by the simple fact that all ideologies can only exist under the condition of ideology-driven information selection ... which is to say, an ideology dominated system always tends to (has to tend to) degeneration and thus, dimensional system reduction. In this sense, the old work of Konrad Lorenz (Nobel Prize in Physiology or Medicine in 1973) about the "Verhausschweinung des Menschen" [D1] (loosely translated one might put it as "the domestication of man") appears in a completely new light. Whatever Lorenz's original intention was when writing his paper during the Nazi terror in the Third Reich and his apparent attempt to support some Nazi race theories, it obviously is a fundamental fact that the reduction of information an entity (any entity, including a human being) considers before making decisions dramatically increases the uncertainty in an inevitable and very principal manner. The number of information bits a given system can hold—as we have been able to show (chapter 5, "The New Abacus," and chapter 6, "The Quantum Gravity Turing Machine")—is equal to the number of dimensions of this very system. The exclusion of information, be it via degeneration, domestication, ideology-governed "political correctness," false "tolerance" or "politeness," dogmatic narratives as in the current COVID-19 crisis, or simply due to system immanent and entity-typical boundaries, reduces the number of information bits the entity can hold and digest and thus, increases the uncertainty. This uncertainty increase is on the lowest quantum and, thus, most fundamental level of the system in question. Therefore:

(a) Any system-inherent attempt of uncertainty reduction is futile. Meaning: A military structure based on ideology (or governed by such) does not have the same decision-making potential as a non-ideology-restricted structure has.
(b) However, so it was shown here, increasing the system by incorporating more information (degrees of freedom, properties, dimensions, thinking outside the space-times of ideology, religion, political correctness, dogmatic rules, etc.) and thus thinking

outside the box (meaning the system, the entity was previously in) is going to lead to success regarding the goal of a reduced uncertainty (see also [4–8, 18]).

[D1] K. Lorenz: Durch Domestikation verursachte Störungen arteigenen Verhaltens. *Zeitschrift für angewandte Psychologie und Charakterkunde* 59, (1940), (1, 2), S. 2–81.

12.3 More Simplification Options

A comprehensive and most holistic study of the non-Klein–Gordon, respectively, non-Laplace outcomes of (7) by not applying condition (23) in connection with a system-readjustment with respect to its degrees of freedom, one can show that vector-function approaches like the Dirac equation can be brought into simpler forms (Laplacians). This finding is important not only with respect to simplifying certain tasks but also with respect to the possibility of reducing the task's overall uncertainty budget. More about this will be presented elsewhere [212].

12.4 Mathematical Psychology as a Weapon

12.4.1 About the "Impossible" Task

Many people are of the opinion that it is impossible to mathematically describe human behavior in a satisfactory manner. Thereby they accept millions of artificial intelligence solutions doing exactly this every second and with respect to almost every aspect of their life. From this fact alone follows that a "Mathematical Psychology" is by no means impossible.

But what if such an approach, which is to say a mathematical description or even a toolbox of models for our "mind," our feelings, and actions, is not just possible, but simply imminent?

Wouldn't it then also be inevitable that such knowledge is used for military purposes?

The answer to both questions is YES, and thus, as a warning, the author here wants to repeat his explanation about the "HOW to do it" question, which he had already given in [18] and which will be the basis of the foundations of a "Mathematical Psychology" as comprehensively outlined in [112].

12.4.2 About How to Do the "Impossible"

It is widely agreed that human beings can be characterized by a list of attributes, which partially entangle with each other ... thereby we mean that the attributes entangle, not the humans. Even though this list may be extremely long and the interaction (or entanglement) of the various attributes might be very complex, in essence we are characterizable by such a list. After all, nothing else is it when we try to—more or less comprehensively and fairly—describe another person or—usually with much less vigor, enthusiasm, and rigorousness—ourselves.

Accepting the fact that humans can be described by a list of entangled attributes, however, still does leave us with the question of how to mathematical handle this list in order to describe processes of action, human interaction, individuality, and other characteristics. Most surprisingly, this work has been already done. Over 100 years ago, it was the great mathematician David Hilbert who found a way to put a set of properties, attributes, degrees of freedom, or—as all these other terms are just coding—**dimensions** into a mathematical formalism, which, in his time, was already known for decades as the one fundamental principle for the determination of the laws of our universe, the Hamilton extremal principle. All truly fundamental natural laws originate from this principle. The only problem is to find the right ingredients to "feed the monster of the Hamilton principle." The essential ingredient, namely, the Lagrange density function, seems to differ from problem to problem and this cannot be when assuming that the Hamilton mechanism should be truly fundamental. Here now came the stroke of genius [2] of Hilbert, who realized that a given set of attributes is nothing else but a set of dimensions and, thus, forms a mathematical space or space-time. The corresponding Lagrange density function is then always the same, namely the so-called Ricci scalar of the very system or space-time. Hilbert applied his calculus on our ordinary 4-dimensional space-time and derived the Einstein field equations and, thus, in just one fluid and conclusive derivation, the complete mathematical basis for the General Theory of Relativity of Albert Einstein [1]. Admittedly, Hilbert never extended his calculus to the space or space-time of attributes characterizing human beings, but the fact that the title of his work [2], in which he derived Einstein's great theory, translates "The Fundaments of Physics" clearly shows us that he saw his approach in a much more general way than just the minimum-principle-origin of Einstein's field equations. Thus, irrespective of the system we are considering, the moment we have its basic attributes and know their entanglement (various interactions), we also immediately know how to place all this knowledge into Hilbert's calculus to come up with a most fundamental and unbiased model for this very system.

Hilbert's creation, de facto, was—no, is—a Theory of Everything [41] or World Formula [3, 18], even though he himself had little chance of fully realizing this. Even in physics, where we now can show that Hilbert's fundamental equation covers both great theories, the General Theory of Relativity and Quantum Theory, the time was not ripe for such a discovery, simply because the mathematical apparatus of Quantum Theory was not fully developed then. While Hilbert brought out his great work in 1915 and knew about the Einstein field equations at the time, the basic quantum equations like the Schrödinger [186, 187], the Klein–Gordon, and the Dirac equation [181] would not follow before the second half of the 1920s.

For convenience, falsifiability, and rigorousness, we have not only presented the Hilbert approach adjusted to our arbitrarily dimensional space-times in the main part of this chapter but also derived the Klein–Gordon equation (one of the main relativistic quantum equations) and its "derivatives," the Schrödinger and the Dirac equation, because it will be our mathematical fundament for the consideration and modeling of psychological processes. With respect to the "thing itself," however, we have to refer to the comprehensive work [112], simply because of the volume of the undertaking.

12.5 War as a Critical State

Already the first few weeks of the 2022 war in Ukraine showed that the alleged predominant Russian military was anything but predominant. Many experts were puzzled and could not explain the situation. In fact, there are many reasons, but one is of special importance with regard to the "stress test" a war (any war) means to—depending on its scale—the participating armies or the whole society. A war, namely, has to be understood as a critical situation or state of the complex system, named society. This criticality is characterized by the absence or—at least—dilution of a characteristic scale [217], which is to say a more or less dominant inter-societal interaction. With such a dominant interaction missing, all scales become important and suddenly the critical situation of the war reveals weak structures which had not been perceptible before. In the above-mentioned conflict, for example, the missing noncommissioned officer structure of the Russian army acted like a weak interface in the command chain and led to severe problems for the Russian forces. These problems were not recognizable outside the critical state the war had created, but near the criticality they summed up to some disastrous defeats for the Russian troops. There were more such weak interface structures within the Russian military complex, but we will not discuss them here.

This effect of weak structures only becoming detectable (measurable) near critical points for a system, will be discussed in one of our next publications [218]. This does not only concern wars but all fields with potentially society-wide criticality (food supply, financial markets, health sector with pandemic situations and so on, social and/or political uproar and so on).

12.6 General Conclusions

The war has already started. It is a war—mainly—of information. There is going to be only one winner and he will take it all.

The moment this one winner will have taken it all, however, Fermi's paradox is going to be resolved.

Appendix A

Derivation of the Metric Klein–Gordon Equation from the Scaled Ricci Scalar R^*

In order to obtain the metric Klein–Gordon equation, we take the scaled Ricci scalar from (1) and demand:

$$0 = 4FF'' + (F')^2(n-6) \quad \Rightarrow \quad F = \begin{cases} \left(f - C_{f0}\right)^{\frac{4}{n-2}} \cdot C_{f1} & n > 2 \\ e^{C_{f0} \cdot f} \cdot C_{f1} & n = 2 \end{cases}, \tag{276}$$

which immediately gives us:

$$\begin{aligned} R^* &= R^*_{\alpha\beta} G^{\alpha\beta} \\ &= \left(\begin{array}{l} \Gamma^{\sigma}_{\alpha\beta,\sigma} g^{\alpha\beta} - \Gamma^{\sigma}_{\beta\sigma,\alpha} g^{\alpha\beta} - \Gamma^{\mu}_{\sigma\alpha} \Gamma^{\sigma}_{\beta\mu} g^{\alpha\beta} + \Gamma^{\sigma}_{\alpha\beta} \Gamma^{\mu}_{\sigma\mu} g^{\alpha\beta} \\ + \dfrac{F'}{F}(1-n)\Delta f + \dfrac{f_{,\alpha} f_{,\beta} g^{\alpha\beta}(1-n)}{4F^2}\left(4FF'' + (F')^2(n-6)\right) \end{array} \right) \frac{1}{F} \\ &= \left(R + \frac{F'}{F}(1-n)\Delta f + \frac{f_{,\alpha} f_{,\beta} g^{\alpha\beta}(1-n)}{4F^2}\left(4FF'' + (F')^2(n-6)\right) \right) \frac{1}{F} \\ &= \left(R + \frac{F'}{F}(1-n)\Delta f \right) \frac{1}{F}; \end{aligned}$$

$$\text{with:} \quad F = F[f]; \quad F' = \frac{\partial F[f]}{\partial f}; \quad F'' = \frac{\partial^2 F[f]}{\partial f^2}. \tag{277}$$

Now we assume the total curvature R^* to vanish and obtain:

$$R^* = 0 \quad \rightarrow \quad 0 = F \cdot R + F' \cdot (1-n) \cdot \Delta f$$

$$\Rightarrow \qquad 0=\begin{cases}\left(f-C_{f0}\right)^{\frac{4}{n-2}}\cdot C_{f1}\left(R+\dfrac{(1-n)}{\left(f-C_{f0}\right)}\cdot\Delta f\right) & n>2\\ e^{C_{f0}\cdot f}\cdot C_{f1}\left(R+C_{f0}\cdot(1-n)\cdot\Delta f\right) & n=2\end{cases}. \tag{278}$$

Thus, in the case of $n > 2$ we always also have the option for a constant (broken symmetry) solution of the kind:

$$0=f-C_{f0}\quad\Rightarrow\quad f=C_{f0}. \tag{279}$$

Otherwise, we have the simple equations:

$$0=\begin{cases}\left(f-C_{f0}\right)\cdot R+(1-n)\cdot\Delta f & n>2\\ R+C_{f0}\cdot(1-n)\cdot\Delta f & n=2\end{cases}. \tag{280}$$

Depending on the curvature term R, which substitutes the classical terms for mass and potential, we can have the classical Klein–Gordon equation or more complex versions of the latter. We also need to point out that potential and mass (or other forms of inertia and fields) can also be obtained via certain (potentially compactified) dimensions when simultaneously choosing suitable structures for the functions f with respect to these add-on dimensions. This was shown in the previous papers by this author and the book [3], but especially and with quite many "practical" examples in [5]. As one of the essential examples, we will give here a complete metrical derivation of the Schrödinger hydrogen atom in appendix I.

The Origin of the Condition $R^* = 0$

A question we need to answer is about the origin of the condition $R^* = 0$. Here we are going to present the corresponding derivations from [18], which, for brevity, are placed under the assumption of simple metrics of type (2).

Assuming the Lagrange matter density L_M and the cosmological constant Λ to vanish, the variation of the Einstein–Hilbert action with a scaled metric $G_{\delta\gamma} = F[f] \cdot g_{\delta\gamma}$ results in:

$$0=\left(\int_{V_G} d^n x\sum_{i=1}^{N}\sqrt{-g\cdot F_i^{\,n}}\left(\begin{pmatrix} R_{\alpha\beta}-\dfrac{1}{2}\cdot R\cdot g_{\alpha\beta}-\dfrac{F_i{}'}{F_i}(n-1)\\ \times\left(f_{i,\alpha\beta}-\Gamma^{\gamma}_{\alpha\beta}f_{i,\gamma}-\dfrac{g_{\alpha\beta}}{2}\Delta f_i\right)\\ +\dfrac{(1-n)}{4F_i^2}\begin{pmatrix}4F_iF_i{}''\\ +F_i{}'F_i{}'(n-6)\end{pmatrix}\left(f_{i,\alpha}f_{i,\beta}-f_{i,\mu}f_{i,\nu}g^{\mu\nu}\dfrac{g_{\alpha\beta}}{2}\right)\end{pmatrix}\delta G^{\alpha\beta}\right)\right). \tag{281}$$

Thereby we have also assumed to have many scaling factors and volumes to perform the integration over (see section 11.4 and following in [18]). This way we will also be able to directly extract a metric Dirac equation as it will be shown in the second part of appendix C of this book.

By splitting up the variation of the scaled metric:

$$0=\left(\int_{V_G} d^n x \sum_{i=1}^{N} \sqrt{-g \cdot F_i^n} \left(\left(\begin{array}{c} R_{\alpha\beta} - \frac{1}{2} \cdot R \cdot g_{\alpha\beta} - \frac{F_i{}'}{F_i}(n-1) \\ \times\left(f_{i,\alpha\beta} - \Gamma^{\gamma}_{\alpha\beta} f_{i,\gamma} - \frac{g_{\alpha\beta}}{2} \Delta f_i \right) \\ + \frac{(1-n)}{4F_i^2}\left(4F_i F_i{}'' + F_i{}' F_i{}'(n-6)\right) \\ \times\left(f_{i,\alpha} f_{i,\beta} - f_{i,\mu} f_{i,\nu} g^{\mu\nu} \frac{g_{\alpha\beta}}{2} \right) \end{array} \right) \left(\begin{array}{c} g^{\alpha\beta} \delta \frac{1}{F_i} \\ + \frac{\delta g^{\alpha\beta}}{F_i} \end{array} \right) \right) \right), \tag{282}$$

we obtain two equations:

$$\begin{aligned} 0 &= \int_{V_G} d^n x \sum_{i=1}^{N} \sqrt{-g \cdot F_i^n} \left(\left(\begin{array}{c} R_{\alpha\beta} - \frac{1}{2} \cdot R \cdot g_{\alpha\beta} - \frac{F_i{}'}{F_i}(n-1) \\ \times\left(f_{i,\alpha\beta} - \Gamma^{\gamma}_{\alpha\beta} f_{i,\gamma} - \frac{g_{\alpha\beta}}{2} \Delta f_i \right) \\ + \frac{(1-n)}{4F_i^2}\left(4F_i F_i{}'' + F_i{}' F_i{}'(n-6)\right) \\ \times\left(f_{i,\alpha} f_{i,\beta} - f_{i,\mu} f_{i,\nu} g^{\mu\nu} \frac{g_{\alpha\beta}}{2} \right) \end{array} \right) g^{\alpha\beta} \delta \frac{1}{F_i} \right) \\ &= \int_{V_G} d^n x \sum_{i=1}^{N} \sqrt{-g \cdot F_i^n} \left(\left(\begin{array}{c} R\left(1-\frac{n}{2}\right) - \frac{F_i{}'}{F_i}(n-1)\left(1-\frac{n}{2}\right)\Delta f_i \\ + \frac{(1-n)}{4F_i^2}\left(4F_i F_i{}'' + F_i{}' F_i{}'(n-6)\right) \\ \times\left(f_{i,\alpha} f_{i,\beta} - f_{i,\mu} f_{i,\nu} g^{\mu\nu} \frac{n}{2} \right) \end{array} \right) \delta \frac{1}{F_i} \right) \\ &= \int_{V_G} d^n x \sum_{i=1}^{N} \sqrt{-g \cdot F_i^n} \left(\left(1-\frac{n}{2}\right) \left(\begin{array}{c} R - \frac{F_i{}'}{F_i}(n-1)\Delta f_i \\ + \frac{f_{i,\alpha} f_{i,\beta}(1-n)}{4F_i^2} g^{\alpha\beta} \left(\begin{array}{c} 4F_i F_i{}'' \\ + F_i{}' F_i{}'(n-6) \end{array} \right) \end{array} \right) \delta \frac{1}{F_i} \right), \end{aligned} \tag{283}$$

$$0=\left(\int_{V_G} d^n x \sum_{i=1}^{N} \sqrt{-g \cdot F_i^n} \left(\left(\begin{matrix} R_{\alpha\beta}-\frac{1}{2}\cdot R\cdot g_{\alpha\beta}-\frac{F_i'}{F_i}(n-1) \\ \times\left(f_{i,\alpha\beta}-\Gamma_{\alpha\beta}^{\gamma} f_{i,\gamma}-\frac{g_{\alpha\beta}}{2}\Delta f_i\right) \\ +\frac{(1-n)}{4F_i^2}\left(\begin{matrix} 4F_iF_i'' \\ +F_i'F_i'(n-6)\end{matrix}\right)\left(f_{i,\alpha}f_{i,\beta}-f_{i,\mu}f_{i,\nu}g^{\mu\nu}\frac{g_{\alpha\beta}}{2}\right)\end{matrix}\right)\frac{\delta g^{\alpha\beta}}{F_i}\right)\right). \tag{284}$$

While we consider the second equation a generalized or quantum Einstein field equation, we are going to use the first one to extract the classical Klein–Gordon equation simply by demanding condition:

$$0=4F_iF_i''+F_i'F_i'(n-6) \tag{285}$$

for the functional wrapper $F_i[f]$. Now, setting the integrand equal to zero, we obtain the following differential equation:

$$0=R-\frac{F_i'}{F_i}(n-1)\Delta f_i+\frac{f_{i,\alpha}f_{i,\beta}(1-n)}{4F_i^2}g^{\alpha\beta}\left(4F_iF_i''+F_i'F_i'(n-6)\right). \tag{286}$$

Incorporation of condition (285) finally gives us:

$$0=R-\frac{F_i'}{F_i}(n-1)\Delta f_i, \tag{287}$$

and thus, (280). The introduction of mass and potential via dimensional entanglement was extensively discussed in chapter 6 of our book [18].

Appendix B

Derivation of the Metric Schrödinger Equation from the Scaled Ricci Scalar R^*

We start from our result of the appendix A and use (280), only that, for the reason of recognition, instead of the symbol f we apply the classical Greek symbol Ψ. In order to be close enough to the classical Schrödinger case, we also assume a Minkowski-like metric time component. We write (280) as follows:

$$0=\begin{cases}\left(\Psi-C_{f0}\right)\cdot R+(1-n)\cdot\left(\Delta_{n-1}\Psi-\dfrac{\partial^2\Psi}{c^2\cdot\partial t^2}\right) & n>2\\ R+C_{f0}\cdot(1-n)\cdot\left(\Delta_{n-1}\Psi-\dfrac{\partial^2\Psi}{c^2\cdot\partial t^2}\right) & n=2.\end{cases} \tag{288}$$

We ignore the n = 2-case and take it that R can be assumed to be linear to f, which is to say that the curvature of space-time is caused by and proportional to the quantum effects residing in it. This gives us:

$$0=\left(\Psi-C_{f0}\right)\cdot R_c\cdot\Psi+(1-n)\cdot\left(\Delta_{n-1}\Psi-\frac{\partial^2\Psi}{c^2\cdot\partial t^2}\right). \tag{289}$$

Assuming the quantum function to be small against the classical physics (an important aspect in Schrödinger's work anyway [186, 187] and also agrees with the practical observation that the space-time curvature is small as we experience our surrounding space as being mainly flat), we discard of the quadratic term, obtaining:

$$\begin{aligned}0&=-C_{f0}\cdot R_c\cdot\Psi+(1-n)\cdot\left(\Delta_{n-1}\Psi-\frac{\partial^2\Psi}{c^2\cdot\partial t^2}\right)\\&=\left[(1-n)\cdot\left(\Delta_{n-1}-\frac{\partial^2}{c^2\cdot\partial t^2}\right)-C_R\right]\Psi.\end{aligned} \tag{290}$$

Then we can separate the time derivative and—for the reason of recognition in connection with Schrödinger's classical approach—we reformulate everything as follows:

$$0=-C_{f0}\cdot R_c\cdot\Psi+(1-n)\cdot\left(\Delta_{n-1}\Psi-\frac{\partial^2\Psi}{c^2\cdot\partial t^2}\right)$$
$$=\left[(1-n)\cdot\left(\Delta_{n-1}-\frac{\partial^2}{c^2\cdot\partial t^2}\right)-C_R\right]\Psi$$
$$\xrightarrow{\left(\Delta_{n-1}-\frac{\partial^2}{c^2\cdot\partial t^2}\right)-\frac{C_R}{(1-n)}=C1\cdot\frac{\partial_t^2}{c^2}+\Delta_{n-1}-M^2}$$
$$\Rightarrow\qquad 0=\left[\Delta-M^2\right]\Psi$$
$$=\left[\left(C1\cdot\frac{\partial_t^2}{c^2}+\underbrace{\frac{1}{\sqrt{g}}\partial_\alpha\sqrt{g}\cdot g^{\alpha\beta}\partial_\beta}_{3D-\Delta-\text{Operator}}\right)-M^2\right]\Psi. \tag{291}$$

Please note that our Greek indices α and β are running only from 1 to $n-1$ and not—as usual—from 0 to $n-1$, because the time or 0-component has been separated. Now we introduce a function $\Psi=\Phi+X$ and demand the following additional condition:

$$\partial_t\Psi=c^2\cdot C2\cdot(\Phi-\mathrm{X}). \tag{292}$$

Together with (291) we obtain:

$$0=-M^2(\Phi+\mathrm{X})+\left(C1\cdot C2\cdot\partial_t(\Phi-\mathrm{X})+\frac{1}{\sqrt{g}}\partial_\alpha\sqrt{g}\cdot g^{\alpha\beta}\partial_\beta(\Phi+\mathrm{X})\right). \tag{293}$$

The following two equations summed up would result in (293):

$$0=-M^2\Phi+\left(C1\cdot C2\cdot\partial_t\Phi+\frac{1}{\sqrt{g}}\partial_\alpha\sqrt{g}\cdot g^{\alpha\beta}\partial_\beta\Phi\right)$$
$$0=-M^2\mathrm{X}+\left(\frac{1}{\sqrt{g}}\partial_\alpha\sqrt{g}\cdot g^{\alpha\beta}\partial_\beta\mathrm{X}-C1\cdot C2\cdot\partial_t\mathrm{X}\right). \tag{294}$$

Comparing with the original Schrödinger equation as given in the form below:

$$\left[-i\cdot\hbar\cdot\partial_t-\frac{\hbar^2}{2m}\Delta_{\text{Schrödinger}}+V_{\text{Schrödinger}}\right]\Psi=0 \tag{295}$$

not only gives us the matter and antimatter solutions (c.f. [3, 41, 186]) but also shows us—as seen and discussed before [18]—that mass m and the potential $V_{\text{Schrödinger}}$ is now been taken on by the metric and the results in hidden (potentially compactified) dimensions, potentially due to entanglement (see [5, 18] for more and the example in appendix I). We also realize that the scalar curvature plays an important role in coding the mass- and potential-term of the classical Schrödinger equation. Thus, a distorted metric and additional dimensions act like an effective potential and/or mass in the Schrödinger approximation (or Schrödinger picture) and vice versa, which is to say: what in classical physics is described as a potential would now become a scalar curvature or (for derivation, see [5]) a set of potentially compactified dimensions, which is providing the necessary interaction. Similarly, we have to formulate for the mass M that what appears as (rest) mass to us is just permanently and locally curved space (showing up in our equations above within the Ricci scalar), whereby the curvature can also be (c.f. [5, 18]) one of the additional dimensions, being potentially entangled ... either among each other, with other "hidden" or compactified dimensions, or with our "visible" dimensions. Disregarding the antimatter solution here, setting the constants $C1$, $C2$ and reshaping the classical Schrödinger equation as:

$$C1\cdot C2=\frac{i\cdot 2\cdot m}{\hbar};\quad X=0;\quad 0=\left[\frac{i\cdot 2\cdot m}{\hbar}\cdot\partial_t+\Delta_{\text{Schrödinger}}-2\cdot V_{\text{Schrödinger}}\right]\Psi \tag{296}$$

gives us the proportionality:

$$\Rightarrow\qquad \left[\Delta_{3D}-2\cdot V_{\text{Schrödinger}}\right]\Psi\triangleq -M^2\Phi+\frac{1}{\sqrt{g}}\partial_\alpha\sqrt{g}\cdot g^{\alpha\beta}\partial_\beta\Phi$$

$$\Rightarrow\qquad \Delta_{3D}\Psi\triangleq\Delta_{3D}\Phi;\quad\Rightarrow V_{\text{Schrödinger}}\triangleq\frac{M^2\Phi}{2\cdot\Psi}\triangleq\frac{C_R}{(1-n)}\frac{\Phi}{2\cdot\Psi}. \tag{297}$$

This way we can now link classical Schrödinger solutions and the corresponding Schrödinger potentials with their metric analogue.

A Schrödinger–Dirac Equation

Using (286), we now add in the Dirac starting point, who took for granted that:

(a) The space-time can be assumed Ricci flat and thus, $R = 0$.

(b) The Laplace operator term in (286) has an eigenvalue M^2, which is just the Klein–Gordon case.

Also setting $F[f] = f$ gives us:

$$0 = f_{i,\alpha} f_{i,\beta} g^{\alpha\beta} + \overbrace{M^2 \cdot \frac{4}{(n-6)}}^{\equiv \mu^2 = \mathbf{\mu} \cdot \mathbf{\mu}} \cdot f_i^2. \tag{298}$$

Again, we introduce a function $f_i = \Psi = \Phi + X$ and demand the following additional condition:

$$\partial_t f_i = \partial_t \Psi = i \cdot c \cdot C2 \cdot (\Phi - \mathrm{X}). \tag{299}$$

After inserting the new function into (298), this results in the following:

$$0 = f_{i,0} f_{i,0} g^{00} + f_{i,\alpha} f_{i,\beta} g^{\alpha\beta} + \overbrace{M^2 \cdot \frac{4}{(n-6)}}^{\equiv \mu^2 = \mathbf{\mu} \cdot \mathbf{\mu}} \cdot f_i^2; \quad \alpha, \beta = \{\not{0}, 1, 2, 3, \ldots\}$$

$$\text{with:} \quad \partial_t f_i = i \cdot c \cdot C2_i \cdot (\Phi_i - \mathrm{X}_i)$$

$$0 = -c^2 \cdot C2_i^2 \cdot (\Phi_i - \mathrm{X}_i)^2 + (\Phi_i + \mathrm{X}_i)_{,\alpha} (\Phi_i + \mathrm{X}_i)_{,\beta} g^{\alpha\beta} + \mathbf{\mu} \cdot \mathbf{\mu} \cdot (\Phi_i + \mathrm{X}_i)^2 \tag{300}$$

$$\Rightarrow \quad \begin{cases} 0 = -c^2 \cdot C2_i^2 \cdot (\Phi_i)^2 + (\Phi_i)_{,\alpha} (\Phi_i)_{,\beta} g^{\alpha\beta} + \mathbf{\mu} \cdot \mathbf{\mu} \cdot (\Phi_i)^2 \\ 0 = -c^2 \cdot C2_i^2 \cdot (-\mathrm{X}_i)^2 + (\mathrm{X}_i)_{,\alpha} (\mathrm{X}_i)_{,\beta} g^{\alpha\beta} + \mathbf{\mu} \cdot \mathbf{\mu} \cdot (\mathrm{X}_i)^2. \end{cases} \tag{301}$$

Now we apply the known relation between the Dirac matrices and the metric tensor, reading:

$$g^{\alpha\beta} \cdot I = \frac{\gamma^\alpha \gamma^\beta + \gamma^\beta \gamma^\alpha}{2} \tag{302}$$

and put (301) into the following form:

$$\Rightarrow \quad \begin{cases} 0 = -c^2 \cdot C2_i^2 \cdot (\Phi_i)^2 + (\Phi_i)_{,\alpha} (\Phi_i)_{,\beta} g^{\alpha\beta} + \mathbf{\mu} \cdot \mathbf{\mu} \cdot (\Phi_i)^2 \\ 0 = -c^2 \cdot C2_i^2 \cdot (-\mathrm{X}_i)^2 + (\mathrm{X}_i)_{,\alpha} (\mathrm{X}_i)_{,\beta} g^{\alpha\beta} + \mathbf{\mu} \cdot \mathbf{\mu} \cdot (\mathrm{X}_i)^2 \end{cases}$$

$$0 = \left(\left(\mathbf{\mu} \cdot \mathbf{\mu} - c^2 \cdot C2_i^2 \right) \cdot (\Phi_i)^2 + (\Phi_i)_{,\alpha} (\Phi_i)_{,\beta} g^{\alpha\beta} \right) \cdot I$$

$$0=\overbrace{\left(\boldsymbol{\mu}\cdot\boldsymbol{\mu}-c^2\cdot C2_i^{\ 2}\right)\cdot I}^{\equiv\frac{\beta^2+\beta^2}{2}\cdot m^2}\cdot\left(\Phi_i\right)^2+\left(\Phi_i\right)_{,\alpha}\left(\Phi_i\right)_{,\beta}g^{\alpha\beta}\cdot I$$

$$\Rightarrow\qquad\left(\Phi_i\right)_{,\alpha}\left(\Phi_i\right)_{,\beta}\left(\frac{\gamma^\alpha\gamma^\beta+\gamma^\beta\gamma^\alpha}{2}\right)+\frac{\beta^2+\beta^2}{2}\cdot m^2\cdot\left(\Phi_i\right)^2=0$$

$$\Rightarrow\qquad\left(\Phi_i\right)_{,\alpha}\left(\Phi_i\right)_{,\beta}\gamma^\alpha\gamma^\beta+\beta^2\cdot m^2\cdot\left(\Phi_i\right)^2=0$$

$$\Rightarrow\qquad\left(\Phi_{i,\alpha}\gamma^\alpha+i\cdot\beta\cdot m\cdot\Phi_i\right)\left(\Phi_{i,\beta}\gamma^\beta-i\cdot\beta\cdot m\cdot\Phi_i\right)=0$$

$$\Rightarrow\qquad\Phi_{i,\alpha}\Phi_{i,\beta}\gamma^\beta\gamma^\alpha+\beta^2\cdot m^2\cdot\Phi_i^{\ 2}=0$$

$$\Rightarrow\qquad\left(\Phi_{i,\alpha}\gamma^\alpha+i\cdot\beta\cdot m\cdot\Phi_i\right)\left(\Phi_{i,\beta}\gamma^\beta-i\cdot\beta\cdot m\cdot\Phi_i\right)=0. \tag{303}$$

$$0=-c^2\cdot C2_i^{\ 2}\cdot\left(-\mathrm{X}_i\right)^2+\left(\mathrm{X}_i\right)_{,\alpha}\left(\mathrm{X}_i\right)_{,\beta}g^{\alpha\beta}+\boldsymbol{\mu}\cdot\boldsymbol{\mu}\cdot\left(\mathrm{X}_i\right)^2$$

$$\Rightarrow\qquad 0=-\left(-c\cdot C2_i\right)^2\cdot\left(\mathrm{X}_i\right)^2+\left(\mathrm{X}_i\right)_{,\alpha}\left(\mathrm{X}_i\right)_{,\beta}g^{\alpha\beta}+\boldsymbol{\mu}\cdot\boldsymbol{\mu}\cdot\left(\mathrm{X}_i\right)^2$$

$$0=\left(\left(\boldsymbol{\mu}\cdot\boldsymbol{\mu}-\left(-c\cdot C2_i\right)^2\right)\cdot\left(\mathrm{X}_i\right)^2+\left(\mathrm{X}_i\right)_{,\alpha}\left(\mathrm{X}_i\right)_{,\beta}g^{\alpha\beta}\right)\cdot I$$

$$0=\overbrace{\left(\boldsymbol{\mu}\cdot\boldsymbol{\mu}-\left(-c\cdot C2_i\right)^2\right)\cdot I}^{\equiv\frac{\beta^2+\beta^2}{2}\cdot M^2}\cdot\left(\mathrm{X}_i\right)^2+\left(\mathrm{X}_i\right)_{,\alpha}\left(\mathrm{X}_i\right)_{,\beta}g^{\alpha\beta}\cdot I$$

$$\Rightarrow\qquad\left(\mathrm{X}_i\right)_{,\alpha}\left(\mathrm{X}_i\right)_{,\beta}\left(\frac{\gamma^\alpha\gamma^\beta+\gamma^\beta\gamma^\alpha}{2}\right)+\frac{\beta^2+\beta^2}{2}\cdot M^2\cdot\left(\mathrm{X}_i\right)^2=0$$

$$\Rightarrow\qquad\left(\mathrm{X}_i\right)_{,\alpha}\left(\mathrm{X}_i\right)_{,\beta}\gamma^\alpha\gamma^\beta+\beta^2\cdot M^2\cdot\left(\mathrm{X}_i\right)^2=0$$

$$\Rightarrow\qquad\left(\mathrm{X}_{i,\alpha}\gamma^\alpha+i\cdot\beta\cdot M\cdot\mathrm{X}_i\right)\left(\mathrm{X}_{i,\beta}\gamma^\beta-i\cdot\beta\cdot M\cdot\mathrm{X}_i\right)=0$$

$$\Rightarrow\qquad\mathrm{X}_{i,\alpha}\mathrm{X}_{i,\beta}\gamma^\beta\gamma^\alpha+\beta^2\cdot M^2\cdot\mathrm{X}_i^{\ 2}=0$$

$$\Rightarrow\qquad\left(\mathrm{X}_{i,\alpha}\gamma^\alpha+i\cdot\beta\cdot M\cdot\mathrm{X}_i\right)\left(\mathrm{X}_{i,\beta}\gamma^\beta-i\cdot\beta\cdot M\cdot\mathrm{X}_i\right)=0. \tag{304}$$

Please note the matrix β is now defined via:

$$\left(\boldsymbol{\mu}\cdot\boldsymbol{\mu}-c^2\cdot C2_i^{\ 2}\right)\cdot I\equiv\frac{\beta^2+\beta^2}{2}\cdot m^2 \tag{305}$$

and:

$$\left(\mathbf{\mu}\cdot\mathbf{\mu}-\left(-c\cdot C2_i\right)^2\right)\cdot I\equiv\frac{\beta^2+\beta^2}{2}\cdot M^2. \tag{306}$$

This gives us two Dirac equations in almost classical form [181] for the two functions Φ and X, namely:

$$0=\Phi_{i,\beta}\gamma^\beta-i\cdot\beta\cdot m\cdot\Phi_i, \tag{307}$$

$$0=\mathrm{X}_{i,\beta}\gamma^\beta-i\cdot\beta\cdot M\cdot\mathrm{X}_i, \tag{308}$$

only that this is not a classical Dirac equation, but a Schrödinger–Dirac equation, where, in contrast to the classical form, the Dirac matrices have dimension $(n-1)\times(n-1)$ and the indices are only running from 1 to $n-1$ and not from 0 to $n-1$.

So, in the Minkowski flat space with dimension $n = 5$, we could directly apply the classical Dirac matrices with just a new indexing:

$$\gamma^1=\begin{pmatrix}1&&&\\&1&&\\&&-1&\\&&&-1\end{pmatrix};\ \gamma^2=\begin{pmatrix}&&&1\\&&1&\\&-1&&\\-1&&&\end{pmatrix}$$

$$\gamma^3=\begin{pmatrix}&&&-i\\&&i&\\&i&&\\-i&&&\end{pmatrix};\ \gamma^4=\begin{pmatrix}&&1&\\&&&-1\\-1&&&\\&1&&\end{pmatrix};\ I=\begin{pmatrix}1&&&\\&1&&\\&&1&\\&&&1\end{pmatrix}, \tag{309}$$

and the index β in (307) and (308) would run from 1 to 4.

Surprise

From the fact that equations (307) and (308) are clearly classical Dirac equations [181] in 4 dimensions with the Dirac matrices (309) and the index β running from 1 to 4, we have to conclude, that Dirac's original work was either founded on a space on—at least—5 dimensions (then *R* could be zero) and not on 4 or on 4 dimensions with a non-vanishing Ricci scalar. We also have to conclude that Dirac's mass, which has to be considered as postulated, in fact, is nothing else (metrically speaking) than the space-time curvature:

$$0=\left[\Delta-M^2\right]\Psi=\left[\left(C1\cdot\frac{\partial_t^2}{c^2}+\underbrace{\frac{1}{\sqrt{g}}\partial_\alpha\sqrt{g}\cdot g^{\alpha\beta}\partial_\beta}_{3D-\Delta-\text{Operator}}\right)-M^2\right]\Psi$$

$$=-C_{f0}\cdot R_c\cdot\Psi+(1-n)\cdot\left(\Delta_{n-1}\Psi-\frac{\partial^2\Psi}{c^2\cdot\partial t^2}\right)$$

$$=\left[(1-n)\cdot\left(\Delta_{n-1}-\frac{\partial^2}{c^2\cdot\partial t^2}\right)-C_R\right]\Psi$$

$$\Rightarrow\qquad \frac{C_{f0}\cdot R_c}{1-n}=M^2$$

$$\Rightarrow\qquad M=\sqrt{\frac{C_{f0}\cdot R_c}{1-n}}\qquad(310)$$

or—alternatively, as derived in the section 6.9.3, "The Metric Origin of Mass," in chapter 6, "The Quantum Gravity Turing Machine"—an additional dimension.

Appendix C

Derivation of the Metric Dirac Equation

In a variety of previous publications, we have shown how the metric Dirac equation can be derived from (1) (e.g., see [3, 5, 15, 16, 17, 18]). Here now we want to revisit our derivation and show a few more options.

Dirac in the Metric Picture and Its Connection to the Classical Quaternion Form

In connection with our scaling or quantum wrapping approach $F[f]$, it should be pointed out that with the condition for f to be a scalar there are still many more options to have vector and tensor functions included, namely, e.g., via:

$$f = h + h_{\alpha} q^{\alpha} + h_{\alpha\beta} q^{\alpha\beta} + \cdots. \tag{311}$$

It needs to be pointed out that also arbitrary spinor forms can be incorporated into (311), which may be written as:

$$f = h + h_{\alpha A} q^{\alpha A} + h_{\alpha\beta A} q^{\alpha\beta A} + h_{\alpha\beta AB} q^{\alpha\beta AB} \cdots. \tag{312}$$

Going back to (1), giving us the Ricci scalar of a scaled metric of type (2)[1], we now assume the Laplace operator to have eigen solutions, which gives us:

$$R^{*} = R^{*}_{\alpha\beta} G^{\alpha\beta}$$

$$= \left(R + \frac{F'}{F}(1-n)\Delta f + \frac{f_{,\alpha} f_{,\beta} g^{\alpha\beta}(1-n)}{4F^2}\left(4FF'' + (F')^2 (n-6)\right)\right)\frac{1}{F}$$

$$\text{with:} \quad F = F[f]; \quad F' = \frac{\partial F[f]}{\partial f}; \quad F'' = \frac{\partial^2 F[f]}{\partial f^2}$$

[1]With respect to general metrics and the derivation of Dirac equation for curved space-times, we refer to [212, 214].

$$\xrightarrow{\Delta f = B^2 f}$$

$$= \left(R + \frac{F'}{F}(1-n)B^2 f + \frac{f_{,\alpha} f_{,\beta} g^{\alpha\beta}(1-n)}{4F^2}\left(4FF'' + (F')^2(n-6)\right)\right)\frac{1}{F}. \tag{313}$$

As $F[f]$ could still be freely chosen, we assume (just as within the derivation of the Schrödinger equation in appendix B) a Ricci scalar R being linearly dependent on f and set:

$$F = f - C_f. \tag{314}$$

This simplifies (313) to:

$$R^* = R^*_{\alpha\beta} G^{\alpha\beta}$$

$$= \left(R + \frac{1}{f - C_f}(1-n)\Delta f + \frac{f_{,\alpha} f_{,\beta} g^{\alpha\beta}(1-n)}{4(f - C_f)^2}(n-6)\right)\frac{1}{f - C_f}. \tag{315}$$

Now we demand:

$$R^* = R^*_{\alpha\beta} G^{\alpha\beta} = 0 \tag{316}$$

and obtain:

$$0 = \frac{4 \cdot R \cdot (f - C_f)^2}{(1-n)\cdot(n-6)} + \frac{4\cdot(f - C_f)}{n-6}\cdot B^2 \cdot f + f_{,\alpha} f_{,\beta} g^{\alpha\beta}$$

$$= \frac{4}{(1-n)\cdot(n-6)}\left(R\cdot\left(f^2 - 2\cdot C_f \cdot f + C_f^2\right) + (1-n)\cdot\left(f^2 - C_f \cdot f\right)\cdot B^2\right) + f_{,\alpha} f_{,\beta} g^{\alpha\beta}$$

$$= \frac{4}{(1-n)\cdot(n-6)}\begin{pmatrix} f^2 \cdot \left(R + (1-n)\cdot B^2\right) \\ -2\cdot f \cdot \left(C_f \cdot \left(R + \frac{(1-n)\cdot B^2}{2}\right)\right) + C_f^2 \cdot R \end{pmatrix} + f_{,\alpha} f_{,\beta} g^{\alpha\beta}$$

$$= \frac{4\cdot\left(R + (1-n)\cdot B^2\right)}{(1-n)\cdot(n-6)}\left(f^2 - 2\cdot f \cdot \frac{C_f \cdot \left(R + \frac{(1-n)\cdot B^2}{2}\right)}{R + (1-n)\cdot B^2} + \frac{C_f^2 \cdot R}{R + (1-n)\cdot B^2}\right) + f_{,\alpha} f_{,\beta} g^{\alpha\beta}$$

$$=\frac{4\cdot\left(R+(1-n)\cdot B^2\right)}{(1-n)\cdot(n-6)}\left(\begin{array}{c}\left(f-\frac{C_f\cdot\left(R+\frac{(1-n)\cdot B^2}{2}\right)}{R+(1-n)\cdot B^2}\right)^2\\-\left(\frac{C_f\cdot\left(R+\frac{(1-n)\cdot B^2}{2}\right)}{R+(1-n)\cdot B^2}\right)^2+\frac{C_f^2\cdot R}{R+(1-n)\cdot B^2}\end{array}\right)+f_{,\alpha}f_{,\beta}g^{\alpha\beta}.$$

(317)

We see that we find no reasonable setting for either B or C_f for which we could make the term in big parentheses in the last line of the equation above a complete square. We therefore assume that there may exist a function *F*[*f*] for which we find a development for (313) resulting in:

$$R^*=0$$
$$=\left(R+\frac{F'}{F}(1-n)B^2f+\frac{f_{,\alpha}f_{,\beta}g^{\alpha\beta}(1-n)}{4F^2}\left(4FF''+(F')^2(n-6)\right)\right)\frac{1}{F}$$

$$\Rightarrow\quad 0=C_{f1}^2\cdot\left(f-C_{f0}\right)^2+f_{,\alpha}f_{,\beta}g^{\alpha\beta};\quad C_{f0}=C_{f0}[n,B,R];\quad C_{f1}=C_{f1}[n,B,R].$$

(318)

Alternatively, as finding a suitable *F*[*f*] seems to be an extremely difficult task, one may also just assume to have an *R*-dependency of the kind:

$$R=R[f]=\left(R_0+R_1\cdot f+R_2\cdot f^2\right)\cdot C_{f0}^2\cdot(n-2)\cdot(n-1).\qquad(319)$$

We also chose $F=e^{f\cdot C_{f0}}\cdot C_{f1}$. This makes (313) with (316) to:

$$0=\frac{R}{(1-n)}+\frac{F'}{F}B^2f+\frac{f_{,\alpha}f_{,\beta}g^{\alpha\beta}}{4F^2}\left(4FF''+(F')^2(n-6)\right)$$

$$\Rightarrow\qquad 0=\frac{4F^2}{4FF''+(F')^2(n-6)}\left(\frac{R}{(1-n)}+\frac{F'}{F}B^2f\right)+f_{,\alpha}f_{,\beta}g^{\alpha\beta}$$

$$0 = \frac{B^2 \cdot C_{f0} \cdot f \cdot (n-1) - R}{C_{f0}^2 \cdot (n-2) \cdot (n-1)} + f_{,\alpha} f_{,\beta} g^{\alpha\beta}$$

$$= f \cdot \frac{B^2}{C_{f0} \cdot (n-2)} - \left(R_0 + R_1 \cdot f + R_2 \cdot f^2\right) + f_{,\alpha} f_{,\beta} g^{\alpha\beta}$$

$$\Rightarrow \qquad R_0 + R_1 \cdot f + R_2 \cdot f^2 - f \cdot \frac{B^2}{C_{f0} \cdot (n-2)} = f_{,\alpha} f_{,\beta} g^{\alpha\beta}$$

$$= R_2 \cdot \left(f^2 + \frac{f}{R_2} \left(R_1 - \frac{B^2}{R_2 \cdot C_{f0} \cdot (n-2)} \right) + \frac{R_0}{R_2} \right) = f_{,\alpha} f_{,\beta} g^{\alpha\beta}. \tag{320}$$

Now we demand:

$$\frac{1}{R_2} \left(R_1 - \frac{B^2}{R_2 \cdot C_{f0} \cdot (n-2)} \right) = 2 \cdot \sqrt{\frac{R_0}{R_2}}, \tag{321}$$

thereby fixing the constant:

$$C_{f0} = \frac{B^2}{R_2 \cdot \left(R_1 - 2 \cdot \sqrt{R_0 \cdot R_2}\right) \cdot (n-2)} \tag{322}$$

and subsequently obtaining:

$$R_2 \cdot \left(f + \sqrt{\frac{R_0}{R_2}} \right)^2 = f_{,\alpha} f_{,\beta} g^{\alpha\beta}. \tag{323}$$

Now we apply our intrinsic structural setting for f as given in (311), restrict it to its simplest form, which is to say:

$$f = h_\lambda q^\lambda. \tag{324}$$

and obtain:

$$R_2 \cdot \left(f + \sqrt{\frac{R_0}{R_2}} \right)^2 = f_{,\alpha} f_{,\beta} g^{\alpha\beta}$$

$$\xrightarrow{f = h_\lambda q^\lambda}$$

$$R_2 \cdot \left(f + \sqrt{\frac{R_0}{R_2}} \right)^2 = f_{,\alpha} f_{,\beta} g^{\alpha\beta} = -h_{\lambda,\alpha} q^\lambda g^{\alpha\beta} h_{\lambda,\beta} q^\lambda$$

$$\xrightarrow{g^{\alpha\beta} = \mathbf{e}^\alpha \cdot \mathbf{e}^\beta} = -h_{\lambda,\alpha} q^\lambda \mathbf{e}^\alpha \cdot \mathbf{e}^\beta h_{\lambda,\beta} q^\lambda$$

$$\xrightarrow{q^\lambda \mathbf{e}^\alpha = \mathbf{Q}^{\lambda\alpha}} = -h_{\lambda,\alpha} \mathbf{Q}^{\lambda\alpha} \cdot \mathbf{Q}^{\lambda\beta} h_{\lambda,\beta}. \tag{325}$$

Please note that in principle, the evaluation should have been:

$$R_2 \cdot \left(f + \sqrt{\frac{R_0}{R_2}} \right)^2 = f_{,\alpha} f_{,\beta} g^{\alpha\beta} = -h_{\lambda,\alpha} q^{\lambda} g^{\alpha\beta} h_{\lambda,\beta} q^{\lambda}$$

$$= -\left(h_{\lambda} q^{\lambda}{}_{,\alpha} + h_{\lambda,\alpha} q^{\lambda} \right) g^{\alpha\beta} \left(h_{\lambda} q^{\lambda}{}_{,\beta} + h_{\lambda,\beta} q^{\lambda} \right)$$

$$= -h_{\lambda} q^{\lambda}{}_{,\alpha} g^{\alpha\beta} h_{\lambda} q^{\lambda}{}_{,\beta} - h_{\lambda,\alpha} q^{\lambda} g^{\alpha\beta} h_{\lambda} q^{\lambda}{}_{,\beta} - h_{\lambda} q^{\lambda}{}_{,\alpha} g^{\alpha\beta} h_{\lambda,\beta} q^{\lambda} - h_{\lambda,\alpha} q^{\lambda} g^{\alpha\beta} h_{\lambda,\beta} q^{\lambda}. \tag{326}$$

However, as we haven't had any need to fix both vectors h_{λ} or q^{λ}, yet, we assume that one is a vector of constants, and by making q^{λ} this constant vector, we obtain (325) instead of the complicated result (326).

Now, only in order to obtain the classical Dirac equation, we approximate the curvature term on the left-hand side of (325) as a square of the linear function of the scalar product f and therefore we can write:

$$R_2 \cdot \left(f + \sqrt{\frac{R_0}{R_2}} \right)^2 \simeq h_{\lambda} q^{\lambda} \mathbf{M} \cdot \mathbf{M} h_{\lambda} q^{\lambda} = -h_{\lambda,\alpha} q^{\lambda} g^{\alpha\beta} h_{\lambda,\beta} q^{\lambda}$$

$$\xrightarrow{g^{\alpha\beta} = \mathbf{e}^{\alpha} \cdot \mathbf{e}^{\beta}} = -h_{\lambda,\alpha} q^{\lambda} \mathbf{e}^{\alpha} \cdot \mathbf{e}^{\beta} h_{\lambda,\beta} q^{\lambda}$$

$$\Rightarrow \qquad \mp h_{\lambda} q^{\lambda} \mathbf{M} = i \cdot h_{\lambda,\alpha} q^{\lambda} \mathbf{e}^{\alpha}$$

$$\Rightarrow \qquad 0 = i \cdot h_{\lambda,\alpha} \mathbf{e}^{\alpha} \pm h_{\lambda} \mathbf{M} \tag{327}$$

$$\Rightarrow \qquad 0 = i \cdot h_{\lambda,\alpha} \mathbf{e}^{\alpha} \pm h_{\lambda} \mathbf{M}. \tag{328}$$

Please note that (328) represents a generalized Dirac equation in an arbitrary number of dimensions. By reducing this general and arbitrarily n-dimensional equation to 4 dimensions, we almost directly recognize the Dirac equation, only that in our—the metric—case, the mass-terms have to be at least a vector, while the Dirac matrices for the derivatives of the function vector h_{λ} (c.f. [181]) are becoming the metric base vector components. In order to obtain the classical Dirac equation in the natural units form:

$$0 = i \cdot \gamma^{\alpha} h_{\lambda,\alpha} \pm h_{\lambda} \cdot m, \tag{329}$$

which we can also write as:

$$0 = i \cdot \gamma^{\alpha} h_{\lambda,\alpha} \pm I \cdot h_{\lambda} \cdot m, \tag{330}$$

and with the Dirac gamma matrices γ^α and the unit matrix I given as follows:

$$\gamma^0=\begin{pmatrix}1&&&\\&1&&\\&&-1&\\&&&-1\end{pmatrix};\ \gamma^1=\begin{pmatrix}&&&1\\&&1&\\&-1&&\\-1&&&\end{pmatrix}$$

$$\gamma^2=\begin{pmatrix}&&&-i\\&&i&\\&i&&\\-i&&&\end{pmatrix};\ \gamma^3=\begin{pmatrix}&&1&\\&&&-1\\-1&&&\\&1&&\end{pmatrix};\ I=\begin{pmatrix}1&&&\\&1&&\\&&1&\\&&&1\end{pmatrix}, \tag{331}$$

we simply need to multiply (328) with a suitable object **C**. This gives us the identities:

$$\mathbf{C}\cdot\mathbf{e}^\alpha=\gamma^\alpha;\quad \mathbf{C}\cdot\mathbf{M}=I\cdot m \tag{332}$$

and provides a connection between the classical quaternion and the metric Dirac equation. As the classical Dirac equation was given in Minkowski coordinates, with base vectors of the kind:

$$\mathbf{e}^0=\begin{pmatrix}i\\0\\0\\0\end{pmatrix};\ \mathbf{e}^1=\begin{pmatrix}0\\1\\0\\0\end{pmatrix};\ \mathbf{e}^2=\begin{pmatrix}0\\0\\1\\0\end{pmatrix};\ \mathbf{e}^3=\begin{pmatrix}0\\0\\0\\1\end{pmatrix}, \tag{333}$$

we could find the corresponding structure of the object **C** via the first equation in (332), and it becomes clear that we need something of the kind:

$$C^\alpha_\beta\cdot\mathbf{e}^\beta=\gamma^\alpha, \tag{334}$$

$$\Rightarrow\begin{cases}C^0_\alpha=\begin{pmatrix}-i&&&\\&1&&\\&&-1&\\&&&-1\end{pmatrix};\ C^1_\alpha=\begin{pmatrix}&&&1\\&&1&\\&-1&&\\i&&&\end{pmatrix}\\ C^2_\alpha=\begin{pmatrix}&&&-i\\&&i&\\&i&&\\-1&&&\end{pmatrix};\ C^3_\alpha=\begin{pmatrix}&&1&\\&&&-1\\i&&&\\&1&&\end{pmatrix}\end{cases}. \tag{335}$$

From here, we deduce that **M** must be something more complex than just a vector as only the following approach gives us what we need:

$$C_\beta^\alpha \cdot \mathbf{M} = I \cdot m \quad \Rightarrow \quad C_\beta^\alpha \cdot M_\lambda^\beta = \delta_\lambda^\alpha \cdot m = I \cdot m. \tag{336}$$

Thereby we need to point out that $h_\lambda q^\lambda \mathbf{M} \cdot \mathbf{M} h_\lambda q^\lambda$ in (327) only is a linear approximation of the complex expression $R_2 \cdot \left(f + \sqrt{\frac{R_0}{R_2}} \right)^2$ and that the classical Dirac equation was just postulated [181], while our new Dirac equation (323) has a fundamental metric and extremal-principle origin. The result after solving (336) with respect to the object M_λ^β reads:

$$\Rightarrow \begin{cases} M_0^\alpha = \begin{pmatrix} i & & & \\ & 1 & & \\ & & -1 & \\ & & & -1 \end{pmatrix}; \quad M_1^\alpha = \begin{pmatrix} & & & -i \\ & & -1 & \\ & 1 & & \\ 1 & & & \end{pmatrix} \\ M_2^\alpha = \begin{pmatrix} & & & -1 \\ & & -i & \\ & -i & & \\ i & & & \end{pmatrix}; \quad M_3^\alpha = \begin{pmatrix} & & -i & \\ & & & 1 \\ 1 & & & \\ & 1 & & \end{pmatrix} \end{cases} \tag{337}$$

and gives us the connection of the classical Dirac equation (329) from [181] with our new metric Dirac equation (323) in its linear approximated form (327). We realize that in the metric picture, the corresponding inertia parameters are not just scalars.

Now we also want to see the outcome without the approximations as applied on the left-hand side of (327)

$$R_2 \cdot \left(f + \sqrt{\frac{R_0}{R_2}} \right)^2 = f_{,\alpha} f_{,\beta} g^{\alpha\beta} = -h_{\lambda,\alpha} q^\lambda g^{\alpha\beta} h_{\lambda,\beta} q^\lambda$$
$$= -h_{\lambda,\alpha} q^\lambda \mathbf{e}^\alpha \cdot \mathbf{e}^\beta h_{\lambda,\beta} q^\lambda. \tag{338}$$

Initially, we realize that there can be a constant (asymmetric or Higgs type) solution via:

$$f + \sqrt{\frac{R_0}{R_2}} = 0. \tag{339}$$

In principle, one might like to avoid a constant curvature term, and by setting $R_0 = 0$, one could just use the approximation from above, only that, this time, meaning with $R_0 = 0$, it is no approximation anymore and we can write:

$$R_2 \cdot \left(f + \sqrt{\frac{R_0}{R_2}} \right)^2 = \xrightarrow{R_0=0} = R_2 \cdot f^2$$

$$\xrightarrow{R_2=\mathbf{M}\cdot\mathbf{M}}$$

$$= h_\lambda q^\lambda \mathbf{M} \cdot \mathbf{M} h_\lambda q^\lambda = -h_{\lambda,\alpha} q^\lambda g^{\alpha\beta} h_{\lambda,\beta} q^\lambda$$

$$\xrightarrow{g^{\alpha\beta}=\mathbf{e}^\alpha\cdot\mathbf{e}^\beta} = -h_{\lambda,\alpha} q^\lambda \mathbf{e}^\alpha \cdot \mathbf{e}^\beta h_{\lambda,\beta} q^\lambda$$

$$\Rightarrow \qquad \mp h_\lambda q^\lambda \mathbf{M} = i \cdot h_{\lambda,\alpha} q^\lambda \mathbf{e}^\alpha$$

$$\Rightarrow \qquad 0 = i \cdot h_{\lambda,\alpha} \mathbf{e}^\alpha \pm h_\lambda \mathbf{M}, \tag{340}$$

obtaining the metric Dirac equation just as shown above.

Without $R_0 = 0$, we can still decompose the left-hand side and get:

$$R_2 \cdot \left(f + \sqrt{\frac{R_0}{R_2}} \right)^2 =$$

$$\xrightarrow{R_2=\mathbf{M}\cdot\mathbf{M}} = \begin{pmatrix} \left(h_\lambda q^\lambda + \sqrt{\frac{R_0}{R_2}} \right) \mathbf{M} \\ \cdot \mathbf{M} \left(h_\lambda q^\lambda + \sqrt{\frac{R_0}{R_2}} \right) \end{pmatrix} = -h_{\lambda,\alpha} q^\lambda g^{\alpha\beta} h_{\lambda,\beta} q^\lambda$$

$$\xrightarrow{g^{\alpha\beta}=\mathbf{e}^\alpha\cdot\mathbf{e}^\beta} = -h_{\lambda,\alpha} q^\lambda \mathbf{e}^\alpha \cdot \mathbf{e}^\beta h_{\lambda,\beta} q^\lambda$$

$$\Rightarrow \qquad \mp \left(h_\lambda q^\lambda + \sqrt{\frac{R_0}{R_2}} \right) \mathbf{M} = i \cdot h_{\lambda,\alpha} q^\lambda \mathbf{e}^\alpha$$

$$\xrightarrow{\sqrt{\frac{R_0}{R_2}} \equiv q^\lambda \Re_\lambda}$$

$$\Rightarrow \qquad 0 = i \cdot h_{\lambda,\alpha} \mathbf{e}^\alpha \pm \left(h_\lambda + \Re_\lambda \right) \mathbf{M}. \tag{341}$$

Here now the new field parameter $\Re_\lambda$ would act as the asymmetric or “Higgs” field. It does not need to be a constant but should be in the case of $h_{\lambda,\alpha} = 0$.

It needs to be pointed out here that the summation over β in (334) is not the only option we have in order to obtain the Dirac gamma matrices. Similarly, we may just perform the evaluation as follows:

$$C_a^{cb} \cdot \mathbf{e}^\beta = C_a^{cb} \cdot E_b{}^\beta \equiv D_a^{c\beta} = \gamma^\beta, \tag{342}$$

$$\Rightarrow \begin{cases} C_a^{c0} = \begin{pmatrix} -i & & & \\ & 1 & & \\ & & -1 & \\ & & & -1 \end{pmatrix}; \quad C_a^{c1} = \begin{pmatrix} & & & 1 \\ & & 1 & \\ & -1 & & \\ i & & & \end{pmatrix} \\ C_a^{c2} = \begin{pmatrix} & & & -i \\ & & i & \\ & i & & \\ -1 & & & \end{pmatrix}; \quad C_a^{c3} = \begin{pmatrix} & & 1 & \\ & & & -1 \\ i & & & \\ & 1 & & \end{pmatrix}. \end{cases} \tag{343}$$

Consequently, we obtain for $\mathbf{M}$:

$$C_a^{cb} \cdot \mathbf{M} = I \cdot m \quad \Rightarrow \quad C_a^{cb} \cdot M_{b\lambda}^a = \delta_\lambda^c \cdot m = I \cdot m. \tag{344}$$

The result after solving (344) with respect to the object $M_{b\lambda}^a$ reads:

$$\Rightarrow \begin{cases} M_{0\lambda}^a = \begin{pmatrix} i & & & \\ & 1 & & \\ & & -1 & \\ & & & -1 \end{pmatrix}; \quad M_{1\lambda}^a = \begin{pmatrix} & & & i \\ & & -1 & \\ & 1 & & \\ 1 & & & \end{pmatrix} \\ M_{2\lambda}^a = \begin{pmatrix} & & & -1 \\ & & -i & \\ & -i & & \\ i & & & \end{pmatrix}; \quad M_{3\lambda}^a = \begin{pmatrix} & & -i & \\ & & & 1 \\ 1 & & & \\ & -1 & & \end{pmatrix}. \end{cases} \tag{345}$$

Here it has to be pointed out, however, that our complex forms of $\mathbf{M}$ are a result from the approximation (327) (first line) and not necessarily a genuine equivalence to real inertia. More discussion will be needed here.

The well-informed reader may ask at this point why we explicitly tried to obtain the Dirac equation from the $f_{,\alpha} f_{,\beta} g^{\alpha\beta}$-operator, while Dirac in [181] only extracted the square root (operational factorization) from the Klein–Gordon equation via the introduction of quaternions. The reason for this excursion lies in the fact that the factorization of the $f_{,\alpha} f_{,\beta} g^{\alpha\beta}$-operator is straightforward and works for all kinds of

symmetric metrics, while this is not the case for the Laplace operator in the Klein–Gordon equation. In [15], sub-section "Dirac's "Luck" and the Flat Space Condition," we have shown that for certain (scale-Ricci-flat) space-times, there is a direct connection between the two operators, which makes our evaluation, even though it does not use Dirac's starting point (the Klein–Gordon equation), above absolutely worthwhile ... if not to say essential!

Please note that the classical Dirac equation is also often given with the so-called β-matrix with $\beta=\begin{pmatrix}1&&&\\&1&&\\&&-1&\\&&&-1\end{pmatrix}$. The Dirac equation then reads:

$$0=f_{i,\beta}\gamma^{\beta}-i\cdot\beta\cdot m\cdot f_i. \tag{346}$$

Insisting on the outcome of the β-matrix in our metrically derived Dirac equation is leading us to the equations (derivation see [16]):

$$C_a^{cb}\cdot\mathbf{M}=\beta\cdot m \;\Rightarrow\; C_a^{cb}\cdot M_{b\lambda}^{a}=\beta\cdot m \tag{347}$$

and gives the following solution with respect to the object $M_{b\lambda}^{a}$:

$$\Rightarrow\quad\left\{\begin{array}{l}M_{0\lambda}^{a}=\begin{pmatrix}i&&&\\&1&&\\&&1&\\&&&1\end{pmatrix};\; M_{1\lambda}^{a}=\begin{pmatrix}&&&i\\&&1&\\&1&&\\1&&&\end{pmatrix}\\ M_{2\lambda}^{a}=\begin{pmatrix}&&&1\\&&i&\\&-i&&\\i&&&\end{pmatrix};\; M_{3\lambda}^{a}=\begin{pmatrix}&&i&\\&&&-1\\1&&&\\&-1&&\end{pmatrix}\end{array}\right. . \tag{348}$$

We saw that with a simple matrix multiplication, we had come from our metric Dirac equation (328) to the classical one (346) (whether with the $\boldsymbol{\beta}$-matrix or I does not matter):

$$0=i\cdot f_{i,\alpha}\mathbf{e}^{\alpha}\pm f_i\mathbf{M}\,/\cdot\mathbf{C}$$

$$\Rightarrow\quad 0=i\cdot\gamma^{\alpha}f_{i,\alpha}+\beta\cdot f_i\cdot m. \tag{349}$$

Thereby, in the case of 4-dimensional Minkowski coordinates, the objects C and M are given via (343) and (348). Our metric Dirac equation (328), on the other hand, just came from a simple condition about the Ricci scalar. In [16], we have shown that this is actually just a by-product of a more complete Einstein–Hilbert variation. Further below in this appendix, we are going to rederive this important finding.

Getting Rid of the Quaternions

Knowing now that the fundamental metric origin of the quaternions, which Dirac had to apply in order to extract the root of the Klein–Gordon equation, and the metric and subsequently the base vectors (see transition from the first to the second line in (327)), we also have the means to potentially avoid these mathematical structures. We are keen to do so, because:

(a) They are kind of nasty ... in a purely practical-mathematical sense, of course.

(b) It is difficult to change the number of dimensions (e.g., the classical Dirac equation is only defined in a 4-dimensional space-time[2]).

In order to achieve this goal, we have to find a different way to separate, respectively decompose, the metric tensor. Instead of applying base vectors for the decomposition, we intend to use vectors. This "vectorial root extraction" or "vector decomposition" was introduced by the author a long time ago. It was motivated by an adventurous story of two Jews in Nazi Germany, who had left a mathematical message inside the gas chamber where they had been killed in 1944 (e.g., introduction part of [3] or the nicely illustrated book [35]).

Apparently, as demonstrated above, there also seems to be an approach which extracts the Dirac operator directly out of the Einstein field equations or the Einstein–Hilbert action and, thus, the General Theory of Relativity With respect to the connection with the classical approach, we refer to the sub-section above and present here only the new technique which does not require the use of quaternions as Dirac needed them.

It is easy to prove that the scalar product of the following vector:

[2]Higher dimensional Dirac equations require higher dimensional Dirac matrices, which can be derived (just like the classical 4-dimensional Dirac matrices) via composition of Pauli matrices [211].

$$V_{\Omega}=\left\{\begin{matrix} a+b+c+d+e,a+b+c+d-e,a+b+c-d+e,a+b+c-d-e, \\ a+b-c+d+e,a+b-c+d-e,a+b-c-d+e,a+b-c-d-e, \\ a-b+c+d+e,a-b+c+d-e,a-b+c-d+e,a-b+c-d-e, \\ a-b-c+d+e,a-b-c+d-e,a-b-c-d+e,a-b-c-d-e \end{matrix}\right\}$$
$$\equiv\left[a\pm b\pm c\pm d\pm e\right]_{\Omega}$$

(350)

gives:

$$V_{\Omega}\cdot V_{\Omega}=a^{2}+b^{2}+c^{2}+d^{2}+e^{2}. \qquad (351)$$

Even the introduction of "virtual" parameters ε can be incorporated into the vector V_{Ω} as follows:

$$V_{\Omega}=\left\{a+b+c,a+b-c,a-b+i\cdot c,a-b-i\cdot c\right\}\equiv\left[a\pm b\pm I\cdot c\right]_{\Omega}$$
$$\text{with:}\quad V_{\Omega}\cdot V_{\Omega}=a^{2}+b^{2}. \qquad (352)$$

We also have the possibility to explicitly demand mixed terms, which is to say, we can give any metric tensor in form of a metric equivalent as V_{Ω}-vector. The "peculiar" thing about the new form only is to be seen in the fact that our vector has too many components. Taking, for example, the possibility of the purely diagonal metric case with the subsequent scalar product (351), we already end up with 2^{n-1} components, which consequently gives more "Dirac" equations than we have metric conditions. Thus, the form (350), as simple and attractive it may appear, cannot be the irreducible form for the vector V_{Ω}. In order to find the latter, we simply have to compose a general vector of the type:

$$V_{\Omega}=\left\{\begin{matrix} a_{0}\cdot a+b_{0}\cdot b+c_{0}\cdot c+d_{0}\cdot d+e_{0}\cdot e+\ldots, \\ a_{0}\cdot a+b_{1}\cdot b+c_{1}\cdot c+d_{1}\cdot d+e_{1}\cdot e+\ldots, \\ a_{0}\cdot a+b_{1}\cdot b+c_{2}\cdot c+d_{2}\cdot d+e_{2}\cdot e+\ldots, \\ a_{0}\cdot a+b_{1}\cdot b+c_{2}\cdot c+d_{3}\cdot d+e_{3}\cdot e+\ldots, \\ a_{0}\cdot a+b_{1}\cdot b+c_{2}\cdot c+d_{3}\cdot d+e_{4}\cdot e+\ldots, \\ a_{0}\cdot a+b_{1}\cdot b+c_{2}\cdot c+d_{3}\cdot d+e_{4}\cdot e+\ldots, \\ \ldots \end{matrix}\right\}$$

$$\equiv\left[a_{j}\cdot a+b_{j}\cdot b+c_{j}\cdot c+d_{j}\cdot d+e_{j}\cdot e+\ldots\right]_{\Omega}. \qquad (353)$$

The constants a_j, b_j, c_j, and so on have to be determined from the line element equation, which we here assume to have the structure:

$$ds^2 = a^2 \cdot dx_0^2 + b^2 \cdot dx_1^2 + c^2 \cdot dx_2^2 + d^2 \cdot dx_3^2 + e^2 \cdot dx_4^2 + \ldots$$
$$+a \cdot b \cdot dx_0 dx_1 + a \cdot c \cdot dx_0 dx_2 + a \cdot d \cdot dx_0 dx_3 + a \cdot e \cdot dx_0 dx_4 + \ldots$$
$$+b \cdot c \cdot dx_1 dx_2 + b \cdot d \cdot dx_1 dx_3 + b \cdot e \cdot dx_1 dx_4 + \ldots$$
$$+c \cdot d \cdot dx_2 dx_3 + c \cdot e \cdot dx_2 dx_4 + \ldots + d \cdot e \cdot dx_3 dx_4 + \ldots \tag{354}$$

with the coordinates x_i, in comparison with the scalar product:

$$V_\Omega \cdot V_\Omega$$
$$= \begin{bmatrix} a_j \cdot a + b_j \cdot b + c_j \cdot c \\ + d_j \cdot d + e_j \cdot e + \cdots \end{bmatrix}_\Omega \cdot \begin{bmatrix} a_j \cdot a + b_j \cdot b + c_j \cdot c \\ + d_j \cdot d + e_j \cdot e + \cdots \end{bmatrix}_\Omega . \tag{355}$$

The resulting equations for the determination of the constants a_j, b_j, c_j, and so on are to be taken from:

$$V_\Omega \cdot V_\Omega$$
$$= \begin{bmatrix} a_j \cdot a + b_j \cdot b + c_j \cdot c \\ + d_j \cdot d + e_j \cdot e + \cdots \end{bmatrix}_\Omega \cdot \begin{bmatrix} a_j \cdot a + b_j \cdot b + c_j \cdot c \\ + d_j \cdot d + e_j \cdot e + \cdots \end{bmatrix}_\Omega$$
$$= a^2 + b^2 + c^2 + d^2 + e^2 + \cdots + a \cdot b + a \cdot c + a \cdot d + a \cdot e + \cdots$$
$$+b \cdot c + b \cdot d + b \cdot e + \cdots + c \cdot d + c \cdot e + \cdots + d \cdot e + \cdots \tag{356}$$

via comparison of coefficients.

The number of coefficients and, thus, the number of "Dirac" equations will now be equal to the number of dimensions.

Let us choose an example in 3 dimensions, with the line element given as:

$$ds^2 = a^2 \cdot dx_0^2 + b^2 \cdot dx_1^2 + c^2 \cdot dx_2^2$$
$$+ a \cdot b \cdot dx_0 dx_1 + a \cdot c \cdot dx_0 dx_2 + b \cdot c \cdot dx_1 dx_2 . \tag{357}$$

The irreducible V_Ω-vector would read:

$$V_\Omega = \{a_0 \cdot a + b_0 \cdot b + c_0 \cdot c, a_0 \cdot a + b_1 \cdot b + c_1 \cdot c, a_0 \cdot a + b_1 \cdot b + c_2 \cdot c\}$$
$$\equiv \begin{bmatrix} a_j \cdot a + b_j \cdot b + c_j \cdot c \end{bmatrix}_\Omega \tag{358}$$

and the subsequent equations for the determination of the constants a_j, b_j, c_j can be written as:

$$a^2 = 3 \cdot a_0^2 \cdot a^2, \quad b^2 = b^2 \cdot \left(2 \cdot b_0^2 + b_1^2\right),$$
$$c^2 = c^2 \cdot \left(b_0^2 + b_1^2 + b_2^2\right), \quad a \cdot b = a \cdot b \cdot 2 \cdot a_0 \cdot \left(2 \cdot b_0 + b_1\right),$$
$$b \cdot c = b \cdot c \cdot 2 \cdot \left(b_0 \cdot c_0 + b_0 \cdot c_2 + b_1 \cdot c_1\right), \quad a \cdot c = a \cdot c \cdot 2 \cdot a_0 \cdot \left(c_0 + c_1 + c_2\right). \tag{359}$$

There are 8 different solutions to the equations above and one of them would just be:

$$a_0 = -\frac{1}{\sqrt{3}},\quad b_0 = -\frac{1}{2}\sqrt{\frac{5}{6}+\sqrt{\frac{2}{3}}},\quad b_1 = \frac{1}{\sqrt{2}} - \frac{1}{2\sqrt{3}}$$

$$c_0 = -\frac{18+\sqrt{6}}{12\sqrt{3}},\quad c_1 = \frac{1}{6}\left(\sqrt{2}-\sqrt{3}\right),\quad c_2 = \frac{1}{6}\sqrt{\frac{7}{2}-\sqrt{6}}. \tag{360}$$

For practical purposes, shorter settings for the V_Ω-vector could be useful:

$$V_\Omega = \left\{a_0 \cdot a + b_0 \cdot b + c_0 \cdot c, b_1 \cdot b + c_1 \cdot c, c_2 \cdot c\right\}$$

$$\equiv \left[a_j \cdot a + b_j \cdot b + c_j \cdot c\right]_\Omega. \tag{361}$$

The corresponding equations for the determination of the constants a_j, b_j, c_j then read:

$$a^2 = a_0^2 \cdot a^2,\quad b^2 = b^2 \cdot \left(b_0^2 + b_1^2\right),\quad c^2 = c^2 \cdot \left(c_0^2 + c_1^2 + c_2^2\right),$$

$$a \cdot b = a \cdot b \cdot 2 \cdot a_0 \cdot b_0,\quad b \cdot c = b \cdot c \cdot 2 \cdot \left(b_0 \cdot c_0 + b_1 \cdot c_1\right),\quad a \cdot c = a \cdot c \cdot 2 \cdot a_0 \cdot c_0 \tag{362}$$

and we obtain somewhat simpler results for the subsequent 8 solutions (here only one example):

$$a_0 = -1,\quad b_0 = -\frac{1}{2},\quad b_1 = -\frac{\sqrt{3}}{2}$$

$$c_0 = -\frac{1}{2},\quad c_1 = -\frac{1}{2\sqrt{3}},\quad c_2 = -\sqrt{\frac{2}{3}}. \tag{363}$$

In 4 dimensions, we have 10 equations for the a_j, b_j, c_j of the form:

$$a^2 = a_0^2 \cdot a^2,\quad b^2 = b^2 \cdot \left(b_0^2 + b_1^2\right),\quad c^2 = c^2 \cdot \left(c_0^2 + c_1^2 + c_2^2\right),$$

$$d^2 = d^2 \cdot \left(d_0^2 + d_1^2 + d_2^2 + d_3^2\right),$$

$$a \cdot b = a \cdot b \cdot 2 \cdot a_0 \cdot b_0,\quad b \cdot c = b \cdot c \cdot 2 \cdot \left(b_0 \cdot c_0 + b_1 \cdot c_1\right),$$

$$b \cdot d = b \cdot d \cdot 2 \cdot \left(b_0 \cdot d_0 + b_1 \cdot d_1\right),\quad a \cdot c = a \cdot c \cdot 2 \cdot a_0 \cdot c_0,$$

$$a \cdot d = a \cdot d \cdot 2 \cdot a_0 \cdot d_0,\quad c \cdot d = c \cdot d \cdot 2 \cdot \left(c_0 \cdot d_0 + c_1 \cdot d_1 + c_2 \cdot d_2\right), \tag{364}$$

when applying the short V_Ω-vector as:

$$V_\Omega = \{a_0 \cdot a + b_0 \cdot b + c_0 \cdot c + d_0 \cdot d, b_1 \cdot b + c_1 \cdot c + d_1 \cdot d, c_2 \cdot c + d_2 \cdot d, d_3 \cdot d\}$$
$$\equiv \left[a_j \cdot a + b_j \cdot b + c_j \cdot c + d_j \cdot d\right]_\Omega. \tag{365}$$

One solution out of the 16 shall be given here:

$$a_0 = -1,\quad b_0 = -\frac{1}{2},\quad b_1 = -\frac{\sqrt{3}}{2}$$
$$c_0 = -\frac{1}{2},\quad c_1 = -\frac{1}{2\sqrt{3}},\quad c_2 = -\sqrt{\frac{2}{3}}$$
$$d_0 = -\frac{1}{2},\quad d_1 = -\frac{1}{2\sqrt{3}},\quad d_2 = -\frac{1}{2\sqrt{6}},\quad d_3 = -\frac{\sqrt{\frac{5}{2}}}{2}. \tag{366}$$

The simplification in the case of purely diagonal metrics is dramatic as the solutions then consist only of 1 and –1. Here just the simplest case:

$$a_0 = 1,\quad b_0 = 0,\quad b_1 = 1,\quad c_0 = 0,\quad c_1 = 0,\quad c_2 = 1$$
$$d_0 = 0,\quad d_1 = 0,\quad d_2 = 0,\quad d_3 = 1. \tag{367}$$

Now we go back to the Ricci scalar of our scale-wrapped metric tensor $G_{\alpha\beta} = F[f]^* g_{\alpha\beta}$ and the resulting equation (1) in the case of $\Delta f = 0$. Taking what we have just derived about the V_Ω-vectors, we can rewrite the right-hand side of (1) as follows:

$$h_{\lambda,\alpha} q^\lambda g^{\alpha\beta} h_{\lambda,\beta} q^\lambda \xrightarrow{g^{\alpha\beta} \to V_\Omega \cdot V_\Omega} = V\left[q^\lambda h_{\lambda,\alpha}\right]_\Omega \cdot V\left[h_{\lambda,\beta} q^\lambda\right]_\Omega$$
$$= V\left[q^\lambda h_{\lambda,\alpha}\right]_\Omega \cdot V\left[h_{\lambda,\beta} q^\lambda\right]_\Omega$$
$$\left[q^\lambda \left(a_j \cdot \partial_0 + b_j \cdot \partial_1 + c_j \cdot \partial_2 + d_j \cdot \partial_3 + \cdots\right) h_\lambda\right]_\Omega \cdot$$
$$\left[q^\lambda \left(a_j \cdot \partial_0 + b_j \cdot \partial_1 + c_j \cdot \partial_2 + d_j \cdot \partial_3 + \cdots\right) h_\lambda\right]_\Omega, \tag{368}$$

which in 4-dimensional space-times and the V_Ω-vectors from (365) results in:

$$h_{\lambda,\alpha} q^\lambda g^{\alpha\beta} h_{\lambda,\beta} q^\lambda \xrightarrow{g^{\alpha\beta} \to V_\Omega \cdot V_\Omega}$$

$$= \begin{Bmatrix} q^\lambda \left(a_0 \cdot \partial_0 + b_0 \cdot \partial_1 + c_0 \cdot \partial_2 + d_0 \cdot \partial_3\right) h_\lambda \\ q^\lambda \left(b_1 \cdot \partial_1 + c_1 \cdot \partial_2 + d_1 \cdot \partial_3\right) h_\lambda \\ q^\lambda \left(c_2 \cdot \partial_2 + d_2 \cdot \partial_3\right) h_\lambda \\ q^\lambda \left(d_3 \cdot \partial_3\right) h_\lambda \end{Bmatrix} \cdot \begin{Bmatrix} q^\lambda \left(a_0 \cdot \partial_0 + b_0 \cdot \partial_1 + c_0 \cdot \partial_2 + d_0 \cdot \partial_3\right) h_\lambda \\ q^\lambda \left(b_1 \cdot \partial_1 + c_1 \cdot \partial_2 + d_1 \cdot \partial_3\right) h_\lambda \\ q^\lambda \left(c_2 \cdot \partial_2 + d_2 \cdot \partial_3\right) h_\lambda \\ q^\lambda \left(d_3 \cdot \partial_3\right) h_\lambda \end{Bmatrix}. \tag{369}$$

Subsequently the complete equation (1) would read:

$$\frac{F[f]^3 \cdot R^*}{C_{N1} \cdot \left(\frac{\partial F[f]}{\partial f}\right)^2 - C_{N2} \cdot F[f] \cdot \frac{\partial^2 F[f]}{\partial f^2}} \xrightarrow{4D(see:[1])} \frac{2 \cdot F[f]^3 \cdot R^*}{3 \cdot \left(\left(\frac{\partial F[f]}{\partial f}\right)^2 - 2 \cdot F[f] \cdot \frac{\partial^2 F[f]}{\partial f^2}\right)}$$

$$= h_{\lambda,\alpha} q^\lambda g^{\alpha\beta} h_{\lambda,\beta} q^\lambda \xrightarrow{g^{\alpha\beta} \to V_\Omega \cdot V_\Omega}$$

$$= \begin{Bmatrix} q^\lambda \left(a_0 \cdot \partial_0 + b_0 \cdot \partial_1 + c_0 \cdot \partial_2 + d_0 \cdot \partial_3\right) h_\lambda \\ q^\lambda \left(b_1 \cdot \partial_1 + c_1 \cdot \partial_2 + d_1 \cdot \partial_3\right) h_\lambda \\ q^\lambda \left(c_2 \cdot \partial_2 + d_2 \cdot \partial_3\right) h_\lambda \\ q^\lambda \left(d_3 \cdot \partial_3\right) h_\lambda \end{Bmatrix} \cdot \begin{Bmatrix} q^\lambda \left(a_0 \cdot \partial_0 + b_0 \cdot \partial_1 + c_0 \cdot \partial_2 + d_0 \cdot \partial_3\right) h_\lambda \\ q^\lambda \left(b_1 \cdot \partial_1 + c_1 \cdot \partial_2 + d_1 \cdot \partial_3\right) h_\lambda \\ q^\lambda \left(c_2 \cdot \partial_2 + d_2 \cdot \partial_3\right) h_\lambda \\ q^\lambda \left(d_3 \cdot \partial_3\right) h_\lambda \end{Bmatrix}. \tag{370}$$

Applying the linearization from the postulated Klein–Gordon equation, respectively the subsequent classical Dirac theory as also used in (327) in order to show the transition from the metric Dirac equation to its classical form, we obtain:

$$\frac{2\cdot F[f]^3\cdot R^*}{3\cdot\left(\left(\frac{\partial F[f]}{\partial f}\right)^2-2\cdot F[f]\cdot\frac{\partial^2 F[f]}{\partial f^2}\right)}\simeq h_\lambda q^\lambda\mathbf{M}\cdot\mathbf{M}h_\lambda q^\lambda=h_{\lambda,\alpha}q^\lambda g^{\alpha\beta}h_{\lambda,\beta}q^\lambda$$

$$\xrightarrow{g^{\alpha\beta}\to V_\Omega\cdot V_\Omega}$$

$$=\begin{Bmatrix}q^\lambda(a_0\cdot\partial_0+b_0\cdot\partial_1+c_0\cdot\partial_2+d_0\cdot\partial_3)h_\lambda\\ q^\lambda(b_1\cdot\partial_1+c_1\cdot\partial_2+d_1\cdot\partial_3)h_\lambda\\ q^\lambda(c_2\cdot\partial_2+d_2\cdot\partial_3)h_\lambda\\ q^\lambda(d_3\cdot\partial_3)h_\lambda\end{Bmatrix}\cdot\begin{Bmatrix}q^\lambda(a_0\cdot\partial_0+b_0\cdot\partial_1+c_0\cdot\partial_2+d_0\cdot\partial_3)h_\lambda\\ q^\lambda(b_1\cdot\partial_1+c_1\cdot\partial_2+d_1\cdot\partial_3)h_\lambda\\ q^\lambda(c_2\cdot\partial_2+d_2\cdot\partial_3)h_\lambda\\ q^\lambda(d_3\cdot\partial_3)h_\lambda\end{Bmatrix}$$

$$\Rightarrow\qquad h_\lambda q^\lambda\mathbf{M}=\begin{Bmatrix}q^\lambda(a_0\cdot\partial_0+b_0\cdot\partial_1+c_0\cdot\partial_2+d_0\cdot\partial_3)h_\lambda\\ q^\lambda(b_1\cdot\partial_1+c_1\cdot\partial_2+d_1\cdot\partial_3)h_\lambda\\ q^\lambda(c_2\cdot\partial_2+d_2\cdot\partial_3)h_\lambda\\ q^\lambda(d_3\cdot\partial_3)h_\lambda\end{Bmatrix}$$

$$\Rightarrow\qquad 0=\begin{Bmatrix}q^\lambda(a_0\cdot\partial_0+b_0\cdot\partial_1+c_0\cdot\partial_2+d_0\cdot\partial_3)h_\lambda\\ q^\lambda(b_1\cdot\partial_1+c_1\cdot\partial_2+d_1\cdot\partial_3)h_\lambda\\ q^\lambda(c_2\cdot\partial_2+d_2\cdot\partial_3)h_\lambda\\ q^\lambda(d_3\cdot\partial_3)h_\lambda\end{Bmatrix}\pm h_\lambda q^\lambda\mathbf{M}\qquad(371)$$

and thus:

$$\Rightarrow\qquad 0=\begin{Bmatrix}(a_0\cdot\partial_0+b_0\cdot\partial_1+c_0\cdot\partial_2+d_0\cdot\partial_3)h_\lambda\\ (b_1\cdot\partial_1+c_1\cdot\partial_2+d_1\cdot\partial_3)h_\lambda\\ (c_2\cdot\partial_2+d_2\cdot\partial_3)h_\lambda\\ (d_3\cdot\partial_3)h_\lambda\end{Bmatrix}\pm h_\lambda\mathbf{M}$$

$$\Rightarrow\qquad 0=\left[\begin{Bmatrix}a_0\cdot\partial_0+b_0\cdot\partial_1+c_0\cdot\partial_2+d_0\cdot\partial_3\\ b_1\cdot\partial_1+c_1\cdot\partial_2+d_1\cdot\partial_3\\ c_2\cdot\partial_2+d_2\cdot\partial_3\\ d_3\cdot\partial_3\end{Bmatrix}\pm\begin{Bmatrix}M_0\\ M_1\\ M_2\\ M_3\end{Bmatrix}\right]h_\lambda.\qquad(372)$$

The latter could be seen as a curvature-linearized metric Dirac equation in vector form.

There are quite some differences to the classical Dirac case:

(A) The functions h_λ need all to be Laplace functions, while in the postulated Dirac equation, the resulting solutions and, thus, the outcoming functions f also fulfill the Klein–Gordon equation (for more see the sub-sections above), which, as we know, gives a 4-dimensional Laplace equation in the case of vanishing masses. Thus, we always have to guarantee the condition $\Delta f = \Delta h_\lambda q^\lambda = 0$ in addition to the metric Dirac equation (372) (or its non-linearized equivalent (370)). An exception from this rule can be made in the case where $F[f]$ is chosen such that we have $\frac{\partial F}{\partial f} \simeq 0$ around the f being considered.

(B) The masses term **M** is a vector and not just a scalar as in the classical Dirac case.

(C) We obtain extremely simple equations in the case of purely diagonal metrics, namely:

$$\Rightarrow \qquad 0 = \left[\begin{Bmatrix} a_0 \cdot \partial_0 \\ b_1 \cdot \partial_1 \\ c_2 \cdot \partial_2 \\ d_3 \cdot \partial_3 \end{Bmatrix} \pm \begin{Bmatrix} M_0 \\ M_1 \\ M_2 \\ M_3 \end{Bmatrix}\right] h_\lambda. \tag{373}$$

(D) With the time t-dependency of a great variety of metrics and subsequent Laplace solutions for these very metrics usually being found of the type $\exp[\pm c{*}C_t{*}x_0] = \exp[\pm c{*}C_t{*}t]$ (c.f. the many examples given above), we have no problem in recognizing the rest mass term $M_t^2 = M_0^2 = \frac{m^2 \cdot c^2}{\hbar^2}$ in the case of the "particle at rest," where we have:

$$\Rightarrow \qquad 0 = \left[\begin{Bmatrix} a_0 \cdot \partial_0 \\ b_1 \cdot \partial_1 \\ c_2 \cdot \partial_2 \\ d_3 \cdot \partial_3 \end{Bmatrix} \pm \begin{Bmatrix} M_0 \\ M_1 = 0 \\ M_2 = 0 \\ M_3 = 0 \end{Bmatrix}\right] h_\lambda. \tag{374}$$

Thus, we are also back with the Dirac matter and antimatter solutions.

(E) Still there are four functions $h_{\lambda\, =\, 0,1,2,3}$ with the classical matter $\exp[-c{*}M_0{*}t]$ and antimatter $\exp[c{*}M_0{*}t]$ solution, plus two more functions, which should code the two classical spin states. Thereby the origin of the latter could either be a residual from the linearization of the curvature term in (371) (first line) or, as demonstrated on many examples in the upper sections, the dimensional entanglement with other degrees of freedom and thus, dimensions of higher order. We also saw in connection with elastic approaches that the spin comes in quite naturally in the case of shear-approaches ([5, 9] and appendix *F* below in this book).

(F) **It should be pointed out, however, that in order to obtain non-vanishing solutions in Cartesian (Minkowski) coordinates—even for the particle at rest $f[...]= f[t]$—we should have either the curvature of our space-time, which is to say $R^{*}\neq 0$, or dimensional entanglement with higher-order dimensions (see sections above). Otherwise, all M_i would vanish and with it (as one can easily deduce from (374)) all h_λ. Of course, one might just demand the derivatives of the scalar product $h^\lambda q^\lambda$ or the scalar itself to become zero, but here we are not interested in such trivial zero-sum solutions.**

This, meaning the fact that the metric Dirac equation requires curvature or entanglement in order to "have something to work with," is in severe contrast to the classical, postulated Dirac theory, which is completely developed on the basis of the flat Minkowski space-time and where no curvature is needed in order to "obtain something." This, however, is a result of the postulation, while, obviously, the metrically derived Dirac equation does not allow for "something to be" without this something also to produce curvature or dimensional entanglement, along the way creating inertia, spin, energy and so on (c.f. the chapters above, especially the appendix of chapter 6, "The Quantum Gravity Turing Machine").

Of course, things are getting much more interesting when moving to certain symmetries such as the spatial spherical one, but for here and now we shall leave this to the interested reader or own later publications (whatever comes first).

So, apparently, there is a way to avoid Dirac's quaternions. It should be pointed out, however, that the introduction of the V_Ω-vectors alone does not do the trick. Looking carefully at equation (371), we realize the combination of the V_Ω-vectors with the vector of constants q^λ residing in there. We could easily assume the two forming one mathematical object like:

$$\begin{Bmatrix} q^\lambda\left(a_0\cdot\partial_0+b_0\cdot\partial_1+c_0\cdot\partial_2+d_0\cdot\partial_3\right)h_\lambda \\ q^\lambda\left(b_1\cdot\partial_1+c_1\cdot\partial_2+d_1\cdot\partial_3\right)h_\lambda \\ q^\lambda\left(c_2\cdot\partial_2+d_2\cdot\partial_3\right)h_\lambda \\ q^\lambda\left(d_3\cdot\partial_3\right)h_\lambda \end{Bmatrix}$$

$$=\begin{Bmatrix} q^\lambda\left(a_0\cdot\partial_0+b_0\cdot\partial_1+c_0\cdot\partial_2+d_0\cdot\partial_3\right) \\ q^\lambda\left(b_1\cdot\partial_1+c_1\cdot\partial_2+d_1\cdot\partial_3\right) \\ q^\lambda\left(c_2\cdot\partial_2+d_2\cdot\partial_3\right) \\ q^\lambda\left(d_3\cdot\partial_3\right) \end{Bmatrix}h_\lambda$$

$$=q^\lambda\begin{Bmatrix} \left(a_0\cdot\partial_0+b_0\cdot\partial_1+c_0\cdot\partial_2+d_0\cdot\partial_3\right) \\ \left(b_1\cdot\partial_1+c_1\cdot\partial_2+d_1\cdot\partial_3\right) \\ \left(c_2\cdot\partial_2+d_2\cdot\partial_3\right) \\ \left(d_3\cdot\partial_3\right) \end{Bmatrix}h_\lambda$$

$$=q^\lambda\begin{Bmatrix} a_0\cdot\partial_0 \\ 0 \\ 0 \\ 0 \end{Bmatrix}h_\lambda+q^\lambda\begin{Bmatrix} b_0\cdot\partial_1 \\ b_1\cdot\partial_1 \\ 0 \\ 0 \end{Bmatrix}h_\lambda+q^\lambda\begin{Bmatrix} c_0\cdot\partial_2 \\ b_1\cdot\partial_2 \\ c_2\cdot\partial_2 \\ 0 \end{Bmatrix}h_\lambda+q^\lambda\begin{Bmatrix} d_0\cdot\partial_3 \\ d_1\cdot\partial_3 \\ d_2\cdot\partial_3 \\ d_3\cdot\partial_3 \end{Bmatrix}h_\lambda$$

$$\equiv\mathbf{Q}^{\lambda\alpha}h_{\lambda,\alpha}, \tag{375}$$

where we are indeed back with some quaternion-like objects. This time, however, we also have a way to decompose these objects and separate the vector of constants q^λ from it. Some special care needs to be taken about curvature and mass terms resulting from entanglement (c.f. [3] and corresponding sections in [5]). The interested reader will find more in papers [15, 16, 17], but as there is still work to do, we shall leave this for later or the interested and mathematically skilled reader.

Treating the Classical Dirac Approach with the New Method

In principle, we should also be able to apply the V_Ω-vector method onto the classical Klein–Gordon equation, thereby following the Dirac path only that we—this time—avoid the quaternions. First, we bring the Minkowski Laplace operator in the V_Ω-vector form:

$$\Delta f=\frac{\left(\sqrt{g}\cdot g^{\alpha\beta}f_{,\beta}\right)_{,\alpha}}{\sqrt{g}}\xrightarrow{g^{\alpha\beta}\to\text{Minkowski}}=\left(g^{\alpha\beta}f_{,\beta}\right)_{,\alpha}$$

$$= \begin{Bmatrix} a_0 \cdot \partial_0 + b_0 \cdot \partial_1 + c_0 \cdot \partial_2 + d_0 \cdot \partial_3 \\ b_1 \cdot \partial_1 + c_1 \cdot \partial_2 + d_1 \cdot \partial_3 \\ c_2 \cdot \partial_2 + d_2 \cdot \partial_3 \\ d_3 \cdot \partial_3 \end{Bmatrix} \cdot \begin{Bmatrix} a_0 \cdot \partial_0 + b_0 \cdot \partial_1 + c_0 \cdot \partial_2 + d_0 \cdot \partial_3 \\ b_1 \cdot \partial_1 + c_1 \cdot \partial_2 + d_1 \cdot \partial_3 \\ c_2 \cdot \partial_2 + d_2 \cdot \partial_3 \\ d_3 \cdot \partial_3 \end{Bmatrix} f. \tag{376}$$

Now we add the mass-vector to the decomposed Klein–Gordon equation:

$$\Rightarrow \qquad 0 = \begin{Bmatrix} a_0 \cdot \partial_0 + b_0 \cdot \partial_1 + c_0 \cdot \partial_2 + d_0 \cdot \partial_3 \\ b_1 \cdot \partial_1 + c_1 \cdot \partial_2 + d_1 \cdot \partial_3 \\ c_2 \cdot \partial_2 + d_2 \cdot \partial_3 \\ d_3 \cdot \partial_3 \end{Bmatrix} f \pm f \cdot \mathbf{M} \tag{377}$$

and, remembering that we can give the scalar function f any arbitrary intrinsic structure, we obtain the V_Ω-vector equivalent of the Dirac equation:

$$0 = \left[\begin{Bmatrix} a_0 \cdot \partial_0 + b_0 \cdot \partial_1 + c_0 \cdot \partial_2 + d_0 \cdot \partial_3 \\ b_1 \cdot \partial_1 + c_1 \cdot \partial_2 + d_1 \cdot \partial_3 \\ c_2 \cdot \partial_2 + d_2 \cdot \partial_3 \\ d_3 \cdot \partial_3 \end{Bmatrix} \pm \begin{Bmatrix} M \\ M \\ M \\ M \end{Bmatrix} \right] q^\lambda h_\lambda$$

$$\Rightarrow \qquad 0 = \left[q^\lambda \begin{Bmatrix} a_0 \cdot \partial_0 + b_0 \cdot \partial_1 + c_0 \cdot \partial_2 + d_0 \cdot \partial_3 \\ b_1 \cdot \partial_1 + c_1 \cdot \partial_2 + d_1 \cdot \partial_3 \\ c_2 \cdot \partial_2 + d_2 \cdot \partial_3 \\ d_3 \cdot \partial_3 \end{Bmatrix} \pm q^\lambda \begin{Bmatrix} M \\ M \\ M \\ M \end{Bmatrix} \right] h_\lambda. \tag{378}$$

Direct Derivation of the Metric Dirac Equation from the Scaled Ricci Scalar R^*

For convenience, we here repeat the derivation of the classical Dirac equation [181] directly from the Einstein–Hilbert action as it was performed in [16]. The Einstein–Hilbert action reads (4):

$$\delta W = 0 = \delta \int_V d^n x \left(\sqrt{-G} \cdot \left(R^* - 2\Lambda + L_M \right) \right). \tag{379}$$

Thereby we assume a simple metric of type (2) and apply the corresponding scaled Ricci scalar R^*:

$$R^* = R^*_{\alpha\beta} G^{\alpha\beta}$$

$$=\begin{pmatrix}\Gamma^{\sigma}_{\alpha\beta,\sigma}g^{\alpha\beta}-\Gamma^{\sigma}_{\beta\sigma,\alpha}g^{\alpha\beta}-\Gamma^{\mu}_{\sigma\alpha}\Gamma^{\sigma}_{\beta\mu}g^{\alpha\beta}+\Gamma^{\sigma}_{\alpha\beta}\Gamma^{\mu}_{\sigma\mu}g^{\alpha\beta}\\+\frac{F'}{F}(1-n)\Delta f+\frac{f_{,\alpha}f_{,\beta}g^{\alpha\beta}(1-n)}{4F^2}\left(4FF''+(F')^2(n-6)\right)\end{pmatrix}\frac{1}{F}$$

$$=\left(R+\frac{F'}{F}(1-n)\Delta f+\frac{f_{,\alpha}f_{,\beta}g^{\alpha\beta}(1-n)}{4F^2}\left(4FF''+(F')^2(n-6)\right)\right)\frac{1}{F}$$

$$\text{with: } F=F[f];\ F'=\frac{\partial F[f]}{\partial f};\ F''=\frac{\partial^2 F[f]}{\partial f^2}, \tag{380}$$

being based upon the scaled metric $G_{\delta\gamma} = F[f] \cdot g_{\delta\gamma}$. Assuming the Lagrange matter density L_M and the cosmological constant Λ to vanish, the variation results in:

$$0=\left(\int_{V_G}d^nx\sum_{i=1}^{N}\sqrt{-g\cdot F_i^n}\left(\begin{pmatrix}R_{\alpha\beta}-\frac{1}{2}\cdot R\cdot g_{\alpha\beta}-\frac{F_i'}{F_i}(n-1)\left(f_{i,\alpha\beta}-\Gamma^{\gamma}_{\alpha\beta}f_{i,\gamma}-\frac{g_{\alpha\beta}}{2}\Delta f_i\right)\\+\frac{(1-n)}{4F_i^2}\begin{pmatrix}4F_iF_i''\\+F_i'F_i'(n-6)\end{pmatrix}\left(f_{i,\alpha}f_{i,\beta}-f_{i,\mu}f_{i,\nu}g^{\mu\nu}\frac{g_{\alpha\beta}}{2}\right)\end{pmatrix}\delta G^{\alpha\beta}\right)\right). \tag{381}$$

Thereby we have also assumed to have many scaling factors and volumes to perform the integration over (for more discussion about this, see [17, 18]). By splitting up the variation of the scaled metric:

$$0=\left(\int_{V_G}d^nx\sum_{i=1}^{N}\sqrt{-g\cdot F_i^n}\left(\begin{pmatrix}R_{\alpha\beta}-\frac{1}{2}\cdot R\cdot g_{\alpha\beta}\\-\frac{F_i'}{F_i}(n-1)\left(f_{i,\alpha\beta}-\Gamma^{\gamma}_{\alpha\beta}f_{i,\gamma}-\frac{g_{\alpha\beta}}{2}\Delta f_i\right)\\+\frac{(1-n)}{4F_i^2}\begin{pmatrix}4F_iF_i''\\+F_i'F_i'(n-6)\end{pmatrix}\begin{pmatrix}f_{i,\alpha}f_{i,\beta}\\-f_{i,\mu}f_{i,\nu}g^{\mu\nu}\frac{g_{\alpha\beta}}{2}\end{pmatrix}\end{pmatrix}\begin{pmatrix}g^{\alpha\beta}\delta\frac{1}{F_i}\\+\frac{\delta g^{\alpha\beta}}{F_i}\end{pmatrix}\right)\right), \tag{382}$$

we obtain two equations:

$$0=\left(\int_{V_G} d^n x \sum_{i=1}^{N}\sqrt{-g\cdot F_i^n}\left(\left(1-\frac{n}{2}\right)\left(\begin{array}{c} R-\frac{F_i{}'}{F_i}(n-1)\Delta f_i \\ +\frac{f_{i,\alpha}f_{i,\beta}(1-n)}{4F_i^2}g^{\alpha\beta}\left(\begin{array}{c}4F_iF_i{}'' \\ +F_i{}'F_i{}'(n-6)\end{array}\right)\end{array}\right)\delta\frac{1}{F_i}\right)\right),$$

(383)

$$0=\left(\int_{V_G} d^n x \sum_{i=1}^{N}\sqrt{-g\cdot F_i^n}\left(\left(\begin{array}{c} R_{\alpha\beta}-\frac{1}{2}\cdot R\cdot g_{\alpha\beta} \\ -\frac{F_i{}'}{F_i}(n-1)\left(f_{i,\alpha\beta}-\Gamma^{\gamma}_{\alpha\beta}f_{i,\gamma}-\frac{g_{\alpha\beta}}{2}\Delta f_i\right) \\ +\frac{(1-n)}{4F_i^2}\left(\begin{array}{c}4F_iF_i{}'' \\ +F_i{}'F_i{}'(n-6)\end{array}\right)\left(\begin{array}{c} f_{i,\alpha}f_{i,\beta} \\ -f_{i,\mu}f_{i,\nu}g^{\mu\nu}\frac{g_{\alpha\beta}}{2}\end{array}\right)\end{array}\right)\frac{\delta g^{\alpha\beta}}{F_i}\right)\right).$$

(384)

While we consider the second equation a generalized or quantum Einstein field equation, we use the first one to derive the classical Dirac equation [181]. Setting the integrand equal to zero, we obtain the following differential equation:

$$0=R-\frac{F_i{}'}{F_i}(n-1)\Delta f_i+\frac{f_{i,\alpha}f_{i,\beta}(1-n)}{4F_i^2}g^{\alpha\beta}\left(4F_iF_i{}''+F_i{}'F_i{}'(n-6)\right). \quad (385)$$

Now we add in the Dirac starting point, where he—Dirac [181]—assumed that:

(a) The space-time can be assumed Ricci flat and thus, $R = 0$.

(b) The Laplace operator term in (385) has an eigenvalue M^2.

Also setting $F[f] = f$ gives us:

$$0=f_{i,\alpha}f_{i,\beta}g^{\alpha\beta}+M^2\cdot\overbrace{\frac{4}{(n-6)}}^{\equiv\mu^2=\mathbf{\mu}\cdot\mathbf{\mu}}\cdot f_i^2. \quad (386)$$

Now we apply the known relation between the Dirac matrices

$$\gamma^0 = \begin{pmatrix} 1 & & & \\ & 1 & & \\ & & -1 & \\ & & & -1 \end{pmatrix}; \quad \gamma^1 = \begin{pmatrix} & & & 1 \\ & & 1 & \\ & -1 & & \\ -1 & & & \end{pmatrix}$$

$$\gamma^2 = \begin{pmatrix} & & & -i \\ & & i & \\ & i & & \\ -i & & & \end{pmatrix}; \quad \gamma^3 = \begin{pmatrix} & & 1 & \\ & & & -1 \\ -1 & & & \\ & 1 & & \end{pmatrix}; \quad I = \begin{pmatrix} 1 & & & \\ & 1 & & \\ & & 1 & \\ & & & 1 \end{pmatrix}, \quad (387)$$

and the metric tensor, reading:

$$g^{\alpha\beta} \cdot I = \frac{\gamma^\alpha \gamma^\beta + \gamma^\beta \gamma^\alpha}{2}, \quad (388)$$

and put (386) into the following form:

$$0 = \left(f_{i,\alpha} f_{i,\beta} g^{\alpha\beta} + \mu^2 \cdot f_i^2 \right) \cdot I$$

$$\Rightarrow \quad f_{i,\alpha} f_{i,\beta} \left(\frac{\gamma^\alpha \gamma^\beta + \gamma^\beta \gamma^\alpha}{2} \right) + \frac{\beta^2 + \beta^2}{2} \cdot m^2 \cdot f_i^2 = 0$$

$$\Rightarrow \quad f_{i,\alpha} f_{i,\beta} \gamma^\alpha \gamma^\beta + \beta^2 \cdot m^2 \cdot f_i^2 = 0$$

$$\Rightarrow \quad \left(f_{i,\alpha} \gamma^\alpha + i \cdot \beta \cdot m \cdot f_i \right) \left(f_{i,\beta} \gamma^\beta - i \cdot \beta \cdot m \cdot f_i \right) = 0$$

$$\Rightarrow \quad f_{i,\alpha} f_{i,\beta} \gamma^\beta \gamma^\alpha + \beta^2 \cdot m^2 \cdot f_i^2 = 0$$

$$\Rightarrow \quad \left(f_{i,\alpha} \gamma^\alpha + i \cdot \beta \cdot m \cdot f_i \right) \left(f_{i,\beta} \gamma^\beta - i \cdot \beta \cdot m \cdot f_i \right) = 0. \quad (389)$$

Thereby we have also used the β-matrix with $\beta = \begin{pmatrix} 1 & & & \\ & 1 & & \\ & & -1 & \\ & & & -1 \end{pmatrix}$.

This gives us the Dirac equation in its classical form:

$$0 = f_{i,\beta} \gamma^\beta - i \cdot \beta \cdot m \cdot f_i. \quad (390)$$

The advantage of the Dirac approach clearly is that here the structure of the equation with the quaternions already assures the satisfaction of the eigenvalue equation for the Laplace operator term, which is to say the Klein–Gordon condition:

$$\Delta f_i = -M^2 \cdot f_i. \tag{391}$$

Seeing the simplicity of the derivation of the Dirac equation with the Einstein–Hilbert action as the starting point, as just performed here and comparing this with Dirac's path, one can only wonder "how on earth this genius came up with this idea of introducing the quaternions."

It is—we have no other word for it—a miracle [16].

Appendix D

Metric Originated Derivation of the Electro-Magnetic Interaction (and Matter) via a Set of Creative Transformations

In order to allow for an easy entrance into the theoretical apparatus, we shall start with the Minkowski flat space metric tensor in Cartesian coordinates t, x, y, z as follows:

$$g_{\alpha\beta}^{\text{flat}} = \begin{pmatrix} -c^2 & 0 & 0 & 0 \\ 0 & 1 & 0 & 0 \\ 0 & 0 & 1 & 0 \\ 0 & 0 & 0 & 1 \end{pmatrix}. \tag{392}$$

This metric solves the Einstein field equations (see [1]) and corresponds to a Ricci scalar of $R = 0$. Now we introduce the following special "transformation" or rather "scaling":

$$G_{\alpha\beta} = g_{\alpha\beta}^{\text{flat}} \cdot F\left[f\left[t,x,y,z\right]\right] = \begin{pmatrix} -c^2 & 0 & 0 & 0 \\ 0 & 1 & 0 & 0 \\ 0 & 0 & 1 & 0 \\ 0 & 0 & 0 & 1 \end{pmatrix} \cdot F\left[f\left[t,x,y,z\right]\right] \tag{393}$$

and evaluate the resulting new Ricci scalar R^*:

$$R^* = \frac{R}{F[f]} + \begin{pmatrix} -\Gamma^{\mu}_{\sigma\alpha}\Gamma^{**\sigma}_{\beta\mu} + \Gamma^{\sigma}_{\alpha\beta}\Gamma^{**\mu}_{\sigma\mu} - \Gamma^{**\mu}_{\sigma\alpha}\Gamma^{\sigma}_{\beta\mu} + \Gamma^{**\sigma}_{\alpha\beta}\Gamma^{\mu}_{\sigma\mu} \\ +\Gamma^{**\sigma}_{\alpha\beta\,,\sigma} - \Gamma^{**\sigma}_{\beta\sigma\,,\alpha} - \Gamma^{**\mu}_{\sigma\alpha}\Gamma^{**\sigma}_{\beta\mu} + \Gamma^{**\sigma}_{\alpha\beta}\Gamma^{**\mu}_{\sigma\mu} \end{pmatrix} \frac{g^{\alpha\beta}}{F[f]}$$

$$\xrightarrow{R=0;\ g_{\alpha\beta}^{\text{flat}}}$$

$$=\frac{1}{F[f]^3}\cdot\begin{pmatrix}\left(C_{N1}\cdot\left(\frac{\partial F[f]}{\partial f}\right)^2-C_{N2}\cdot F[f]\cdot\frac{\partial^2 F[f]}{\partial f^2}\right)\cdot\left(\tilde{\nabla}_g f\right)^2\\ -C_{N2}\cdot F[f]\cdot\frac{\partial F[f]}{\partial f}\cdot\Delta_g f\end{pmatrix}$$

$$=\frac{1}{F[f]^3}\cdot\left(\begin{pmatrix}\frac{3}{2}\cdot\left(\frac{\partial F[f]}{\partial f}\right)^2\\ -3\cdot F[f]\cdot\frac{\partial^2 F[f]}{\partial f^2}\end{pmatrix}\cdot\left(\tilde{\nabla}_g f\right)^2-3\cdot F[f]\cdot\frac{\partial F[f]}{\partial f}\cdot\Delta_g f\right)$$

$$=\frac{3}{2\cdot F[f]^3}\cdot\begin{pmatrix}\begin{pmatrix}\left(\frac{\partial F[f]}{\partial f}\right)^2\\ -2\cdot F[f]\cdot\frac{\partial^2 F[f]}{\partial f^2}\end{pmatrix}\cdot\begin{bmatrix}-\frac{(\partial_t f)^2}{c^2}+(\partial_x f)^2\\ +(\partial_y f)^2+(\partial_z f)^2\end{bmatrix}\\ -2\cdot F[f]\cdot\frac{\partial F[f]}{\partial f}\cdot\left[-\frac{\partial_t^2}{c^2}+\partial_x^2+\partial_y^2+\partial_z^2\right]f\end{pmatrix}. \quad (394)$$

Thereby we have used:

$$\Gamma^{*\gamma}_{\alpha\beta}=\frac{g^{\gamma\sigma}}{2\cdot F[f]}\left(\left[F[f]\cdot g_{\sigma\alpha}\right]_{,\beta}+\left[F[f]\cdot g_{\sigma\beta}\right]_{,\alpha}-\left[F[f]\cdot g_{\alpha\beta}\right]_{,\sigma}\right)$$

$$=\frac{g^{\gamma\sigma}}{2\cdot F[f]}\left(\left[F[f]\cdot g_{\sigma\alpha}\right]_{,\beta}+\left[F[f]\cdot g_{\sigma\beta}\right]_{,\alpha}-\left[F[f]\cdot g_{\alpha\beta}\right]_{,\sigma}\right)$$

$$=\frac{g^{\gamma\sigma}}{2}\left(g_{\sigma\alpha,\beta}+g_{\sigma\beta,\alpha}-g_{\alpha\beta,\sigma}\right)+\frac{g^{\gamma\sigma}}{2\cdot F[f]}\begin{pmatrix}F[f]_{,\beta}\cdot g_{\sigma\alpha}+F[f]_{,\alpha}\cdot g_{\sigma\beta}\\ -F[f]_{,\sigma}\cdot g_{\alpha\beta}\end{pmatrix}$$

$$=\Gamma^{\gamma}_{\alpha\beta}+\frac{g^{\gamma\sigma}}{2\cdot F[f]}\begin{pmatrix}F[f]_{,\beta}\cdot g_{\sigma\alpha}+F[f]_{,\alpha}\cdot g_{\sigma\beta}\\ -F[f]_{,\sigma}\cdot g_{\alpha\beta}\end{pmatrix}$$

$$\equiv\Gamma^{\gamma}_{\alpha\beta}+\Gamma^{**\gamma}_{\alpha\beta}. \quad (395)$$

The reader will find a more comprehensive elaboration about the motivation of our special transformation in section 3.3, "The Ricci Scalar Quantization," of [3] and the previous chapters of this book. However, it is quite entertaining to follow this simple trial and observe its amazing evolution into something rather unexpected (at least if taking its origin as a metric solution to the Einstein field equations) out of our "wrapper-transformation" (393). As here the result is the essential aspect and as the various derivations are to be found in the book [3], we will not go too much into detail as this only spoils the reading fun and the punch line.

The symbol $\tilde{\nabla}_g$ in (394) denotes a first-order differential operator similar to the Nabla-operator in the metric $g_{\alpha\beta}$. The symbols C_{Ni} stand for constants, which only depend on the number of dimensions. Please note that the transformation (393) is just the simplest form of a general approach like:

$$G_{\alpha\beta} = F\left[f\left[t,x,y,z\right]\right]_{\alpha\beta}^{ij} g_{ij} \rightarrow G_{\alpha\beta} = F\left[f\left[t,x,y,z\right]\right] \cdot \delta_{\alpha}^{i}\delta_{\beta}^{j} g_{ij}. \qquad (396)$$

Without loss of generality, we can now demand that the first term in parenthesis in (394) would be zero, which is to say, we choose our arbitrary function F such that we have:

$$C_{N1} \cdot \left(\frac{\partial F\left[f\right]}{\partial f}\right)^2 - C_{N2} \cdot F\left[f\right] \cdot \frac{\partial^2 F\left[f\right]}{\partial f^2} = 0. \qquad (397)$$

For instance, in 4 dimensions this would always be the case for $F[f] = f^2$, giving us the Klein–Gordon-like equation:

$$\left[\Delta_g + \frac{R^* \cdot f^4}{f \cdot C_{N2} \cdot 2 \cdot f}\right] f - \left[\Delta_g + \frac{R^* \cdot f^2}{2}\right] f = 0; \quad \Delta_{g\text{-Coordinates}} = \Delta_g. \qquad (398)$$

Setting the solution $F[f] = f^2$ into our transformation starting point (393), however, one would directly obtain an all-scale quantum-dominated metric solution for $G_{\alpha\beta}$. This is in total contrast to the daily experiences that quantum effects are only important in smaller scales. In order to overcome this problem, we evaluate the general solution of condition (397) in 4 dimensions and find:

$$F\left[f\right] = C_1 \cdot f + \frac{C_1^2 \cdot f^2}{4 \cdot C_2} + C_2. \qquad (399)$$

Thereby the constants C_i are arbitrary.

Repetition: The Higgs Field as a Mere By-Product

For entertainment and further motivation, it should be pointed out here that the functional wrapper $F[f]$ from (399) could also be adapted as follows:

$$F[f]=C_1\cdot f^2+\frac{C_1^2\cdot f^4}{4\cdot C_2}+C_2. \tag{400}$$

With arbitrary constants C_i, this assures the appearance of linear Laplace operator terms according to the condition (397) for the resulting Ricci scalar of the transformed metric in (396) as follows:

$$R^*=\frac{1}{F[f]^3}\cdot\left(-C_{N2}\cdot F[f]\cdot\frac{\partial F[f]}{\partial f}\cdot\Delta_g f^2\right)$$
$$\xrightarrow{4D}$$
$$R^*\cdot\frac{\left(2\cdot C_2+C_1\cdot f^2\right)^3}{8\cdot C_2\cdot C_1}=-3\cdot\Delta_g f^2. \tag{401}$$

Now we assume f to be a constant, which automatically leads to the simple equation:

$$R^*\cdot\frac{\left(2\cdot C_2+C_1\cdot f^2\right)^3}{8\cdot C_2\cdot C_1}=0. \tag{402}$$

Also assuming that the Ricci scalar curvature R^* shall be proportional to f^n, with an arbitrary exponent $n \geq 0$, we obtain the familiar trivial solution of $f_0 = 0$. However, from (402) we also obtain the non-trivial ground states:

$$f^2=-\frac{2\cdot C_2}{C_1}. \tag{403}$$

As the constants C_i are arbitrary, we could simply set them as follows:

$$2\cdot C_2=-\mu^2;\quad C_1=2\cdot\lambda. \tag{404}$$

This gives us the additional ground state solutions directly in the classical Higgs field style [200], namely:

$$\left(f_{1,2}\right)^2=\frac{\mu^2}{2\cdot\lambda}\quad\Rightarrow\quad f_{1,2}=\pm\frac{\mu}{\sqrt{2\cdot\lambda}}. \tag{405}$$

The measured value for the $f_{1,2}$ is known to be [185]:

$$\left|f_{1,2}\right|=\frac{\mu}{\sqrt{2\cdot\lambda}}=\frac{246\,\text{GeV}}{\sqrt{2}\cdot c^2}. \tag{406}$$

Rewriting (401) with the use of (404) as:

$$R^*\cdot\frac{\left(-\mu^2+2\cdot\lambda\cdot f^2\right)^3}{8\cdot\mu^2\cdot 2\cdot\lambda}=f^n\cdot\frac{\left(2\cdot\lambda\cdot f^2-\mu^2\right)^3}{16\cdot\mu^2\cdot\lambda}=3\cdot\Delta_g f^2 \tag{407}$$

gives us the total Higgs–Ricci-curvature connection. It also, automatically, gives us a curvature value R^* at the ground state, which is:

$$R^* \sim \left|f_{1,2}\right|^n = \left(\frac{\mu}{\sqrt{2\cdot\lambda}}\right)^n = \left(\frac{246\,\text{GeV}}{\sqrt{2}\cdot c^2}\right)^n. \tag{408}$$

Now, with the Higgs mass m_H known to be 125 GeV/c^2 and the fact that we have [185]:

$$m_H = \sqrt{2\cdot\mu^2} = \sqrt{4\cdot\lambda\cdot\left(f_{1,2}\right)^2}, \tag{409}$$

we can obtain:

$$\lambda \simeq 0.13;\quad \mu \simeq 88.8\ \text{GeV/c}^2. \tag{410}$$

Back to the Main Part of this Appendix

Please note that we obtain the simple case $F[f] = f^2$, which, e.g., was first applied in [120], with the condition:

$$C_1^2 = 2\cdot C_2 \tag{411}$$

and the subsequent limit procedure $C_2 \to 0$ as follows:

$$\Rightarrow F[f] = \sqrt{2\cdot C_2}\ f + f^2 + C_2 \xrightarrow[C_2\to 0]{\lim} F[f] = f^2. \tag{412}$$

Using (399) instead of $F = f^2$ in (393) gives us:

$$R^* = \frac{1}{F[f]^3}\cdot\left(\begin{array}{c}\left(C_{N1}\cdot\left(\frac{\partial F[f]}{\partial f}\right)^2 - C_{N2}\cdot F[f]\cdot\frac{\partial^2 F[f]}{\partial f^2}\right)\cdot\left(\tilde{\nabla}_g f\right)^2 \\ -C_{N2}\cdot F[f]\cdot\frac{\partial F[f]}{\partial f}\cdot\Delta_g f\end{array}\right)$$

$$= -C_{N2}\cdot\frac{\frac{\partial F[f]}{\partial f}\cdot\Delta_g f}{F[f]^2} \xrightarrow{4D} = -3\cdot\frac{\left(C_1 + \frac{C_1^2\cdot f}{2\cdot C_2}\right)}{\left(C_1\cdot f + \frac{C_1^2\cdot f^2}{4\cdot C_2} + C_2\right)^2}\cdot\Delta_g f. \tag{413}$$

Immediately we see that in the many cases with $R^* = 0$, we would obtain classical quantum Klein–Gordon equations. The Taylor expansion for small f (which is to say at $f = 0$) for the general case in the second line in (413) also gives us equations leading to these classical equations. For instance, if only considering up to linear f as follows:

$$\frac{R^*}{3}\cdot\frac{\left(C_1\cdot f+\frac{C_1^2\cdot f^2}{4\cdot C_2}+C_2\right)^2}{C_1+\frac{C_1^2\cdot f}{2\cdot C_2}}+\Delta_g f=0$$

$$\xrightarrow{\text{small } f}$$

$$\frac{R^*}{3}\cdot\left(\frac{C_2^2}{C_1}+\frac{3}{2}\cdot C_2\cdot f+\ldots\right)+\Delta_g f=0 \tag{414}$$

and assuming constant R^*, we already have the typical structure of the Klein–Gordon equation plus a constant term $\frac{R^*}{3}\cdot\frac{C_2^2}{C_1}$. Such a constant would still not influence the principal structural character of our resulting quantum equations because a simple transformation of:

$$f^*=\frac{C_2^2}{C_1}+\frac{3}{2}\cdot C_2\cdot f;\quad \Delta_g f^*=\Delta_g f \tag{415}$$

does give us back the ordinary Klein–Gordon equation with:

$$\frac{R^*}{3}\cdot f^*+\Delta_g f^*=0. \tag{416}$$

The resulting transformation in 4 dimensions would give the following quantum metric:

$$G_{\alpha\beta}=g_{\alpha\beta}\cdot F[f]=g_{\alpha\beta}\cdot\left(C_1\cdot f+\frac{C_1^2\cdot f^2}{4\cdot C_2}+C_2\right). \tag{417}$$

This clearly allows for the classical physics or Einstein situation with quantum effects becoming dominant only at small scales and thus, a dominating constant part C_2 at bigger scales. As a result, one might separate as follows in 4 dimensions:

$$G_{\alpha\beta}=\overbrace{\overbrace{g_{\alpha\beta}\cdot C_2}^{\text{class. GTR}}+\overbrace{g_{\alpha\beta}\cdot\left(C_1\cdot f\cdot\left(1+\frac{C_1\cdot f}{4\cdot C_2}\right)\right)}^{\text{quantum metric}}}^{\text{Quantum Gravity in } 4D}. \tag{418}$$

Thereby it does not come as a surprise that we have already found so many cases where f has to be determined by classical quantum equations (see [3–18] and references in [3]), respectively where our equations give similar or even equal solutions. Normalization of the quantum gravity metric $G_{\alpha\beta}$ requires division by C_2 and gives:

$$G_{\alpha\beta}^{\text{norm}}=\frac{G_{\alpha\beta}}{C_2}=\overbrace{\underbrace{g_{\alpha\beta}}_{\text{class. GTR}}+\overbrace{g_{\alpha\beta}\cdot\left(\frac{C_1}{C_2}\cdot f\cdot\left(1+\frac{C_1\cdot f}{4\cdot C_2}\right)\right)}^{\text{quantum metric}}}^{\text{Quantum Gravity in }4D}. \tag{419}$$

Now we apply the recipe outlined above to our simple flat space example in order to see how the transformation can bring about matter.

We saw that by taking the metric (392), using rule (397) with the subsequent solution (399), we obtain the new Ricci scalar curvature:

$$R^{*}=-3\cdot\frac{\left(C_1+\frac{C_1^2\cdot f}{2\cdot C_2}\right)}{\left(C_1\cdot f+\frac{C_1^2\cdot f^2}{4\cdot C_2}+C_2\right)^2}\cdot\Delta_g f. \tag{420}$$

Now we demand that the Ricci scalar should be a conserved quantity and as we have $R = 0$ with the flat space metric (392), we shall also demand $R^* = 0$. This gives us the simplest equation:

$$0=\Delta_g f=\left[\partial_x^2+\partial_y^2+\partial_z^2-\frac{\partial_t^2}{c^2}\right]f. \tag{421}$$

Just for an increase of generality, we extend (392) as follows (A, B, D are constants):

$$g_{\alpha\beta}^{\text{flat}}=\begin{pmatrix}-c^2 & 0 & 0 & 0\\ 0 & A^2 & 0 & 0\\ 0 & 0 & B^2 & 0\\ 0 & 0 & 0 & D^2\end{pmatrix}, \tag{422}$$

which makes (421) to:

$$0=\Delta_g f=\left[\frac{\partial_x^2}{A^2}+\frac{\partial_y^2}{B^2}+\frac{\partial_z^2}{D^2}-\frac{\partial_t^2}{c^2}\right]f. \tag{423}$$

Applying the separation approach $f[t,x,y,z] = T[t]*X[x]*Y[y]*Z[z]$ gives us the solutions:

$$T[t]=C_{t1}\cdot\cos\left[c\cdot C_t\cdot t\right]+C_{t2}\cdot\sin\left[c\cdot C_t\cdot t\right], \tag{424}$$

$$X[x]=C_{x1}\cdot\cos\left[A\cdot C_x\cdot x\right]+C_{x2}\cdot\sin\left[A\cdot C_x\cdot x\right], \tag{425}$$

$$Y[y]=C_{y1}\cdot\cos\left[B\cdot C_y\cdot y\right]+C_{y2}\cdot\sin\left[B\cdot C_y\cdot y\right], \tag{426}$$

$$Z[z]=C_{z1}\cdot\cos\left[D\cdot C_z\cdot z\right]+C_{z2}\cdot\sin\left[D\cdot C_z\cdot z\right] \tag{427}$$

and the following characteristic equation:

$$0=C_x^2+C_y^2+C_z^2-C_t^2. \tag{428}$$

The matter coded with this solution for R^* can now directly be obtained via the Einstein field equations with matter, which reads:

$$R^{*\alpha\beta}-\frac{1}{2}R^*\cdot g^{\alpha\beta}+\Lambda\cdot g^{\alpha\beta}=-\kappa\cdot T^{\alpha\beta}. \tag{429}$$

Here we have: $R^{\alpha\beta}$, $T^{\alpha\beta}$ the Ricci- and the energy momentum tensor, respectively, while the parameters Λ and κ are constants (usually called cosmological and coupling constant, respectively). As we had per demand $R^* = R = 0$ and as we also assume the cosmological constant to be equal to zero, we obtain the following identity:

$$R^{*\alpha\beta}=-\kappa\cdot T^{\alpha\beta}. \tag{430}$$

Thus, in order to find out what kind of matter our transformation (393) has created, we simply need to evaluate the corresponding Ricci tensor of the transformed metric $G_{\alpha\beta}$. In order to keep the presentation general[1] and as the evaluation of the derivatives is simple, we give the Ricci tensor with the function f. But for simplicity, we choose $F[f]$ to be $F[f] = (f + C_1)$:

$$R^{*00}=\frac{(A\cdot B\cdot c)^2\left(f_{,z}\right)^2+D^2\left(A^2c^2\left(f_{,y}\right)^2+B^2\left(\begin{array}{l}c^2\left(f_{,x}\right)^2+3A^2\left(f_{,t}\right)^2\\-2A^2\left(C_1+f\right)f_{,t,t}\end{array}\right)\right)}{(A\cdot B\cdot D)^2c^4\left(C_1+f\right)^6}. \tag{431}$$

$$R^{*11}=\frac{-(A\cdot B\cdot c)^2\left(f_{,z}\right)^2+D^2\left(\begin{array}{l}-A^2c^2\left(f_{,y}\right)^2+B^2\left(\begin{array}{l}3c^2\left(f_{,x}\right)^2\\-2c^2\left(C_1+f\right)f_{,x,x}\end{array}\right)\\+A^2\left(f_{,t}\right)^2\end{array}\right)}{(c\cdot B\cdot D)^2A^4\left(C_1+f\right)^6}. \tag{432}$$

[1]This comes in handy the moment we want to exploit the additive character of our metric solutions. Thereby the additivity is assured by the means of condition . See further below in the main text!

$$R^{*22}=\frac{-(A\cdot B\cdot c)^{2}\left(f_{,z}\right)^{2}+D^{2}\begin{pmatrix}3A^{2}c^{2}\left(f_{,y}\right)^{2}-2A^{2}c^{2}\left(C_{1}+f\right)f_{,y,y}\\-B^{2}c^{2}\left(f_{,x}\right)^{2}+B^{2}A^{2}\left(f_{,t}\right)^{2}\end{pmatrix}}{(c\cdot A\cdot D)^{2}B^{4}\left(C_{1}+f\right)^{6}}. \tag{433}$$

$$R^{*33}=\frac{3(A\cdot B\cdot c)^{2}\left(f_{,z}\right)^{2}+D^{2}\begin{pmatrix}-A^{2}c^{2}\left(f_{,y}\right)^{2}-B^{2}c^{2}\left(f_{,x}\right)^{2}+B^{2}A^{2}\left(f_{,t}\right)^{2}\\+2\left(C_{1}+f\right)\left(A^{2}c^{2}f_{,y,y}+B^{2}\left(c^{2}f_{,x,x}-A^{2}f_{,t,t}\right)\right)\end{pmatrix}}{(c\cdot A\cdot D)^{2}B^{4}\left(C_{1}+f\right)^{6}}. \tag{434}$$

$$R^{*01}=\frac{2\left(-2f_{,x}f_{,t}+\left(C_{1}+f\right)f_{,t,x}\right)}{A^{2}c^{2}\left(C_{1}+f\right)^{6}};\quad R^{*02}=\frac{2\left(-2f_{,y}f_{,t}+\left(C_{1}+f\right)f_{,t,y}\right)}{B^{2}c^{2}\left(C_{1}+f\right)^{6}}$$

$$R^{*03}=\frac{2\left(-2f_{,z}f_{,t}+\left(C_{1}+f\right)f_{,t,z}\right)}{D^{2}c^{2}\left(C_{1}+f\right)^{6}};\quad R^{*12}=\frac{2\left(2f_{,y}f_{,x}-\left(C_{1}+f\right)f_{,x,y}\right)}{A^{2}B^{2}\left(C_{1}+f\right)^{6}}$$

$$R^{*13}=\frac{2\left(2f_{,z}f_{,x}-\left(C_{1}+f\right)f_{,x,z}\right)}{A^{2}D^{2}\left(C_{1}+f\right)^{6}};\quad R^{*23}=\frac{2\left(2f_{,y}f_{,z}-\left(C_{1}+f\right)f_{,z,y}\right)}{B^{2}D^{2}\left(C_{1}+f\right)^{6}}. \tag{435}$$

Previously we had discussed corresponding solutions in spherical [137] and cylindrical [138] geometries and it is clear that we have obtained photons or photonic forms. What is more, with condition (397) we have automatically achieved the additive character for all our solutions (424) to (428), allowing the application of simple integral transform methods (in fact, it is just the scaling option and a bit of distortion we needed so far) in order to construct almost arbitrary photonic matter forms.

Without applying the technique of superposition of our solution (424) to (428), however, the subsequent photonic solution would be a plane structure with no changes along the x-axis. It was shown in [41, 53] how such a structure can be localized also in lateral directions via standard integral transform methods and that the subsequent solutions do fulfill the Maxwell equations. The connection with the energy momentum tensor for the electromagnetic field is therefore achieved via (430). The electrostatic and magnetic vector fields **E** and **B** with components E_i and B_i form the energy momentum tensor in a contravariant form as follows:

$$\left(T^{\alpha\beta}\right)=\begin{pmatrix} \frac{1}{2}\cdot(\mathbf{E}\cdot\mathbf{E}+\mathbf{B}\cdot\mathbf{B}) & (\mathbf{E}\times\mathbf{B})^{T} \\ \mathbf{E}\times\mathbf{B} & \frac{1}{2}\cdot(\mathbf{E}\cdot\mathbf{E}+\mathbf{B}\cdot\mathbf{B})\cdot\delta_{ik}-E_iE_k-B_iB_k \end{pmatrix}$$

$$\delta_{ik}=\begin{pmatrix} 1 & 0 & 0 \\ 0 & 1 & 0 \\ 0 & 0 & 1 \end{pmatrix}. \tag{436}$$

Please note that in SI-unites this tensor reads:

$$\left(T^{\alpha\beta}\right)=\begin{pmatrix} \dfrac{\varepsilon_0\cdot\mathbf{E}\cdot\mathbf{E}+\dfrac{\mathbf{B}\cdot\mathbf{B}}{\mu_0}}{2} & c\cdot\varepsilon_0\cdot(\mathbf{E}\times\mathbf{B})^{T} \\ c\cdot\varepsilon_0\cdot\mathbf{E}\times\mathbf{B} & \dfrac{\varepsilon_0\cdot\mathbf{E}\cdot\mathbf{E}+\dfrac{\mathbf{B}\cdot\mathbf{B}}{\mu_0}}{2}\cdot\delta_{ik}-\varepsilon_0\cdot E_iE_k-\dfrac{B_iB_k}{\mu_0} \end{pmatrix}. \tag{437}$$

The constants ε_0, μ_0 are the electric and the magnetic field constant, respectively.

For brevity, we shall leave it to the mathematically skilled and interested reader to work out the detailed connection of the two classical fields with our quantum metric distortion. For illustration, however, we just consider the energy density of the electromagnetic field given via the component T^{00} and, thus, obtain:

$$R^{*00}=\frac{(A\cdot B\cdot c)^2\left(f_{,z}\right)^2+D^2\left(A^2c^2\left(f_{,y}\right)^2+B^2\left(\begin{array}{c}c^2\left(f_{,x}\right)^2+3A^2\left(f_{,t}\right)^2\\-2A^2\left(C_1+f\right)f_{,t,t}\end{array}\right)\right)}{(A\cdot B\cdot D)^2c^4\left(C_1+f\right)^6}$$

$$=-\kappa\cdot T^{00}=-\kappa\cdot\frac{\varepsilon_0\cdot\mathbf{E}\cdot\mathbf{E}+\dfrac{\mathbf{B}\cdot\mathbf{B}}{\mu_0}}{2}$$

$$=-\frac{\kappa}{2}\cdot\left(\varepsilon_0\cdot\left(E_1^2+E_2^2+E_3^2\right)+\frac{B_1^2+B_2^2+B_3^2}{\mu_0}\right). \tag{438}$$

In fact, we find quadratic terms on both sides of the equation, and it appears kind of attractive to seek the connections between the classical E- and B-fields and the quantum metric distortion in the following way:

$$\frac{\left(f_{,\mathrm{x}}\right)^2+C_{tx}\left(3\left(f_{,\mathrm{t}}\right)^2-2\left(C_1+f\right)f_{,\mathrm{t,t}}\right)}{\left(C_1+f\right)^6}=-\frac{\kappa_x}{2}\cdot\left(\varepsilon_0\cdot E_1^2+\frac{B_1^2}{\mu_0}\right)$$

$$\frac{\left(f_{,\mathrm{y}}\right)^2+C_{ty}\left(3\left(f_{,\mathrm{t}}\right)^2-2\left(C_1+f\right)f_{,\mathrm{t,t}}\right)}{\left(C_1+f\right)^6}=-\frac{\kappa_y}{2}\cdot\left(\varepsilon_0\cdot E_2^2+\frac{B_2^2}{\mu_0}\right)$$

$$\frac{\left(f_{,\mathrm{z}}\right)^2+C_{tz}\left(3\left(f_{,\mathrm{t}}\right)^2-2\left(C_1+f\right)f_{,\mathrm{t,t}}\right)}{\left(C_1+f\right)^6}=-\frac{\kappa_z}{2}\cdot\left(\varepsilon_0\cdot E_3^2+\frac{B_3^2}{\mu_0}\right). \quad (439)$$

Thus, obviously we have created photonic matter from a flat space vacuum solution to the Einstein field equations by means of a simple metric transformation. The scalar Ricci curvature was thereby kept unchanged, which is to say $R = R^*$. The equations necessary to solve this were the classical quantum Klein–Gordon equations, which we directly obtain from the $R = R^*$-condition.

Appendix E

The Fundamental Connection to Materials Science

It was shown in a variety of papers by this author (e.g., [24] to [26] and [201]) that one major problem of mathematically dealing with time- and stress-dependent material properties results from the space-time dependency of these very properties. It was even shown with first-principles methods [25] that the usual time-functional approach, which expands the ordinary linear theory of elasticity to an effective (global) time dependency, simply cannot work in a general manner because its basic assumption of material homogeneity does not hold. The integral (space or time integral) effective laws applied to overcome this problem, on the other hand, are always only valid in connection with a certain group of experiments, respectively deformations and stress situations or processes.

The problem, so it was shown in [25], seems to result from the restriction of the material constitutive laws to linearity, homogeneity, and the time dependency only being functionality (add-on like or grafted) incorporated.

This author believes that the incorporation of time and non-linearity as true additional degrees of freedom provides quite some potential for mirroring the natural behavior much better than the current approaches can ever do. Degrees of freedom, however, are just dimensions and so we end up with elastic material space-times instead of the usual linear elastic spaces being (time-functionally) made time dependent.

Now it was shown by the author that any kind of process, a certain space-time is subjected to, including those which add energy, mass, or any other kinds of "stress," can be described by the transformation of the kind (5) (e.g., [3]).

Naturally, this can also be done to a material space-time and the learned material scientist will immediately realize the connection, because we have the following analogy:

$$G_{\alpha\beta} = F\left[f\left[t,x,y,z\right]\right]_{\alpha\beta}^{ij} g_{ij} \xrightarrow{\text{simplest}} G_{\alpha\beta} = F\left[f\left[t,x,y,z\right]\right] \cdot \delta_{\alpha}^{i}\delta_{\beta}^{j} g_{ij}$$

$$\uparrow\downarrow$$

$$\varepsilon_{\alpha\beta} = E_{\alpha\beta}^{ij}\sigma_{ij} \xrightarrow{\text{simplest}} \varepsilon_{\alpha\beta} = E \cdot \delta_{\alpha}^{i}\delta_{\beta}^{j}\sigma_{ij}. \tag{440}$$

This connection not only provides us with quite some interesting analogy between the cosmologic space-time and elasticity but also gives us access to the mathematical apparatus from both sides of the wall. For instance, while there is still a great struggle in dealing with non-homogeneous Young's moduli, there are methods to deal with general $F[...]$ in (5). As mathematically, however, there is no difference to (440), we immediately see one advantage in forming this connection. The explicit time dependency in F of (5) and the classical problems in treating creep in (440) are adding another point.

Just in order to give an example that there are also situations where materials science can help to dramatically simplify certain problems in General Theory of Relativity, we want to consider a complex diagonal metric g_{ij}. It can be shown that there often exists a transformation:

$$G_{\alpha\beta} = F_{\alpha\beta}^{ij} g_{ij} \xrightarrow{\text{isotropic}} G_{\alpha\beta} = \delta_{j}^{i} \cdot \Psi\left[t,x,y,z\right], \tag{441}$$

which renders the resulting metric completely isotropic. The skilled material scientist will have no difficulty in finding that the transformation necessary to achieve this does have the following structure:

$$F_{\alpha\beta}^{ij} = \begin{pmatrix} 1 & v & v & v & & & & \vdots & & \\ v & 1 & v & v & & & & \vdots & & \\ v & v & 1 & v & & & \cdots & 0 & \cdots & \\ v & v & v & 1 & & & & \vdots & & \\ & & & & 1+v & & & & & \\ & & & & & 1+v & & & & \\ & & \vdots & & & & 1+v & & & \\ & \cdots & 0 & \cdots & & & & 1+v & & \\ & & \vdots & & & & & & 1+v & \\ & & & & & & & & & 1+v \end{pmatrix}. \tag{442}$$

Thereby we have used the symmetries of the transformation (being the same for F and Young's modulus tensor) and applied the Klein-Voigt presentation for shortage. The reader will already have recognized the typical form of a Young's modulus tensor, only that it is not in 3, but in 4 material dimensions. Thus, it automatically includes different time behavior at each position in space, giving a much greater flexibility to describe real material problems.

Now, with the connection established, we might like to ask how to model and understand processes (transformations) our 4-dimensional material space-time is subjected to. Well, the answer is not too difficult if we just turn our attention to the energy momentum tensor and observe what happens to it when something has been done to our material. The corresponding mathematical apparatus was given in [3, 141].

Thereby it was worked out how one has to connect solutions from the general approach (396) (or (440) if we are already thinking in material space-times) to the resulting energy momentum tensor and its changes. As shown in the previous appendix (see also [3] and [141], section 2, "An Unusual Introduction," sub-section 2.1, "Derivation of Electro-Magnetic Interaction (and Matter) via a Set of Creative Transformations"), we will not need anything else but the field tensor of the electromagnetic interaction to close the circle. The electrostatic and magnetic vector fields **E** and **B** with components E_i and B_i, on the other hand, form the energy momentum tensor in contravariant form, which was already given above as follows:

$$\left(T^{\alpha\beta}\right)=\begin{pmatrix} \frac{1}{2}\cdot(\mathbf{E}\cdot\mathbf{E}+\mathbf{B}\cdot\mathbf{B}) & (\mathbf{E}\times\mathbf{B})^{T} \\ \mathbf{E}\times\mathbf{B} & \frac{1}{2}\cdot(\mathbf{E}\cdot\mathbf{E}+\mathbf{B}\cdot\mathbf{B})\cdot\delta_{ik}-E_iE_k-B_iB_k \end{pmatrix}$$

$$\delta_{ik}=\begin{pmatrix} 1 & 0 & 0 \\ 0 & 1 & 0 \\ 0 & 0 & 1 \end{pmatrix}. \tag{443}$$

Now some critical readers might ask what—on earth—the electromagnetic field tensors have to do with the strains and stresses in general materials. After all, materials consist of atoms.

This author answers with a counter question, namely:

Well, and what holds these atoms together? What is the field acting between them?

Yes, it is a field completely based on electromagnetic interaction, and as such, the E-B-form to better and more holistically describe, interpret, and understand the inner workings of materials just seems to fit perfectly.

Thereby it might come in very handy, especially in connection with questions regarding spin-glasses, spin-bio-molecules, and all sorts of constructive solutions regarding quantum computer concepts, that we also already have a Dirac-based approach.

It should explicitly be noted, however, that the restriction to the E-B energy momentum tensor only applies for "ordinary," which is to say earthly materials, as has been usually treated by classical materials science. Pure photonic matter, strong or weak force dominated matter (like atom cores), degenerated matter due to huge gravity fields, or even black hole material [13] cannot be treated that way.

Appendix F

The Linear Elastic Space-Time

Evaluating the quantum Einstein field equations from [9] and comparing our peculiarly simple result in 2 dimensions, in 6 dimensions, and in n dimensions (peculiar if one considers the complexity of the general starting point (17) and its even more complicated result) with the fundamental equation of equilibrium for the linear elastic space free of body forces [202, 203], one finds a great similarity.

Why is this?

Does it mean that our space-time is linearly elastic, and what do all those many possibilities of various stress, strain, and displacement fields then mean with respect to—let's say—elementary particle physics?

Well, let's see.

We take the generalized Einstein–Hilbert action from (18) and (19) and concentrate on the case $p = 4$, $n = 6$ where, just as elaborated in [5, 9, 18], we assume 2 dimensions to be curled up or compactified and f only depends on the remaining 4 dimensions. It was also shown in [5] that the compactified dimensions can then bring about inertia.

Giving f an internal structure like:

$$f = Q^{\gamma} \phi_{\gamma}, \tag{444}$$

where we assume Q^{γ} to be a vector of constants, and inserting it into (17) (assuming a metric of constants) leads us to:

$$0 = \left(11 \cdot \left(g^{\kappa\alpha} g^{\lambda\beta} + g^{\lambda\alpha} g^{\kappa\beta}\right) - 14 \cdot g^{\kappa\lambda} g^{\alpha\beta}\right) f_{,\alpha\beta}$$

$$\Rightarrow \qquad 0 = \left(14 \cdot g^{\kappa\lambda} g^{\alpha\beta} - 11 \cdot \left(g^{\kappa\alpha} g^{\lambda\beta} + g^{\lambda\alpha} g^{\kappa\beta}\right)\right) Q^{\gamma} \phi_{\gamma,\alpha\beta}$$

$$= Q^{\gamma} \left(14 \cdot g^{\kappa\lambda} g^{\alpha\beta} - 11 \cdot \left(g^{\kappa\alpha} g^{\lambda\beta} + g^{\lambda\alpha} g^{\kappa\beta}\right)\right) \phi_{\gamma,\alpha\beta}$$

$$\Rightarrow \qquad 0 = 14 \cdot g^{\kappa\lambda} Q^{\gamma} \phi_{\gamma,\beta}{}^{\beta} - 11 \cdot \left(Q^{\gamma} \phi_{\gamma}{}^{,\kappa\lambda} + Q^{\gamma} \phi_{\gamma}{}^{,\lambda\kappa} \right). \tag{445}$$

Comparison with the fundamental (load free) equation of linear elasticity, which in the isotropic case with Poisson's ratio μ and the displacement vector h_γ reads[1]:

$$0 = 2 \cdot (1 - 2 \cdot \mu) h_{\gamma,\beta}{}^{\beta} - \left(h_{\gamma}{}^{,\gamma\lambda} + h_{\gamma}{}^{,\gamma\kappa} \right), \tag{446}$$

directly leads us not only to a great variety of possible solutions but also to some rather interesting analogies. One of these solutions, which was already discussed here, is the pure shear stress solution with the following example (in Cartesian coordinates[2]):

$$h_\gamma = \left\{ -\partial_y, \partial_x, 0 \right\} \psi \quad \text{with:} \quad \psi_{,\beta}{}^{\beta} = 0. \tag{447}$$

Applied onto (445), we may set:

$$h_\gamma = D_\gamma^\sigma \phi_\sigma = \left\{ 0, -\partial_y, \partial_x, 0 \right\} \psi \quad \text{with:} \quad \psi_{,\beta}{}^{\beta} = 0. \tag{448}$$

Shear Fields = Spin Fields?

It seems that this type of solution to (445) provides the characteristics of a spin field. Thus, what a spin is in elementary particle physics would become a shear field in quantum Einstein field equations. We apply this approach as follows:

$$\begin{aligned} 0 &= 14 \cdot g^{\kappa\lambda} Q^{\gamma} \phi_{\gamma,\beta}{}^{\beta} - 11 \cdot \left(Q^{\gamma} \phi_{\gamma}{}^{,\kappa\lambda} + Q^{\gamma} \phi_{\gamma}{}^{,\lambda\kappa} \right) \\ &= 14 \cdot g^{\kappa\lambda} Q^{\gamma} \phi_{\gamma,\beta}{}^{\beta} - 11 \cdot \left(q^{\kappa} h_{\gamma}{}^{,\gamma\lambda} + q^{\lambda} h_{\gamma}{}^{,\gamma\kappa} \right) \end{aligned} \tag{449}$$

and directly see that, because of the structure of our solution:

$$h_\gamma = \pm \left\{ \begin{matrix} \{-\partial_x, \partial_t, 0, 0\}, \{-\partial_y, 0, \partial_t, 0\}, \{-\partial_z, 0, 0, \partial_t\} \\ \{0, -\partial_y, \partial_x, 0\}, \{0, -\partial_z, 0, \partial_x\}, \{0, 0, -\partial_z, \partial_y\} \end{matrix} \right\} \psi; \quad Q^{\gamma} \phi_{\gamma,\beta}{}^{\beta} = 0, \tag{450}$$

we automatically get:

$$14 \cdot g^{\kappa\lambda} \overbrace{Q^{\gamma} \phi_{\gamma,\beta}{}^{\beta}}^{=0} - 11 \cdot \left(\overbrace{q^{\kappa} h_{\gamma}{}^{,\gamma\lambda}}^{=0} + \overbrace{q^{\lambda} h_{\gamma}{}^{,\gamma\kappa}}^{=0} \right) = \ \Rightarrow \ = 0. \tag{451}$$

---

[1]It should be noted that in the original equation (e.g., [202]), the non-Laplacian term has a positive sign, but this is not of interest to the shear solutions we are going to discuss here.

[2]Please note that in Cartesian coordinates, the difference between co- and contravariant tensor forms vanishes.

Quark-Like 1/3 and 2/3 Charges?

Now we extend this approach a bit and apply it as follows:

$$\begin{aligned} 0 &= 14 \cdot g^{\kappa\lambda} Q^{\gamma\sigma} \phi_{\gamma\sigma,\beta}{}^{\beta} - 11 \cdot \left(Q^{\gamma\sigma} \phi_{\gamma\sigma}{}^{,\kappa\lambda} + Q^{\gamma\sigma} \phi_{\gamma\sigma}{}^{,\lambda\kappa} \right) \\ &= 14 \cdot g^{\kappa\lambda} Q^{\gamma\sigma} \phi_{\gamma\sigma,\beta}{}^{\beta} - 11 \cdot \left(q^{\kappa\sigma} h_{\gamma\sigma}{}^{,\gamma\lambda} + q^{\lambda\sigma} h_{\gamma\sigma}{}^{,\gamma\kappa} \right) \end{aligned} \quad (452)$$

or even:

$$\begin{aligned} 0 &= 14 \cdot g^{\kappa\lambda} Q^{\gamma\sigma\mu} \phi_{\gamma\sigma\mu,\beta}{}^{\beta} - 11 \cdot \left(Q^{\gamma\sigma\mu} \phi_{\gamma\sigma\mu}{}^{,\kappa\lambda} + Q^{\gamma\sigma\mu} \phi_{\gamma\sigma\mu}{}^{,\lambda\kappa} \right) \\ &= 14 \cdot g^{\kappa\lambda} Q^{\gamma\sigma\mu} \phi_{\gamma\sigma\mu,\beta}{}^{\beta} - 11 \cdot \left(q^{\kappa\sigma\mu} h_{\gamma\sigma\mu}{}^{,\gamma\lambda} + q^{\lambda\sigma\mu} h_{\gamma\sigma\mu}{}^{,\gamma\kappa} \right). \end{aligned} \quad (453)$$

Then, as shown in [3], more such forms of shear-like solutions can be found easily via the following recipe:

$$h_\gamma[x_k] = \left\{ \dots, A_1 \overbrace{\frac{\partial^2 \psi}{\partial \xi_2 \partial \xi_3}}^{pos\ \xi_1}, \dots, A_2 \overbrace{\frac{\partial^2 \psi}{\partial \xi_1 \partial \xi_3}}^{pos\ \xi_2}, \dots, A_3 \overbrace{\frac{\partial^2 \psi}{\partial \xi_1 \partial \xi_2}}^{pos\ \xi_3}, \dots \right\}$$

$$h_\gamma[x_k] = \left\{ \dots, A_1 \overbrace{\frac{\partial^3 \psi}{\partial \xi_2 \partial \xi_3 \partial \zeta_4}}^{pos\ \xi_1}, \dots, A_2 \overbrace{\frac{\partial^3 \psi}{\partial \xi_1 \partial \xi_3 \partial \xi_4}}^{pos\ \xi_2}, \dots, A_3 \overbrace{\frac{\partial^3 \psi}{\partial \xi_1 \partial \xi_2 \partial \xi_4}}^{pos\ \xi_3}, \dots, A_4 \overbrace{\frac{\partial^3 \psi}{\partial \zeta_1 \partial \xi_2 \partial \xi_3}}^{pos\ \xi_4} \dots \right\}$$

$$\dots$$

$$\forall (\dots,\dots) = 0; \quad \sum_{\forall k} A_k = 0. \quad (454)$$

In a Minkowski space-time with $c = 1$ and the coordinates t, x, y, and z, we could have the following types of such solutions:

$$\begin{aligned} &h_\gamma = \pm \left\{ \frac{\partial}{\partial z}, 0, 0, -\frac{\partial}{\partial t} \right\} \psi; \quad h_\gamma = \pm \left\{ \frac{\partial}{\partial y}, 0, -\frac{\partial}{\partial t}, 0 \right\} \psi; \quad h_\gamma = \pm \left\{ \frac{\partial \psi}{\partial x}, -\frac{\partial}{\partial t}, 0, 0 \right\} \psi \\ &h_\gamma = \pm \left\{ 0, \frac{\partial}{\partial z}, 0, -\frac{\partial}{\partial x} \right\} \psi; \quad h_\gamma = \pm \left\{ 0, \frac{\partial}{\partial y}, -\frac{\partial}{\partial x}, 0 \right\} \psi; \quad h_\gamma = \pm \left\{ 0, 0, \frac{\partial}{\partial z}, -\frac{\partial}{\partial y} \right\} \psi, \end{aligned} \quad (455)$$

$$\begin{aligned} &h_\gamma = \left\{ A_1 \frac{\partial^2}{\partial x \partial y}, A_2 \frac{\partial^2}{\partial t \partial y}, A_3 \frac{\partial^2}{\partial x \partial t}, 0 \right\} \psi; \quad h_\gamma = \left\{ A_1 \frac{\partial^2}{\partial x \partial z}, A_2 \frac{\partial^2}{\partial t \partial z}, 0, A_3 \frac{\partial^2}{\partial x \partial t} \right\} \psi \\ &h_\gamma = \left\{ A_1 \frac{\partial^2}{\partial z \partial y}, 0, A_2 \frac{\partial^2}{\partial t \partial z}, A_3 \frac{\partial^2}{\partial z \partial t} \right\} \psi; \quad h_\gamma = \left\{ 0, A_1 \frac{\partial^2}{\partial z \partial y}, A_2 \frac{\partial^2}{\partial x \partial z}, A_3 \frac{\partial^2}{\partial x \partial y} \right\} \psi. \end{aligned} \quad (456)$$

$$h_\gamma = \left\{ A_1 \frac{\partial^3}{\partial x \partial y \partial z}, A_1 \frac{\partial^3}{\partial t \partial y \partial z}, A_1 \frac{\partial^3}{\partial x \partial t \partial z}, A_4 \frac{\partial^3}{\partial x \partial y \partial t} \right\} \psi; \quad \sum_{\forall k} A_k = 0. \quad (457)$$

Thereby we find that the last solution in (456) with a setting of the kind $A_1 = 1/3, A_2 = 1/3$, and $A_3 = -2/3$ shows a peculiar closeness to the charges of quarks, the building blocks of hadrons and baryons.

Hydrostatic Particle Fields

As said above, comparison of (445) with the fundamental (load free) equation of linear elasticity, which in the isotropic case with Poisson's ratio μ and the displacement vector h_γ reads:

$$0 = 2\cdot(1-2\cdot\mu)h_{\gamma,\beta}{}^{\beta} - \left(h_{\gamma}{}^{;\gamma\lambda} + h_{\gamma}{}^{;\gamma\kappa}\right), \tag{458}$$

directly leads us to not only a great variety of possible solutions but also to some rather interesting analogies. Another interesting solution could be the pure normal stress (or hydrostatic) solution with the example (in Cartesian coordinates):

$$h_\gamma = \left\{\partial_x, \partial_y, \partial_z\right\}\psi \quad \text{with:} \quad \psi_{,\beta}{}^{\beta} = 0. \tag{459}$$

Applied onto (445), we may set:

$$h_\gamma = D_\gamma^\sigma \phi_\sigma = \begin{cases} \left\{\partial_t, \partial_x, \partial_y, \partial_z\right\}\psi \quad \text{with:} \quad \psi_{,k}{}^{k} = 0; \quad k = 0,1,2,3 \\ \left\{\partial_t, \partial_x, \partial_y, 0\right\}\psi \quad \text{with:} \quad \psi_{,k}{}^{k} = 0; \quad k = 0,1,2 \\ \left\{\partial_t, \partial_x, 0, \partial_z\right\}\psi \quad \text{with:} \quad \psi_{,k}{}^{k} = 0; \quad k = 0,1,3 \\ \left\{\partial_t, 0, \partial_y, \partial_z\right\}\psi \quad \text{with:} \quad \psi_{,k}{}^{k} = 0; \quad k = 0,2,3 \\ \left\{\partial_t, \partial_x, 0, 0\right\}\psi \quad \text{with:} \quad \psi_{,k}{}^{k} = 0; \quad k = 0,1 \\ \left\{\partial_t, 0, 0, \partial_z\right\}\psi \quad \text{with:} \quad \psi_{,k}{}^{k} = 0; \quad k = 0,3 \\ \left\{\partial_t, 0, \partial_z, 0\right\}\psi \quad \text{with:} \quad \psi_{,k}{}^{k} = 0; \quad k = 0,2 \\ \left\{0, \partial_x, \partial_y, \partial_z\right\}\psi \quad \text{with:} \quad \psi_{,k}{}^{k} = 0; \quad k = 1,2,3 \\ \left\{0, \partial_x, \partial_y, 0\right\}\psi \quad \text{with:} \quad \psi_{,k}{}^{k} = 0; \quad k = 1,2 \\ \left\{0, \partial_x, 0, \partial_z\right\}\psi \quad \text{with:} \quad \psi_{,k}{}^{k} = 0; \quad k = 1,3 \\ \left\{0, 0, \partial_y, \partial_z\right\}\psi \quad \text{with:} \quad \psi_{,k}{}^{k} = 0; \quad k = 2,3. \end{cases} \tag{460}$$

The charm of this type of solution to (445) could be seen in the fact that it provides the characteristics of a spherical pressure field (spherical in 4, 3, or 2 dimensions). This could mean a charge, mass, or something. Thus, perhaps, what might be a certain charge in elementary particle physics would become a symmetric normal pressure field in the quantum Einstein field equations. Thus, picking—for instance—the spatial solution:

$$h_\gamma = D_\gamma^\sigma \phi_\sigma = \{0, \partial_x, \partial_y, \partial_z\}\psi \quad \text{with:} \quad \psi_{,k}{}^{k} = 0; \quad k = 1,2,3, \tag{461}$$

we would obtain a stable (time-independent) homogeneous pressure field around the spatial point {0,0,0}. The stresses and strains it eradiates into its neighborhood might be registered there as some kind of force fields and their origin would then be a charged particle, which sends out the field.

Time-Planes and Time-Layers

Another elastic solution, which appears to be of cosmological character, is the plane boundary solution with the example (in Cartesian coordinates):

$$h_\gamma = z \cdot \{\partial_x, \partial_y, \partial_z\}\psi + (1 - 4 \cdot \mu)\{0,0,\partial_z\}\psi \quad \text{with:} \quad \psi_{,\beta}{}^{\beta} = 0. \tag{462}$$

Please note that in the classical literature of elasticity (e.g., [202]), where we have (458) with a positive sign as:

$$0 = 2 \cdot (1 - 2 \cdot \mu) h_{\gamma,\beta}{}^{\beta} + \left(h_\gamma{}^{,\gamma\lambda} + h_\gamma{}^{,\gamma\kappa}\right), \tag{463}$$

solution (462) reads:

$$h_\gamma = z \cdot \{\partial_x, \partial_y, \partial_z\}\psi - (3 - 4 \cdot \mu)\{0,0,\partial_z\}\psi \quad \text{with:} \quad \psi_{,\beta}{}^{\beta} = 0. \tag{464}$$

Applied onto (445), we may set:

$$f = Q^\gamma \phi_\gamma \Leftrightarrow h_\gamma = D_\gamma^\sigma \phi_\sigma = t \cdot \{\partial_x, \partial_x, \partial_y, \partial_z\}\psi + \left(1 - 4 \cdot \frac{9}{11}\right)\{\partial_t, 0,0,0\}\psi$$

$$\text{with:} \quad \psi_{,\beta}{}^{\beta} = 0 \tag{465}$$

and obtain the solution to (445) via:

$$\Rightarrow \quad 0 = 14 \cdot g^{\kappa\lambda} Q^\gamma \phi_{\gamma,\beta}{}^{\beta} - 11 \cdot \left(Q^\gamma \phi_\gamma{}^{,\kappa\lambda} + Q^\gamma \phi_\gamma{}^{,\lambda\kappa}\right)$$

$$= \left(1 - 2 \cdot \frac{9}{11}\right) \cdot g^{\kappa\lambda} q^\beta \phi_{\gamma,\beta}{}^{\gamma} - Q^\kappa \phi_\gamma{}^{,\gamma\lambda}. \tag{466}$$

The importance of this type of solution to (445) here could be seen in the fact that it provides the characteristics of the fundamental elastic solution to an impact or indent on the plane $t = 0$. Especially in connection with the other solutions we already presented here, we would obtain a great variety of indent solutions. Particles (and other stuff, usually been named matter) may just be seen as defects [204], which are the remainders of the hypothetic primordial impact.

As shown elsewhere [24–26, 205, 206], this type of solution also allows the introduction of layered structures, which are in this case layers of time. However, as the type of solution (465) can be easily generalized to layers within all dimensions:

$$f = Q^{\gamma}\phi_{\gamma} \Leftrightarrow h_{\gamma} = \begin{cases} t\cdot\left\{\partial_x,\partial_x,\partial_y,\partial_z\right\}\psi + \left(1-4\cdot\frac{9}{11}\right)\left\{\partial_t,0,0,0\right\}\psi \\ x\cdot\left\{\partial_x,\partial_x,\partial_y,\partial_z\right\}\psi + \left(1-4\cdot\frac{9}{11}\right)\left\{0,\partial_x,0,0\right\}\psi \\ y\cdot\left\{\partial_x,\partial_x,\partial_y,\partial_z\right\}\psi + \left(1-4\cdot\frac{9}{11}\right)\left\{0,0,\partial_y,0\right\}\psi \\ z\cdot\left\{\partial_x,\partial_x,\partial_y,\partial_z\right\}\psi + \left(1-4\cdot\frac{9}{11}\right)\left\{0,0,0,\partial_z\right\}\psi \end{cases}$$

$$\text{with: } \psi_{,\beta}{}^{\beta} = 0 \tag{467}$$

or even be used for the introduction of contacting and / or layered manifolds via (here we only give one example in 3D and 2D; for more complex solutions, the reader is referred to the books of Fabrikant [207, 208], for instance):

$$h_{\gamma} = \begin{cases} x\cdot\left\{0,\partial_x,\partial_y,\partial_z\right\}\psi + C_{\psi}\cdot\left\{0,\partial_x,0,0\right\}\psi \\ y\cdot\left\{0,\partial_x,\partial_y,\partial_z\right\}\psi + C_{\psi}\cdot\left\{0,0,\partial_y,0\right\}\psi \\ z\cdot\left\{0,\partial_x,\partial_y,\partial_z\right\}\psi + C_{\psi}\cdot\left\{0,0,0,\partial_z\right\}\psi \end{cases}$$

$$\text{with: } \psi_{,k}{}^{k} = 0; \quad k = 1,2,3$$

$$h_{\gamma} = \begin{cases} x\cdot\left\{0,\partial_x,\partial_y,0\right\}\psi + C_{\psi}\cdot\left\{0,\partial_x,0,0\right\}\psi \\ y\cdot\left\{0,\partial_x,\partial_y,0\right\}\psi + C_{\psi}\cdot\left\{0,0,\partial_y,0\right\}\psi \end{cases}$$

$$\text{with: } \psi_{,k}{}^{k} = 0; \quad k = 1,2, \tag{468}$$

we end up with a space-time brim-full with "elastic options."

One may even use the analogies found here for an easy way to actually illustrate features of our universe as stress, strain, displacement, and deformation energy fields [209] (c.f. cover picture of [9] and the figures in the next sub-section).

Thus: Our Universe—Not an Accident, Not an Event, but Perhaps an Indent?

There have been hypotheses about the Big Bang just being an impact event. Thereby it was assumed that some colliding super-branes started what we realize as our universe.

Now, by reevaluating the Einstein–Hilbert action for a scaled metric, thereby deriving a set of quantum Einstein field equations, we found solutions which seem to support this idea.

Our universe could indeed just be the deformation field of some kind of "mechanical contact," impact, or indent.

... and the surface at which the contact "was, is, and will be" originated "was, is, and will be" $t = 0$, the beginning of time.

The charm of this idea is not just the fact that it provides the features the original hypotheses were aiming for anyway, but that it also explains time as the propagation of the shockwave, which is still pushing through our universe, if not to say, which is making our universe.

Yes, if this hypothesis takes shape and gets more supporting evidence, it may be a shock to some people to find themselves being the result of an impact and its subsequent shockwave, but at least this scenario would assure everyone and everything to make an impact as it, he, or she actually is impact material.

Brief Discussion Regarding Our Peculiar Results for "Poisson's Ratio"

In the two sections above, we found settings for the quantum Einstein field equations, which resulted in equations similar to the governing equations in elasticity, only that here our equations sported a funny Poisson's ratio $\mu > 0.5$. Such a Poisson's ratio would be equivalent to a negative compression module and means that space-time, if under pressure, would expand and not contract.

We also find some strange behavior when putting the space under certain loading situations. In Figs. F.1 to F.3 below, we illustrate this in the case of some pulling forces being active on planar surface parts within the space. Such loading situations are completely hypothetical and only shown here in order to demonstrate the strange effect of the high μ. The colors code the total shear of von Mises stress of the spatial space.

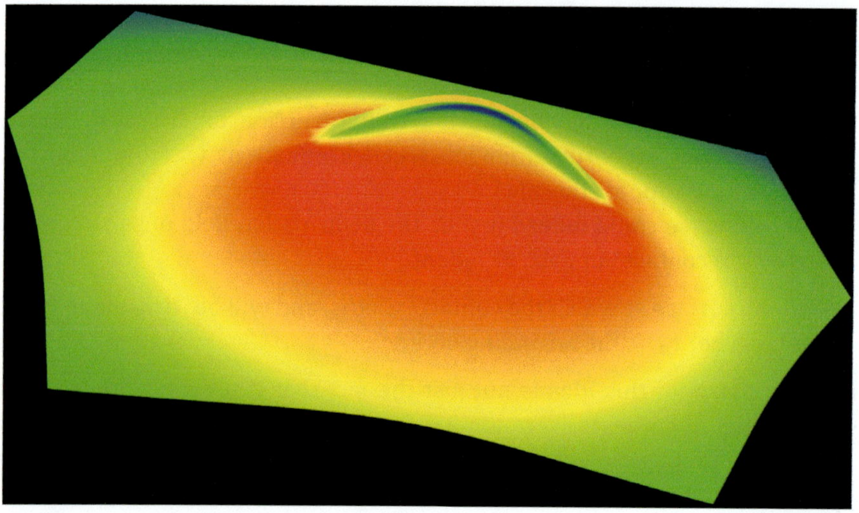

Figure F.1 A stress field from a pulling force in space-time according to the elastic solution of the quantum Einstein field equations, evaluated with the software FilmDoctor [209] using equation (449) with solution (468).

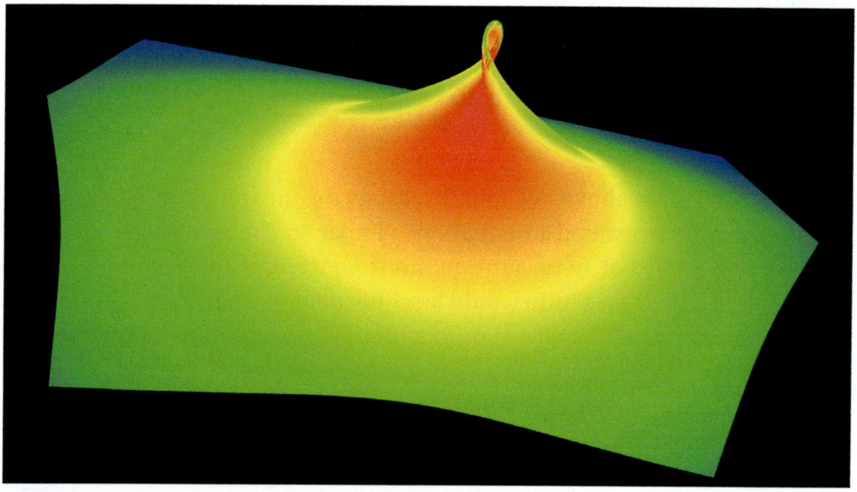

Figure F.2 A stress field from a pulling force in space-time according to the elastic solution of the quantum Einstein field equations, evaluated with the software FilmDoctor [209] using equation (449) with solution (468).

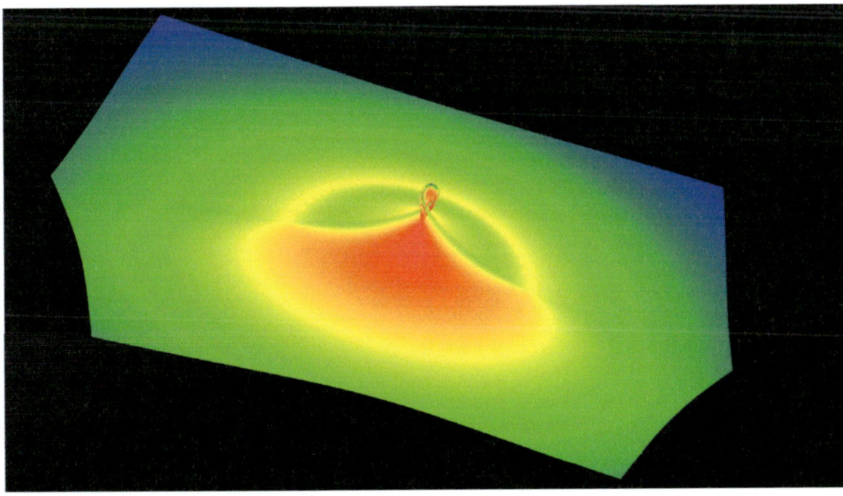

Figure F.3 A stress field from a pulling force in space-time according to the elastic solution of the quantum Einstein field equations, evaluated with the software FilmDoctor [209] using equation (449) with solution (468).

Appendix G

Restoration of the Classical Einstein Field Equations and Where Does Einstein's and Hilbert's Matter Come from?

First, we want to restore the classical Einstein field equations from the equations we have obtained from the scaled metric via Hilbert's variation of the Einstein–Hilbert action. This resulted in the following two equations for metrics of the type (2)[1] (e.g., see section 11.2, "Quantum Gravity Equations," in chapter 11, "The Quantum Gravity Enigma Machine"):

$$\begin{aligned}
&1.\quad 0=\begin{pmatrix} \left(R+\frac{F'}{F}(1-n)\Delta f\right)\left(1-\frac{n}{2}\right) \\ +\frac{f_{,\alpha}f_{,\beta}(1-n)}{4F^2}g^{\alpha\beta}\left(4FF''+F'F'(n-6)\right) \\ \times\left(1-\frac{n}{2}\right)+\kappa T_{\alpha\beta}g^{\alpha\beta}+\Lambda\cdot F\cdot n \end{pmatrix} \\
&2.\quad 0=\begin{pmatrix} R_{\alpha\beta}-R\cdot\frac{g_{\alpha\beta}}{2} \\ +\frac{F'}{F}(1-n)\left(f_{,\alpha\beta}-\Gamma^{\gamma}_{\alpha\beta}f_{,\gamma}-\frac{g_{\alpha\beta}}{2}\Delta f\right) \\ +\frac{(1-n)}{4F^2}\begin{pmatrix} 4FF'' \\ +F'F'(n-6)\end{pmatrix}\begin{pmatrix} f_{,\alpha}f_{,\beta} \\ -f_{,\mu}f_{,\nu}g^{\mu\nu}\frac{g_{\alpha\beta}}{2}\end{pmatrix} \\ +\kappa T_{\alpha\beta}+\Lambda\cdot F\cdot g_{\alpha\beta} \end{pmatrix},
\end{aligned} \tag{469}$$

---

[1]Regarding general metrics, please see appendix I.

which originated via a split-up of the variation term $\delta G^{\alpha\beta}$ from:

$$\delta W = 0$$

$$= \int_V d^n x \left(\sqrt{-g \cdot F^n} \left(\begin{array}{c} R_{\alpha\beta} - R \cdot \frac{g_{\alpha\beta}}{2} \\ + \frac{F'}{F}(1-n)\left(f_{,\alpha\beta} - \Gamma^{\gamma}_{\alpha\beta} f_{,\gamma} - \frac{g_{\alpha\beta}}{2} \Delta f \right) \\ + \frac{(1-n)}{4F^2} \left(\begin{array}{c} 4FF'' \\ +F'F'(n-6) \end{array} \right) \left(\begin{array}{c} f_{,\alpha} f_{,\beta} \\ -f_{,\mu} f_{,\nu} g^{\mu\nu} \frac{g_{\alpha\beta}}{2} \end{array} \right) \\ +\kappa T_{\alpha\beta} + \Lambda \cdot F \cdot g_{\alpha\beta} \end{array} \right) \delta G^{\alpha\beta} \right). \tag{470}$$

Now we simply need to assume an F to be a constant and obtain:

$$\delta W = 0$$

$$= \int_V d^n x \left(\sqrt{-g \cdot F^n} \left(\begin{array}{c} R_{\alpha\beta} - R \cdot \frac{g_{\alpha\beta}}{2} \\ \overbrace{\left\{ \begin{array}{c} + \frac{F'}{F}(1-n)\left(f_{,\alpha\beta} - \Gamma^{\gamma}_{\alpha\beta} f_{,\gamma} - \frac{g_{\alpha\beta}}{2} \Delta f \right) \\ + \frac{(1-n)}{4F^2} \left(\begin{array}{c} 4FF'' \\ +F'F'(n-6) \end{array} \right) \left(\begin{array}{c} f_{,\alpha} f_{,\beta} \\ -f_{,\mu} f_{,\nu} g^{\mu\nu} \frac{g_{\alpha\beta}}{2} \end{array} \right) \end{array} \right\}}^{=0} \\ +\kappa T_{\alpha\beta} + \Lambda \cdot F \cdot g_{\alpha\beta} \end{array} \right) \delta G^{\alpha\beta} \right)$$

$$\Rightarrow \quad = \int_V d^n x \left(\sqrt{-g \cdot F^n} \left(R_{\alpha\beta} - R \cdot \frac{g_{\alpha\beta}}{2} + \kappa T_{\alpha\beta} + \Lambda \cdot F \cdot g_{\alpha\beta} \right) \delta G^{\alpha\beta} \right)$$

$$\Rightarrow \quad 0 = R_{\alpha\beta} - R \cdot \frac{g_{\alpha\beta}}{2} + \kappa T_{\alpha\beta} + \Lambda \cdot F \cdot g_{\alpha\beta}. \tag{471}$$

As desired, we have obtained the classical Einstein field equations.

Discovering the classical field equation structure (c.f. (381)) among our ensemble of options to decompose the Einstein–Hilbert variation (see [13–18]), one might also want to demonstrate the connection to the Einstein–Hilbert matter approach. This way we would finally be able to understand the nature of matter or—at least—the potential nature of Hilbert's and Einstein's postulated and artificially introduced matter terms [1, 2].

For the reason of brevity and simplicity, we apply condition (regarding the generalized results when avoiding this simplification see [17]):

$$4F_iF_i'' + F_i'F_i'(n-6) = 0 \tag{472}$$

and start with the assumption of one global metric $g_{\alpha\beta}$, but an ensemble of F_i and f_i, leading us to:

$$0 = \left(\begin{array}{c} \int\limits_{V_G} d^n x \sqrt{-G} \left(\left(R_{\alpha\beta} - \frac{1}{2} \cdot R \cdot g_{\alpha\beta} \right) \delta G^{\alpha\beta} \right) \\ + \frac{1}{2} \cdot \sum_{i=1}^{N} \int\limits_{V_{Qi}} d^n x \sqrt{-g \cdot F^n} \, \frac{F'}{F} (n-1) \left(\Delta f \cdot g_{\alpha\beta} - 2 \cdot \overbrace{\left(f_{,\alpha\beta} - \Gamma^{\gamma}_{\alpha\beta} f_{,\gamma} \right)}^{\equiv f_{;\alpha\beta}} \right) \delta G^{\alpha\beta} \end{array} \right)$$

$$\Rightarrow \quad = \int\limits_{V_G} d^n x \sqrt{-g} \left(\sum_{i=1}^{N} \sqrt{F_i^n} \cdot \left(R_{\alpha\beta} - \frac{1}{2} \cdot R \cdot g_{\alpha\beta} + \frac{(n-1)}{2} \cdot \frac{F_i'}{F_i} \left(\Delta f_i \cdot g_{\alpha\beta} - 2 \cdot f_{i:\alpha\beta} \right) \right) \delta G^{\alpha\beta} \right). \tag{473}$$

This gives us the classical non-vacuum field equations via:

$$0 = \sum_{i=1}^{N} \sqrt{F_i^n} \cdot \left(R_{\alpha\beta} - \frac{1}{2} \cdot R \cdot g_{\alpha\beta} \right) + \sum_{i=1}^{N} \sqrt{F_i^n} \cdot \left(\frac{(n-1)}{2} \cdot \frac{F_i'}{F_i} \left(\Delta f_i \cdot g_{\alpha\beta} - 2 \cdot f_{i:\alpha\beta} \right) \right)$$

$$\Rightarrow \quad 0 = R_{\alpha\beta} - \frac{1}{2} \cdot R \cdot g_{\alpha\beta} + \frac{\sum_{i=1}^{N} \sqrt{F_i^n} \cdot \left(\frac{(n-1)}{2} \cdot \frac{F_i'}{F_i} \left(\Delta f_i \cdot g_{\alpha\beta} - 2 \cdot f_{i:\alpha\beta} \right) \right)}{\sum_{i=1}^{N} \sqrt{F_i^n}}$$

$$= R_{\alpha\beta} - \frac{1}{2} \cdot R \cdot g_{\alpha\beta} + \frac{\sum_{i=1}^{N} \sqrt{F_i^n} \cdot \left(\frac{(n-1)}{2} \cdot \frac{F_i'}{F_i} \left(\Delta f_i \cdot g_{\alpha\beta} - 2 \cdot \left(f_{,\alpha\beta} - \Gamma^{\gamma}_{\alpha\beta} f_{,\gamma} \right) \right) \right)}{\sum_{i=1}^{N} \sqrt{F_i^n}} \tag{474}$$

and the last term in the last line above gives us the quantum or scale equivalent of the Einstein–Hilbert matter term, respectively the energy momentum tensor:

$$\Rightarrow \quad \kappa T_{\alpha\beta} = \frac{\sum_{i=1}^{N}\sqrt{F_i^n}\cdot\left(\frac{(n-1)}{2}\cdot\frac{F_i'}{F_i}\left(\Delta f_i\cdot g_{\alpha\beta}-2\cdot f_{i;\alpha\beta}\right)\right)}{\sum_{i=1}^{N}\sqrt{F_i^n}}$$

$$= \frac{\sum_{i=1}^{N}\sqrt{F_i^n}\cdot\left(\frac{(n-1)}{2}\cdot\frac{F_i'}{F_i}\left(\Delta f_i\cdot g_{\alpha\beta}-2\cdot\left(f_{,\alpha\beta}-\Gamma_{\alpha\beta}^{\gamma}f_{,\gamma}\right)\right)\right)}{\sum_{i=1}^{N}\sqrt{F_i^n}}. \tag{475}$$

Now, for completeness, here we also want to present the result without the simplification $4FF'' + F'F'(n-6) = 0$. This renders the variation (470), which with the functional "multiplets" F_i looks like:

$$0 = \begin{pmatrix} \int_{V_G} d^n x\sqrt{-G}\left(\left(R_{\alpha\beta}-\frac{1}{2}\cdot R\cdot g_{\alpha\beta}\right)\delta G^{\alpha\beta}\right) \\ +\frac{1}{2}\cdot\sum_{i=1}^{N}\int_{V_{Qi}} d^n x\sqrt{-g\cdot F^n}\,\frac{F'}{F}(n-1)\left(\Delta f\cdot g_{\alpha\beta}-2\cdot\left(f_{,\alpha\beta}-\Gamma_{\alpha\beta}^{\gamma}f_{,\gamma}\right)\right)\delta G^{\alpha\beta} \end{pmatrix}$$

$$\Rightarrow \quad = \int_{V_G} d^n x\sqrt{-g}\left(\sum_{i=1}^{N}\sqrt{F_i^n}\cdot\left(R_{\alpha\beta}-\frac{1}{2}\cdot R\cdot g_{\alpha\beta}+\frac{(n-1)}{2}\cdot\frac{F_i'}{F_i}\left(\Delta f_i\cdot g_{\alpha\beta}-2\cdot f_{i;\alpha\beta}\right)\right)\delta G^{\alpha\beta}\right), \tag{476}$$

as follows:

$$0 = \begin{pmatrix} \int_{V_G} d^n x\sqrt{-G}\left(\left(R_{\alpha\beta}-\frac{1}{2}\cdot R\cdot g_{\alpha\beta}\right)\delta G^{\alpha\beta}\right) \\ +\frac{1}{2}\cdot\sum_{i=1}^{N}\int_{V_{Qi}} d^n x\sqrt{-g\cdot F^n}\begin{pmatrix} \frac{F'}{F}(n-1)\left(\Delta f\cdot g_{\alpha\beta}-2\cdot\left(f_{,\alpha\beta}-\Gamma_{\alpha\beta}^{\gamma}f_{,\gamma}\right)\right) \\ +\frac{(1-n)}{2F_i^2}\begin{pmatrix} F_iF_i'' \\ +F_i'F_i'(n-6)\end{pmatrix}\begin{pmatrix} f_{i,\alpha}f_{i,\beta} \\ -f_{i,\mu}f_{i,\nu}g^{\mu\nu}\frac{g_{\alpha\beta}}{2}\end{pmatrix}\end{pmatrix}\delta G^{\alpha\beta} \end{pmatrix}$$

$$\Rightarrow \quad = \int_{V_G} d^n x\sqrt{-g}\left(\sum_{i=1}^{N}\sqrt{F_i^n}\cdot\begin{pmatrix} R_{\alpha\beta}-\frac{1}{2}\cdot R\cdot g_{\alpha\beta}+\frac{(n-1)}{2} \\ \times\begin{pmatrix} \frac{F_i'}{F_i}\left(\Delta f_i\cdot g_{\alpha\beta}-2\cdot f_{i;\alpha\beta}\right) \\ +\frac{1}{2F_i^2}\begin{pmatrix} F_iF_i'' \\ +F_i'F_i'(n-6)\end{pmatrix}\begin{pmatrix} f_{i,\alpha}f_{i,\beta} \\ -f_{i,\mu}f_{i,\nu}g^{\mu\nu}\frac{g_{\alpha\beta}}{2}\end{pmatrix}\end{pmatrix}\end{pmatrix}\delta G^{\alpha\beta}\right). \tag{477}$$

and gives us the matter in the general form:

$$T_{\alpha\beta}=\frac{\sum_{i=1}^{N}\sqrt{F_i^n}\cdot\left(\frac{(n-1)}{2}\cdot\left(\frac{F_i'}{F_i}\begin{pmatrix}\Delta f_i\cdot g_{\alpha\beta}\\-2\cdot f_{i;\alpha\beta}\end{pmatrix}+\frac{1}{2F_i^2}\begin{pmatrix}F_iF_i''\\+F_i'F_i'(n-6)\end{pmatrix}\begin{pmatrix}f_{i,\alpha}f_{i,\beta}\\-f_{i,\mu}f_{i,\nu}g^{\mu\nu}\frac{g_{\alpha\beta}}{2}\end{pmatrix}\right)\right)}{\kappa\cdot\sum_{i=1}^{N}\sqrt{F_i^n}}. \tag{478}$$

Appendix H

A Few Words about Spin 1/2, 3/2, 5/2, and so on

In the following, we want to investigate potential half-spin solutions for the associated Legendre polynomials, which we needed in various forms within this book.

Here, as an example, we intend to consider the simple form:

$$Y_l^m[v] = C_{Pv} \cdot P_l^m[\cos[v]] + C_{Qv} \cdot Q_l^m[\cos[v]]. \tag{479}$$

Observing the solution $Y[v]$ more closely, we find that there exist singularity-free spin $l = n/2$-solutions for $m = -n/2$ with $n = 1,3,5,7, \ldots$ in the case of $C_{Pv} \neq 0$, $C_{Qv} = 0$ and for $m = +n/2$ in the case of $C_{Pv} = 0$, $C_{Qv} \neq 0$. Figures H.1 and H.2 illustrate the corresponding distributions within the domain of definition for the angle v from 0 to π.

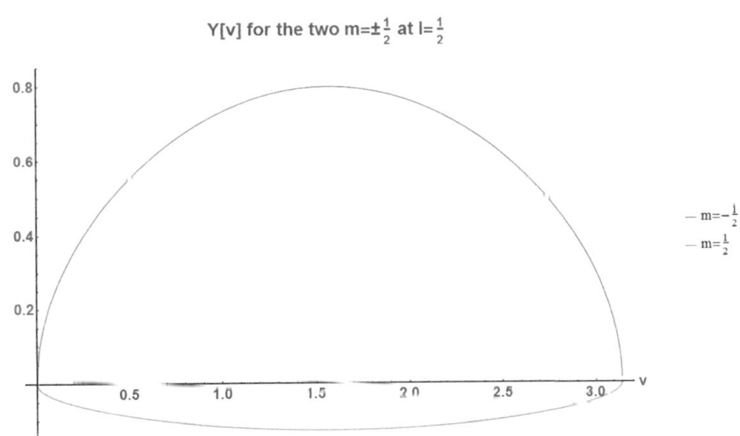

Figure H.1 Spin $l = 1/2$ situation with the two possible spin states $m = \pm 1/2$ according to our angular quantum gravity solution (247). For better illustration and comparability, we divided the "Q-Legendres" by 10.

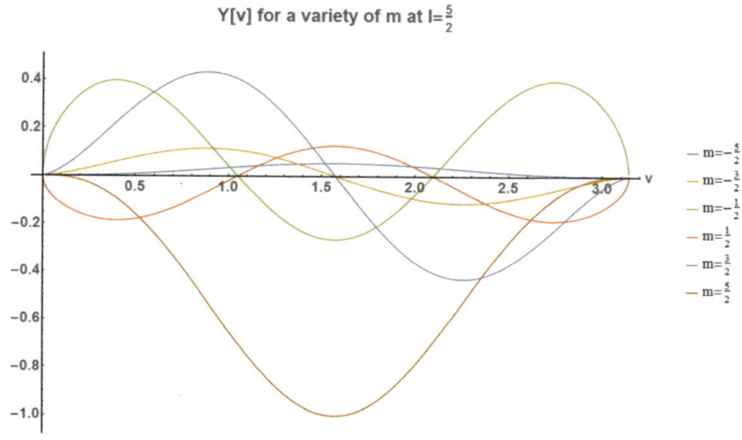

Figure H.2 Spin $l = 5/2$ situation with the five possible spin states $m = \pm\{1/2,3/2,5/2\}$ according to our angular quantum gravity solution (247). Regarding the evaluation, see text. For better illustration, we divided the "Q-Legendres" by 10.

The general solution to the partial differential equation resulting from the Schrödinger hydrogen problem [186, 187, 194] (see also appendix I) can be given as follows:

$$f_{n,l,m}[t,r,\vartheta,\varphi]=g[t]\cdot\Psi_{n,l,m}[r,\vartheta,\varphi]=g[t]\cdot e^{i\cdot m\cdot\varphi}\cdot P_l^m[\cos\vartheta]\cdot R_{n,l}[r]$$

$$=\left(C_1\cdot\cos\left[c\cdot C_{tt}\cdot t\right]+C_2\cdot\sin\left[c\cdot C_{tt}\cdot t\right]\right)\cdot N\cdot \mathrm{e}^{-\rho/2}\cdot\rho^l\cdot L_{n-l-1}^{2l+1}[\rho]\cdot Y_l^m[\vartheta,\varphi]$$

$$\rho=\frac{2\cdot r}{n\cdot a_0};\quad N=\sqrt{\left(\frac{2}{n\cdot a_0}\right)^3\frac{(n-l-1)!}{2\cdot n\cdot(n+l)!}}. \tag{480}$$

It totally suffices here to consider the classical problems in order to also discuss the angular aspects of (206) as the results for these coordinates in both cases are the same functional dependencies.

For nostalgic reasons, we used the symbol "n" (here not the dimension) as the so-called main quantum number. All quantum numbers n, l and the parameter a_0 depend on the constants E, A_i in (206) and have to satisfy certain quantum conditions in order to obtain singularity-free solutions for $f[\ldots]$.

Regarding the conditions for the quantum numbers n, l, and m, we not only have the usual:

$$\{n,\mathrm{l},\mathrm{m}\}\in\mathbb{Z};\quad n\geq 0;\quad l<n;\quad -l\leq m\leq +l \tag{481}$$

but also found the suitable solutions for the half-spin forms as discussed above and derived in Figs. H.1 and H.2. The corresponding main quantum numbers for half-spin l-numbers with l = 1/2, 3/2, ... are simply (just as before with the integers) $n = l + 1 = 3/2, 5/2, 7/2, \ldots$.

It should explicitly be noted, however, that the usual spherical harmonics are inapplicable in cases of half-spin. For $\{n,l,m\} = \{1/2,3/2,5/2,7/2,\ldots\}$, the wave function (480) has to be adapted as follows:

$$f_{n,l,m}[t,r,\vartheta,\varphi] = \mathrm{e}^{\pm i\cdot c\cdot C_t\cdot t}\cdot N\cdot \mathrm{e}^{-\rho/2}\cdot \rho^l\cdot L_{n-l-1}^{2l+1}[\rho]\cdot Z_m[\varphi]\cdot\begin{Bmatrix} P_l^{m<0}[\cos\vartheta] \\ Q_l^{m>0}[\cos\vartheta]\end{Bmatrix}$$

$$= \mathrm{e}^{\pm i\cdot c\cdot C_t\cdot t}\cdot N\cdot \mathrm{e}^{-\rho/2}\cdot \rho^l\cdot L_{n-l-1}^{2l+1}[\rho]\cdot\begin{Bmatrix}\cos[m\cdot\varphi] \\ \sin[m\cdot\varphi]\end{Bmatrix}\cdot\begin{Bmatrix} P_l^{m<0}[\cos\vartheta] \\ Q_l^{m>0}[\cos\vartheta]\end{Bmatrix}. \quad (482)$$

Thereby it was elaborated elsewhere [140] that, in fact, the sin- and the cos-functions seem to make the Pauli exclusion [210] and not the "+" and "–" of m. However, in order to have the usual Fermionic statistic, we may simply define as follows:

$$f_{n,l,m}[t,r,\vartheta,\varphi]$$

$$= e^{\pm i\cdot c\cdot C_t\cdot t}\cdot N\cdot e^{-\rho/2}\cdot \rho^l\cdot L_{n-l-1}^{2l+1}[\rho]\cdot\begin{Bmatrix}\sin[m\cdot\varphi]_{m<0} \\ \cos[m\cdot\varphi]_{m>0}\end{Bmatrix}\cdot\begin{Bmatrix} P_l^{m<0}[\cos\vartheta] \\ Q_l^{m>0}[\cos\vartheta]\end{Bmatrix}. \quad (483)$$

As derived in [140], the resolution of the degeneration with respect to half-spin requires a break of the symmetry, which we achieved by introducing elliptical geometry instead of the spherical one here. Seen experimentally, this is just equivalent to the introduction of an electromagnetic (polarization) field.

Thus, we have metrically derived a fairly general "hydrogen atom." In addition to the Schrödinger structure, our form also sports a time-dependent factor clearly showing the options for matter and antimatter via the ±-sign in the $g[t]$-function. We also found the half-spin states and were able to resolve the spin-degeneration via a simple symmetry break by switching from spherical to elliptical coordinates (see [140] regarding the evaluation). A few illustrations of the half-spin solutions are given in chapter 8, "The Spinor," in this book and [18], section 11.5.

Appendix I

The Metric Schrödinger Hydrogen

Abstract

In this appendix, we are going to derive the Schrödinger hydrogen equation [I1, I2] by using the metric concept of Hilbert's variational integrals [I3], which are based on Hamilton's extremal principles. While Hilbert originally only wanted to derive the Einstein field equations [I4], he, in fact, wrote down a world formula, which also contained all fundamental quantum equations [I5, I6, I7].

It was already shown in the previous chapters of this book how both theories, Einstein's General Theory of Relativity and Quantum Theory, can be extracted from Hilbert's original work, and as an example we will now derive the Schrödinger hydrogen equation.

It will be shown that for this most fundamental derivation of the H-Atom, interestingly, a space or space-time of 7 dimensions is required. Of course, higher numbers of dimensions would also allow one to construct H-atoms (and may even be useful with respect to the metric incorporation of mass or other inertia), but 7 seems to be the irreducible number. This fact alone is an amazing finding. For convenience, we are going to repeat and generalize the fundamental starting point of our derivation.

References for Abstract

I1. E. Schrödinger, Quantisierung als Eigenwertproblem (erste Mitteilung). *Ann. Phys.*, 384(4), 1926, 361–376.

I2. E. Schrödinger, Quantisierung als Eigenwertproblem (zweite Mitteilung). *Ann. Phys.*, 384(6), 1926, 489–527.

13. D. Hilbert, Die Grundlagen der Physik, Teil 1, *Göttinger Nachrichten*, 27, 1915, 395–407.
14. A. Einstein, Grundlage der allgemeinen Relativitätstheorie, *Ann. Phys.*, 49(ser. 4), 1916, 769–822.
15. N. Schwarzer, *The World Formula: A Late Recognition of David Hilbert's Stroke of Genius*, Jenny Stanford Publishing, 2022, ISBN: 9789814877206.
16. N. Schwarzer, *The Math of Body, Soul and the Universe*, Jenny Stanford Publishing, 2023, ISBN 9789814968249.
17. N. Schwarzer, *The Theory of Everything—Quantum and Relativity Is Everywhere—A Fermat Universe*, Jenny Stanford Publishing, 2020, ISBN-10: 9814774472.

A System as Ensemble of Properties

We are assuming that every system, from the elementary particle via a "primitive" or even complex life-form to the whole universe, can be considered an ensemble of potentially entangled properties. These properties are the degrees of freedom of the system and as such they are also the dimensions of that very system forming a space or space-time. The system can mathematically be described via its metric when being of Riemannian character. This most general description of almost any system of practical relevance now allows us to investigate the system when simply assuming that it has to follow certain and very universal laws. The most fundamental law we could think of right now would be a Hamilton extremal principle and this directly brings us to our starting point.

Fundamental Starting Point

Our starting point shall be the Hamilton principle in form of the Einstein–Hilbert action [2]:

$$\delta W = 0 = \delta \int_V d^n x \left(\sqrt{-g} \cdot \Phi_R [R] \right) = \delta \int_V d^n x \left(\sqrt{-g} \cdot \left(R - 2\Lambda + L_M \right) \right). \quad (484)$$

Here R gives the Ricci scalar and g denotes the determinant of the metric tensor $g_{\alpha\beta}$. The term L_M stands for the matter Lagrange density, which was postulated by both Hilbert and Einstein and Λ gives the cosmological constant. Assuming that we can extract a functional factor $F[f]$, with f being a function of all coordinates, from the metric tensor in the following way:

$$G_{\alpha\beta} = g_{\alpha\beta} \cdot F[f], \tag{485}$$

it was shown in [3] that we do obtain not only the Einstein field equations [2] but also the most important classical quantum equations (see also [18]).

Here, we only briefly repeat the essentials.

Setting (485) into (484), we obtain the following Einstein–Hilbert action [18]:

$$\delta W = 0$$

$$= \delta \int_V d^n x \left(\sqrt{-g \cdot F^n} \times \left(\left(\begin{pmatrix} R - \frac{1}{2F} \begin{pmatrix} 2F_{,\alpha\beta}(n-1)g^{\alpha\beta} + 2\Gamma^a_{\alpha\beta}F_{,a}g^{\alpha\beta} \\ -F_{,\alpha}g^{ab}g_{\beta b,a}g^{\alpha\beta} - F_{,\beta}g^{ab}g_{\alpha b,a}g^{\alpha\beta} \\ -n\Gamma^d_{\alpha\beta}F_{,d}g^{\alpha\beta} + \frac{n}{2}F_{,d}g^{cd}g_{ab,c}g^{ab} \end{pmatrix} \\ -\frac{F_{,\alpha} \cdot F_{,\beta}}{4F^2} g^{\alpha\beta}((n-6)(n-1)) \end{pmatrix} \frac{1}{F} \right) \\ -2\Lambda + L_M \right) \right). \tag{486}$$

Thereby, in order to keep things simple and still general enough, one can often restrict the metric selection to diagonal metrics with the following properties:

$$g_{ij} = \begin{pmatrix} g_{00} & \cdots & 0 \\ \vdots & \ddots & \vdots \\ 0 & \cdots & g_{n-1n-1} \end{pmatrix}; \quad g_{jj,j} = 0, \tag{487}$$

which gives an extended Hilbert integral:

$$\delta W = 0$$

$$= \delta \int_V d^n x \left(\sqrt{-g \cdot F^n} \times \left(\left(\begin{pmatrix} R + \frac{F'}{F}(1-n)\Delta f \\ + \frac{f_{,\alpha}f_{,\beta}g^{\alpha\beta}(1-n)}{4F^2}\left(4FF'' + (F')^2(n-6)\right) \end{pmatrix} \frac{1}{F} \right) \\ -2\Lambda + L_M \right) \right). \tag{488}$$

In this appendix, however, we will not use this simplification. After performing the variation in (486), we obtain the following quantum gravity field equations:

$$0=\left(\begin{array}{c}\boxed{R_{\alpha\beta}-\frac{g_{\alpha\beta}}{2}R}+\frac{g_{\alpha\beta}}{2}\left(\begin{array}{c}\frac{1}{2F}\left(\begin{array}{c}2F_{,ij}(n-1)g^{ij}+2\Gamma_{ij}^{a}F_{,a}g^{ij}\\-F_{,i}g^{ab}g_{jb,a}g^{ij}-F_{,j}g^{ab}g_{ib,a}g^{ij}\\-n\Gamma_{ij}^{d}F_{,d}g^{ij}+\frac{n}{2}F_{,d}g^{cd}g_{ab,c}g^{ab}\end{array}\right)\\+\frac{F_{,i}\cdot F_{,j}}{4F^{2}}g^{ij}\left((n-6)(n-1)\right)\end{array}\right)\\-\frac{1}{2F}\left(\begin{array}{c}F_{,\alpha\beta}(n-2)+F_{,ab}g_{\alpha\beta}g^{ab}\\+F_{,a}g^{ab}\left(g_{\beta b,\alpha}-g_{\beta\alpha,b}\right)-F_{,\alpha}g^{ab}g_{\beta b,a}-F_{,\beta}g^{ab}g_{\alpha b,a}\\+F_{,d}g^{cd}\frac{1}{2}n\left(\frac{2}{n}g_{\alpha c,\beta}-g_{\alpha c,\beta}-g_{\beta c,\alpha}+g_{\alpha\beta,c}+\frac{1}{n}g_{\alpha\beta}g_{ab,c}g^{ab}\right)\\-\frac{1}{2F}\left(F_{,\alpha}\cdot F_{,\beta}(3n-6)+g_{\alpha\beta}F_{,c}F_{,d}g^{cd}(4-n)\right)\end{array}\right)\\+\kappa T_{\alpha\beta}+\Lambda\cdot F\cdot g_{\alpha\beta}\end{array}\right).\quad(489)$$

Immediately, we recognize the classical Einstein field equations (see the box in the equation above) inside our new ones. It was shown in [3, 18] and the previous chapters in this book that the scaling terms in (489) give the Klein–Gordon, Dirac, and Schrödinger equations (regarding the latter see also [41] and compare with [186, 187], for instance). Under the variational integral, the integrand (489) has to be multiplied with the variated metric tensor $\delta G^{\alpha\beta}=\delta\left(\frac{g^{\alpha\beta}}{F}\right)$ and therefore could be split up into two equations:

$$0=\left(\begin{array}{c}\boxed{R_{\alpha\beta}-\frac{g_{\alpha\beta}}{2}R}+\frac{g_{\alpha\beta}}{2}\left(\begin{array}{c}\frac{1}{2F}\left(\begin{array}{c}2F_{,ij}(n-1)g^{ij}+2\Gamma_{ij}^{a}F_{,a}g^{ij}\\-F_{,i}g^{ab}g_{jb,a}g^{ij}-F_{,j}g^{ab}g_{ib,a}g^{ij}\\-n\Gamma_{ij}^{d}F_{,d}g^{ij}+\frac{n}{2}F_{,d}g^{cd}g_{ab,c}g^{ab}\end{array}\right)\\+\frac{F_{,i}\cdot F_{,j}}{4F^{2}}g^{ij}\left((n-6)(n-1)\right)\end{array}\right)\\-\frac{1}{2F}\left(\begin{array}{c}F_{,\alpha\beta}(n-2)+F_{,ab}g_{\alpha\beta}g^{ab}\\+F_{,a}g^{ab}\left(g_{\beta b,\alpha}-g_{\beta\alpha,b}\right)-F_{,\alpha}g^{ab}g_{\beta b,a}-F_{,\beta}g^{ab}g_{\alpha b,a}\\+F_{,d}g^{cd}\frac{1}{2}n\left(\frac{2}{n}g_{\alpha c,\beta}-g_{\alpha c,\beta}-g_{\beta c,\alpha}+g_{\alpha\beta,c}+\frac{1}{n}g_{\alpha\beta}g_{ab,c}g^{ab}\right)\\-\frac{1}{2F}\left(F_{,\alpha}\cdot F_{,\beta}(3n-6)+g_{\alpha\beta}F_{,c}F_{,d}g^{cd}(4-n)\right)\end{array}\right)\\+\kappa T_{\alpha\beta}+\Lambda\cdot F\cdot g_{\alpha\beta}\end{array}\right)\delta G^{\alpha\beta}$$

$$
=\begin{pmatrix}
\boxed{R_{\alpha\beta}-\frac{g_{\alpha\beta}}{2}R}+\frac{g_{\alpha\beta}}{2}\begin{pmatrix}
\frac{1}{2F}\begin{pmatrix}
2F_{,ij}(n-1)g^{ij}+2\Gamma^{a}_{ij}F_{,a}g^{ij}\\
-F_{,i}g^{ab}g_{jb,a}g^{ij}-F_{,j}g^{ab}g_{ib,a}g^{ij}\\
-n\Gamma^{d}_{ij}F_{,d}g^{ij}+\frac{n}{2}F_{,d}g^{cd}g_{ab,c}g^{ab}
\end{pmatrix}\\
+\frac{F_{,i}\cdot F_{,j}}{4F^{2}}g^{ij}\left((n-6)(n-1)\right)
\end{pmatrix}\\
-\frac{1}{2F}\begin{pmatrix}
F_{,\alpha\beta}(n-2)+F_{,ab}g_{\alpha\beta}g^{ab}\\
+F_{,a}g^{ab}\left(g_{\beta b,\alpha}-g_{\beta\alpha,b}\right)-F_{,\alpha}g^{ab}g_{\beta b,a}-F_{,\beta}g^{ab}g_{\alpha b,a}\\
+F_{,d}g^{cd}\frac{1}{2}n\left(\frac{2}{n}g_{\alpha c,\beta}-g_{\alpha c,\beta}-g_{\beta c,\alpha}+g_{\alpha\beta,c}+\frac{1}{n}g_{\alpha\beta}g_{ab,c}g^{ab}\right)\\
-\frac{1}{2F}\left(F_{,\alpha}\cdot F_{,\beta}(3n-6)+g_{\alpha\beta}F_{,c}F_{,d}g^{cd}(4-n)\right)
\end{pmatrix}\\
+\kappa T_{\alpha\beta}+\Lambda\cdot F\cdot g_{\alpha\beta}
\end{pmatrix}
\begin{pmatrix}
g^{\alpha\beta}\delta\left(\frac{1}{F}\right)\\
+\frac{\delta g^{\alpha\beta}}{F}
\end{pmatrix},
\tag{490}
$$

$$
0=\left(\left(1-\frac{n}{2}\right)\left(R-\begin{pmatrix}
\frac{1}{2F}\begin{pmatrix}
2F_{,ij}(n-1)g^{ij}+2\Gamma^{a}_{ij}F_{,a}g^{ij}\\
-F_{,i}g^{ab}g_{jb,a}g^{ij}-F_{,j}g^{ab}g_{ib,a}g^{ij}\\
-n\Gamma^{d}_{ij}F_{,d}g^{ij}+\frac{n}{2}F_{,d}g^{cd}g_{ab,c}g^{ab}
\end{pmatrix}\\
+\frac{F_{,i}\cdot F_{,j}}{4F^{2}}g^{ij}\left((n-6)(n-1)\right)
\end{pmatrix}\right)+\Lambda\cdot F\cdot n\right)\delta\left(\frac{1}{F}\right)
$$

$$
0=\begin{pmatrix}
R_{\alpha\beta}-\frac{g_{\alpha\beta}}{2}R+\frac{g_{\alpha\beta}}{2}\begin{pmatrix}
\frac{1}{2F}\begin{pmatrix}
2F_{,ij}(n-1)g^{ij}+2\Gamma^{a}_{ij}F_{,a}g^{ij}\\
-F_{,i}g^{ab}g_{jb,a}g^{ij}-F_{,j}g^{ab}g_{ib,a}g^{ij}\\
-n\Gamma^{d}_{ij}F_{,d}g^{ij}+\frac{n}{2}F_{,d}g^{cd}g_{ab,c}g^{ab}
\end{pmatrix}\\
+\frac{F_{,i}\cdot F_{,j}}{4F^{2}}g^{ij}\left((n-6)(n-1)\right)
\end{pmatrix}\\
-\frac{1}{2F}\begin{pmatrix}
F_{,\alpha\beta}(n-2)+F_{,ab}g_{\alpha\beta}g^{ab}\\
+F_{,a}g^{ab}\left(g_{\beta b,\alpha}-g_{\beta\alpha,b}\right)-F_{,\alpha}g^{ab}g_{\beta b,a}-F_{,\beta}g^{ab}g_{\alpha b,a}\\
+F_{,d}g^{cd}\frac{1}{2}n\left(\frac{2}{n}g_{\alpha c,\beta}-g_{\alpha c,\beta}-g_{\beta c,\alpha}+g_{\alpha\beta,c}+\frac{1}{n}g_{\alpha\beta}g_{ab,c}g^{ab}\right)\\
-\frac{1}{2F}\left(F_{,\alpha}\cdot F_{,\beta}(3n-6)+g_{\alpha\beta}F_{,c}F_{,d}g^{cd}(4-n)\right)
\end{pmatrix}\\
+\kappa T_{\alpha\beta}+\Lambda\cdot F\cdot g_{\alpha\beta}
\end{pmatrix}\frac{\delta g^{\alpha\beta}}{F}.
\tag{491}
$$

While we consider the second equation of the pair in (491) generalized or quantum Einstein field equations, we used the first one to derive the classical quantum equations (Klein–Gordon, Dirac, and Schrödinger) for a generally curved space-time. Demanding the "gravity" to be weak as follows:

$$\delta G^{\alpha\beta}=g^{\alpha\beta}\cdot\delta_0+\overbrace{g^{ab}\delta_{ab}^{\alpha\beta}}^{\text{Gravity}}\xrightarrow{\forall\,\delta_{ab}^{\alpha\beta}\ll\delta_0}=g^{\alpha\beta}\cdot\delta_0, \tag{492}$$

we are left with the scalar and purely metrically obtained quantum equation (the first equation in (491)), which, when excluding the trivial case $n = 2$, reads:

$$0=\left(R-\left(\frac{1}{2F}\begin{pmatrix}2F_{,ij}(n-1)g^{ij}+2\Gamma_{ij}^{a}F_{,a}g^{ij}\\-F_{,i}g^{ab}g_{jb,a}g^{ij}-F_{,j}g^{ab}g_{ib,a}g^{ij}\\-n\Gamma_{ij}^{d}F_{,d}g^{ij}+\frac{n}{2}F_{,d}g^{cd}g_{ab,c}g^{ab}\end{pmatrix}+\frac{F_{,i}\cdot F_{,j}}{4F^{2}}g^{ij}\left((n-6)(n-1)\right)\right)\right)+\Lambda\cdot F\cdot n. \tag{493}$$

From there, which is to say, based on the latter equation, all classically important quantum equations can be obtained. As an example, here we derive the case of the Schrödinger hydrogen atom [186, 187].

The Metrically Obtained Schrödinger Hydrogen Atom

Taking the metric of a space-time with dimensions $t, r, \vartheta, \varphi, \varphi_1, \varphi_2$, and φ_3 of the following form:

$$g_{\alpha\beta}=f\left[t,r,\vartheta,\varphi,\varphi_3\right]^{\frac{4}{5}}\cdot\begin{pmatrix}-c^2&0&0&0&0&0&0\\0&1&0&0&0&0&0\\0&0&r^2&0&0&0&0\\0&0&0&r^2\cdot\sin^2\vartheta&0&0&0\\0&0&0&0&r^{-p-1}&0&0\\0&0&0&0&0&r^{p}&0\\0&0&0&0&0&0&r\end{pmatrix};\quad p^2+p+1=0$$

$$f\left[t,r,\vartheta,\varphi,\varphi_3\right]=f_{\text{Schrödinger}}(r,\vartheta,\varphi)\cdot\begin{pmatrix}\left(C_{tc}\cos\left(c\cdot\sqrt{-Q_n}\cdot t\right)+C_{ts}\sin\left(c\cdot\sqrt{-Q_n}\cdot t\right)\right)\\\times\left(C_{\varphi c}\cos\left(C_{\varphi_3}\cdot y\right)+C_{\varphi s}\sin\left(C_{\varphi_3}\cdot y\right)\right)\end{pmatrix}, \tag{494}$$

assuming weak gravity according to (492) and feeding everything into the extended Einstein field equations (489), being contracted with the contravariant metric tensor in accordance with condition (492), directly leads us to the following scalar quantum gravity equation for the function $f_{\text{Schrödinger}}(r,\vartheta,\varphi)$:

$$\left[-\Delta_{\text{Schrödinger}}-\overbrace{\frac{C_{\varphi_3}^2}{r}}^{V_{\text{Schrödinger }H\text{-atom}}}\right]f_{\text{Schrödinger}}(r,\vartheta,\varphi)=Q_n f_{\text{Schrödinger}}(r,\vartheta,\varphi)$$

$$\text{with: } C_{\varphi_3}^2=\frac{q^2\cdot\mu}{2\pi\varepsilon_0\hbar^2};\quad Q_n=E_n\frac{2\cdot\mu}{\hbar^2};\quad \mu=\frac{m_e m_p}{m_e+m_p}. \tag{495}$$

With $\hbar$, μ, q, and ε_0 denoting the reduced Planck constant, the reduced mass (with mass of proton m_p and electron m_e), the unit charge, and the permittivity constant in vacuum, respectively, the well-known corresponding solution can be given as follows [186, 187, 194]:

$$f_{\text{Schrödinger}}(r,\vartheta,\varphi)=\Psi_{n,l,m}[r,\vartheta,\varphi]=e^{i\cdot m\cdot\varphi}\cdot P_l^m[\cos\vartheta]\cdot R_{n,l}[r]$$

$$=\sqrt{\left(\frac{2}{n\cdot a_0}\right)^3\frac{(n-l-1)!}{2\cdot n\cdot(n+l)!}}\cdot e^{-\rho/2}\cdot\rho^l\cdot L_{n-l-1}^{2l+1}[\rho]\cdot Y_l^m[\vartheta,\varphi];\quad \rho=\frac{2\cdot r}{n\cdot a_0}. \tag{496}$$

The constant a_0 denotes the Bohr radius with:

$$a_0=\frac{4\cdot\pi\cdot\varepsilon_0\cdot\hbar^2}{m_e\cdot q^2}=5.292\cdot10^{-11}\,\text{meter}. \tag{497}$$

The functions P, L, and Y denote the associated Legendre function, the Laguerre polynomials, and the spherical harmonics, respectively.

Regarding the classical discussion of the solution, especially the quantization via the quantum numbers n, l, and m, we have to refer to the textbook literature (e.g., [194, 216]).

Here we are only interested in the influence of the potential- and energy-related other dimensions t, φ_1, φ_2, and φ_3 on the character of the solution (bonded or non-bonded states).

As we require negative E_n for the bonded states, we see that changing the approach (494) to pure oscillations in time as follows:

$$f\left[t,r,\vartheta,\varphi,\varphi_3\right]=f_{\text{Schrödinger}}\left(r,\vartheta,\varphi\right)\cdot\begin{pmatrix}\left(C_{tc}\cos\left(c\cdot\sqrt{Q_n}\cdot t\right)+C_{ts}\sin\left(c\cdot\sqrt{Q_n}\cdot t\right)\right)\\ \times\left(C_{\varphi c}\cos\left(C_{\varphi_3}\cdot y\right)+C_{\varphi s}\sin\left(C_{\varphi_3}\cdot y\right)\right)\end{pmatrix} \tag{498}$$

does give us exactly what we need. A hyperbolic approach in time:

$$f\left[t,r,\vartheta,\varphi,\varphi_3\right]=f_{\text{Schrödinger}}\left(r,\vartheta,\varphi\right)\cdot\begin{pmatrix}\left(C_{tc}\cosh\left(c\cdot\sqrt{Q_n}\cdot t\right)+C_{ts}\sinh\left(c\cdot\sqrt{Q_n}\cdot t\right)\right)\\ \times\left(C_{\varphi c}\cos\left(C_{\varphi_3}\cdot y\right)+C_{\varphi s}\sin\left(C_{\varphi_3}\cdot y\right)\right)\end{pmatrix}, \tag{499}$$

on the other hand, gives us continuous (non-bonded) solutions.

We see that we are able to construct the metric situation for the Schrödinger hydrogen problem in a 7-dimensional space-time.

References

1. A. Einstein, Grundlage der allgemeinen Relativitätstheorie, *Ann. Phys.*, 1916, 49 ser. 4, 769–822.
2. D. Hilbert, Die Grundlagen der Physik, Teil 1, *Göttinger Nachrichten*, 1915, 27, 395–407.
3. N. Schwarzer, *The World Formula: A Late Recognition of David Hilbert's Stroke of Genius*, 2022, Jenny Stanford Publishing, ISBN: 9789814877206.
4. N. Schwarzer, *Societons and Ecotons—The Photons of the Human Society—Control Them and Rule the World*, Part 1 of "Medical Socio-Economic Quantum Gravity," self-published, Amazon Digital Services, 2020, Kindle.
5. N. Schwarzer, *Masses and the Infinity Options Principle—Can We Explain the 3-Generations and the Quantized Mass Problem?*, Part 5 of "Medical Socio-Economic Quantum Gravity," self-published, Amazon Digital Services, 2020, Kindle.
6. N. Schwarzer, *Humanitons—The Intrinsic Photons of the Human Body—Understand Them and Cure Yourself*, Part 2 of "Medical Socio-Economic Quantum Gravity," self-published, Amazon Digital Services, 2020, Kindle.
7. N. Schwarzer, *Mastering Human Crises with Quantum-Gravity-Based but Still Practicable Models—First Measure: SEEING and UNDERSTANDING the WHOLE*, Part 3 of "Medical Socio-Economic Quantum Gravity," self-published, Amazon Digital Services, 2020, Kindle.
8. N. Schwarzer, *Self-Similar Quantum Gravity—How Infinity Brings Simplicity*, Part 4 of "Medical Socio-Economic Quantum Gravity," self-published, Amazon Digital Services, 2020, Kindle.
9. N. Schwarzer, *Towards Quantum Einstein Field Equations*, Part 7 of "Medical Socio-Economic Quantum Gravity," self-published, Amazon Digital Services, December 2020, Kindle

10. N. Schwarzer, *The 3 Generations of Elementary Particles*, Part 6 of "Medical Socio-Economic Quantum Gravity," self-published, Amazon Digital Services, December 2020, Kindle.

11. N. Schwarzer, *Anaphylactic Shock Attack Strategy—The Assault You Don't Realize until It Is TOO Late*, Part 10a of "Medical Socio-Economic Quantum Gravity," self-published, Amazon Digital Services, 2021, Kindle.

12. N. Schwarzer, *The Quantum Einstein Field Equations in Their Simplest Form*, Part 7a of "Medical Socio-Economic Quantum Gravity, self-published, Amazon Digital Services, March 2021, Kindle.

13. N. Schwarzer, *The Quantum Black Hole*, Part 7b of "Medical Socio-Economic Quantum Gravity," self-published, Amazon Digital Services, April 2021, Kindle.

14. N. Schwarzer, *Quantum Cosmology—A Simple Answer to the Flatness Riddle*, Part 7c of "Medical Socio-Economic Quantum Gravity," self-published, Amazon Digital Services, April 2021, Kindle.

15. N. Schwarzer, *The Metric Dirac Equation Revisited and the Geometry of Spinors*, Part 8 of "Medical Socio-Economic Quantum Gravity, self-published, Amazon Digital Services, April 2021, Kindle.

16. N. Schwarzer, *The Dirac Miracle*, Part 8a of "Medical Socio-Economic Quantum Gravity," self-published, Amazon Digital Services, May 2021, Kindle.

17. N. Schwarzer, *My Horcruxes—A Curvy Math to Salvation*, Part 9 of "Medical Socio-Economic Quantum Gravity," self-published, Amazon Digital Services, 2021, Kindle.

18. N. Schwarzer, *The Math of Body, Soul and the Universe*, 2023, Jenny Stanford Publishing, ISBN 9789814968249.

19. M. A. Gilchrist, A. Sasaki, Modeling host-parasite coevolution: a nested approach based on mechanistic models, *J. Theor. Biol.*, 2002, 218, 289–308.

20. M. E. Hochberg, M. Van Baalen, Antagonistic coevolution over productivity gradients, *Am. Nat.*, 1998, 152, 620–634.

21. J. Jaremo, J. Tuomi, P. Nilsson, T. Lennartsson, Plant adaptation to herbivory: mutualistic versus antagonistic coevolution, *Oikos*, 1999 84, 313–320.

22. J. Knisken, M. D. Rausher, Two modes of host-enemy coevolution, *Popul. Ecol.*, 2001, 43, 3–14.

23. N. Schwarzer, *Einstein Had It, but He Did Not See It—Part LXVIII: Most Fundamental Tools for Optimum Decision-Making Based on Quantum Gravity*, self-published, Amazon Digital Services, 2018, Kindle.

24. N. Schwarzer, Scale invariant mechanical surface optimization applying analytical time dependent contact mechanics for layered structures, Chapter 22 in *Applied Nanoindentation in Advanced Materials*, Atul Tiwari (editor), Sridhar

Natarajan (co-editor), 2017, ISBN: 978-1-119-08449-5; www.wiley.com/WileyCDA/WileyTitle/productCd-1119084490.html.

25. N. Schwarzer, From interatomic interaction potentials via Einstein field equation techniques to time dependent contact mechanics, *Mater. Res. Express*, 2014, 1, 1, http://dx.doi.org/10.1088/2053-1591/1/1/015042.
26. N. Schwarzer, Completely analytical tools for the next generation of surface and coating optimization, *Coatings*, 2014, 4, 263–291; doi:10.3390/coatings4020263.
27. N. Schwarzer, *Science Riddles—Riddle No. 8: Could the Schwarzschild Metric Contain Its Own Quantum Solution?*, self-published, Amazon Digital Services, 2018, Kindle.
28. N. Schwarzer, *Einstein Had It, but He Did Not See It—Part XXXIX: EQ or The Einstein Quantum Computer*, self-published, Amazon Digital Services, 2018, Kindle.
29. N. Schwarzer, *The Einstein Quantum Computer—Mathematical Principle and Transition to the Classical Discrete and Quantum Computer Design*, self-published, Amazon Digital Services, 2018, Kindle.
30. N. Schwarzer, *Is There an Ultimate, Truly Fundamental and Universal Computer Machine?*, self-published, Amazon Digital Services, 2019, Kindle.
31. A. M. Turing, On computable numbers, with an application to the Entscheidungsproblem, *Proceedings of the London Mathematical Society*, 1936, 2 (published 1937), 42, 230–265, doi:10.1112/plms/s2-42.1.230.
32. M. M. Vopsen, The information catastrophe, *AIP Adv.*, 2020, 10, 085014, https://doi.org/10.1063/5.0019941.
33. T. Bodan, *7 Days—How to Explain the World to My Dying Child*, self-published, BoD Classic, 2021, ISBN 9783752639728.
34. T. Bodan, *The Eighth Day—Holocaust and the World's Biggest Mysteries: The Other Final Solution—with Mathematical Elaborations* (English Edition), self-published, Amazon Digital Services, 2015, Kindle.
35. T. Bodan, *The Eighth Day—Two Jews against The Third Reich.*, illustrated version, self-published, BoD Classic, 2021, ISBN 9783753417257.
36. N. Schwarzer, T. Bodan, *Sherlock, Watson, Einstein—Part 1: The Mystery of Entanglement and the Spooky Action at a Distance*, self-published, Amazon Digital Services, 2018, Kindle.
37. N. Schwarzer, T. Bodan, *Sherlock, Watson, Stalin—Part 2: The Hell of Gender, Merkel, Communism and the Dictatorship of Parasites*, self-published, Amazon Digital Services, 2018, Kindle.
38. T. Bodan, *EU vs. Britain—The Other Monkey Trial*, self-published, Amazon, 2017, ASIN: 1521304793.

39. T. Bodan, *Grexit, Eurokrise und Flüchtlinge—Vom dummverkauften Souverän zum mitbestimmenden Bürger—Oder: Warum unsere Politiker versagen müssen ja sogar versagen wollen*, self-published, Amazon Digital Services, 2015, Kindle.

40. N. Schwarzer, *How Julius and His Syllables Managed to Talk with Aliens*, self-published, Amazon Digital Services, 2019, Kindle.

41. N. Schwarzer, *The Theory of Everything—Quantum and Relativity Is Everywhere—A Fermat Universe*, 2020, Jenny Stanford Publishing, ISBN-10: 9814774472.

42. N. Schwarzer, *How Einstein Gives Dirac, Klein–Gordon and Schrödinger*. self-published, Amazon Digital Services, 2016, Kindle.

43. N. Schwarzer, *Our Universe, Nothing but an Intelligent Zero?—The Dark Lord's Zero-Sum- & God's Non-Zero-Sum-Game*, self-published, Amazon Digital Services, 2016, Kindle.

44. N. Schwarzer, *Quantized GTR—Understanding the "Friedmanns"?*," self-published, Amazon Digital Services, 2016, Kindle.

45. N. Schwarzer, *Quantum Gravity Monster*, self-published, Amazon Digital Services, 2016, Kindle.

46. N. Schwarzer, *Entanglement: Einstein-Rosen-Bridges = Einstein-Podolsky-Rosen-Joints?*, self-published, Amazon Digital Services, 2016, Kindle.

47. N. Schwarzer, *Particles—Spin, Quark Confinement, the Three Generation-Problem and Matter Antimatter Asymmetry*, self-published, Amazon Digital Services, 2016, Kindle.

48. N. Schwarzer, *Time, Mass and Higgs—How the Three Things Emerge from the Sublevel in a Theory of Everything*, self-published, Amazon Digital Services, 2016, Kindle.

49. N. Schwarzer, *Understanding Time*, self-published, Amazon Digital Services, 2016, Kindle.

50. N. Schwarzer, *Understanding the Electron?—Part I*, self-published, Amazon Digital Services, 2016, Kindle.

51. N. Schwarzer, *General Quantization of Smooth Spaces—From ħ to Plancktensor & Planckfunction*, self-published, Amazon Digital Services, 2016, Kindle.

52. N. Schwarzer, *Understanding the Electron II*, self-published, Amazon Digital Services, 2016, Kindle.

53. N. Schwarzer, *The Photon*, self-published, Amazon Digital Services, 2016, Kindle.

54. N. Schwarzer, *The ANTI-GENDER Proof: Mathematical Proof Against One of the Most Stupid and Dangerous Ideologies, Disguising Itself as "Science"*, self-published, Amazon Digital Services, 2018, Kindle.

55. N. Schwarzer, *Recipe to Quantize the General Theory of Relativity*, self-published, Amazon Digital Services, 2016, Kindle.

56. N. Schwarzer, *The Einstein Oscillator in 1D*, self-published, Amazon Digital Services, 2017, Kindle.

57. N. Schwarzer, *Quantized Schwarzschild*, self-published, Amazon Digital Services, 2017, Kindle.

58. N. Schwarzer, *The Cosmic Atom & Quantum Friedmann Equations: The Quantum States of Friedmann–Lemaître–Robertson–Walker (FLRW) Universes and the 4 Friedmann Equations*, self-published, Amazon Digital Services, 2017, Kindle.

59. N. Schwarzer, *Quantum Tribology—Part I: Theory*, self-published, Amazon Digital Services, 2017, Kindle.

60. T. Bodan, N. Schwarzer, *Quantum Economy*, self-published, Amazon Digital Services, 2017, Kindle.

61. N. Schwarzer, *Quantized Relativized Theology—Where is God?*, self-published, Amazon Digital Services, 2017, Kindle.

62. N. Schwarzer, *Humanized Artificial Intelligence—Spiritual Computers—Make Them Believe and They'll Start to Think*, self-published, Amazon Digital Services, 2017, Kindle.

63. N. Schwarzer, *Einstein Already Had It, but He Did Not See It—Part 0: The Discarded Term from the Einstein–Hilbert-Action*, self-published, Amazon Digital Services, 2017, Kindle.

64. N. Schwarzer, *Einstein Had It, but He Did Not See It—Part I: How the Quantum Theory Resides in the Einstein–Hilbert-Action*, self-published, Amazon Digital Services, 2017, Kindle.

65. N. Schwarzer, *How Can We Measure the Size of a Thought*, self-published, Amazon Digital Services, 2020, Kindle (short version on: www.worldformulaapps.com/wp-content/uploads/2019/12/What-is-the-Size-of-a-Thought.pdf).

66. N. Schwarzer, *Einstein Had It, but He Did Not See It—Part II: String Cosmology*, self-published, Amazon Digital Services, 2017, Kindle.

67. N. Schwarzer, *Einstein Had It, but He Did Not See It—Part III: The Impossible Black-Hole Singularity*, self-published, Amazon Digital Services, 2017, Kindle.

68. N. Schwarzer, *Einstein Had It, but He Did Not See It—Part IV: Sixty e-Foldings*, self-published, Amazon Digital Services, 2017, Kindle.

69. N. Schwarzer, *Einstein Had It, but He Did Not See It—Part V: Amendment to the Heisenberg Uncertainty Principle*, self-published, Amazon Digital Services, 2017, Kindle.

70. N. Schwarzer, *Einstein Had It, but He Did Not See It—Part VI: The Photon and the Flat Space*, self-published, Amazon Digital Services, 2017, Kindle.

71. N. Schwarzer, *Einstein Had It, but He Did Not See It—Part VII: Konrad Zuse's Computing Universe*, self-published, Amazon Digital Services, 2017, Kindle.
72. N. Schwarzer, *Einstein Had It, but He Did Not See It—Part VIII: Derivation of the Speed of Light and Other Funny Limits*, self-published, Amazon Digital Services, 2017, Kindle.
73. N. Schwarzer, *Einstein Einstein Had It, but He Did Not See It—Part IX: Fermat's Last Theorem inside the Einstein–Hilbert-Action?*, self-published, Amazon Digital Services, 2017, Kindle.
74. N. Schwarzer, *Einstein Had It, but He Did Not See It—Part XI: Behind the Event Horizon*, self-published, Amazon Digital Services, 2017, Kindle.
75. N. Schwarzer, *Einstein Had It, but He Did Not See It—Part XIII: Why Equality is Always Hostile to Life*, self-published, Amazon Digital Services, 2019, Kindle.
76. N. Schwarzer, *Einstein Had It, but He Did Not See It—Part XIV: Braneworks or a Brany Universe*, self-published, Amazon Digital Services, 2017, Kindle.
77. N. Schwarzer, "*Einstein Had It, but He Did Not See It—Part XV: The Genetic String-Code of the Universe*, self-published, Amazon Digital Services, 2017, Kindle.
78. N. Schwarzer, *Einstein Had It, but He Did Not See It—Part XVI: An Inner Kerr Solution*, self-published, Amazon Digital Services, 2017, Kindle.
79. N. Schwarzer, *Einstein Had It, but He Did Not See It—Part XVII: A Funny Form of Vacuum*, self-published, Amazon Digital Services, 2017, Kindle.
80. N. Schwarzer, *Einstein Had It, but He Did Not See It—Part XVIII: A Few Matter Solutions*, self-published, Amazon Digital Services, 2017, Kindle.
81. N. Schwarzer, *Einstein Had It, but He Did Not See It—Part XIX: The Cosmologic Neutralizer*, self-published, Amazon Digital Services, 2017, Kindle.
82. N. Schwarzer, *Einstein Had It, but He Did Not See It—Part XX: Higher Order Covariant Variation of the Einstein–Hilbert-Action*, self-published, Amazon Digital Services, 2017, Kindle.
83. N. Schwarzer, *Einstein Had It, but He Did Not See It—Part XXI: A Very Simple Theory of Everything*, self-published, Amazon Digital Services, 2018, Kindle.
84. N. Schwarzer, *Einstein Had It, but He Did Not See It—Part XXII: Finis Fabulae de Foraminibus Atris—Recipe to Evaluate Black Hole Matter*, self-published, Amazon Digital Services, 2018, Kindle.
85. N. Schwarzer, *Einstein Had It, but He Did Not See It—Part XXIII: A Bit Cosmology*, self-published, Amazon Digital Services, 2018, Kindle.
86. N. Schwarzer, *Einstein Had It, but He Did Not See It—Part XXIV: A Variety of Solutions and the Dirac-Schwarzschild-Particle*, self-published, Amazon Digital Services, 2018, Kindle.

87. N. Schwarzer, *Einstein Had It, but He Did Not See It—Part XXV: The Vacuum Friedmann Cosmos and the Origin of a Quantal World*, self-published, Amazon Digital Services, 2018, Kindle.
88. N. Schwarzer, *Einstein Had It, but He Did Not See It—Part XXVI: The Nature of SPIN 1/2*, self-published, Amazon Digital Services, 2018, Kindle.
89. N. Schwarzer, *Einstein Had It, but He Did Not See It—Part XXVII: The Quantum Transformer Always Already Was inside the General Theory of Relativity*, self-published, Amazon Digital Services, 2018, Kindle.
90. N. Schwarzer, *Einstein Had It, but He Did Not See It—Part XXVIII: ¿ Anti Gravity ?, self-published*, Amazon Digital Services, 2018, Kindle.
91. N. Schwarzer, *Einstein Had It, but He Did Not See It—Part XXIX: Space Strain Invariants and the Evaluation of Mass*, self-published, Amazon Digital Services, 2018, Kindle.
92. N. Schwarzer, *Einstein Had It, but He Did Not See It—Part XXX: ¿ Neutrino Oscillations Directly Evaluated from the Einstein-Field-Equations ?*, self-published, Amazon Digital Services, 2018, Kindle.
93. N. Schwarzer, *Einstein Had It, but He Did Not See It—Part XXXI: A Cosmologic Pairwise Entanglement of Dimensions and the Holographic Principle*, self-published, Amazon Digital Services, 2018, Kindle.
94. N. Schwarzer, *Einstein Had It, but He Did Not See It—Part XXXII: Enough to Play With—More Quantum Friedmann Solutions*, self-published, Amazon Digital Services, 2018, Kindle.
95. N. Schwarzer, *Einstein Had It ... Part XXXIII: Elementary Particle Universes (without Spin and Shear Components)*, self-published, Amazon Digital Services, 2018, Kindle.
96. N. Schwarzer, *Einstein Had It, but He Did Not See It—Part XXXIV: Is This the Electron?*, self-published, Amazon Digital Services, 2018, Kindle.
97. N. Schwarzer, *Einstein Had It ... Part XXXV: The 2-Body Problem and the GTR-Origin of Mass*, self-published, Amazon Digital Services, 2018, Kindle.
98. N. Schwarzer, *Einstein Had It ... Part XXXVI: The Classical and Principle Misinterpretation of the Einstein-Field-Equations AND How It Might Be Done Correctly*, self-published, Amazon Digital Services, 2018, Kindle.
99. N. Schwarzer, *Einstein Had It, but He Did Not See It—Part XXXVII: Most Simple Singularity-Free "Black Holes"*, self-published, Amazon Digital Services, 2018, Kindle.
100. N. Schwarzer, *Einstein Had It, but He Did Not See It—Part XXXVIII: The Einstein Hydrogen Atom*, self-published, Amazon Digital Services, 2018, Kindle.

101. N. Schwarzer, *Einstein Had It, but He Did Not See It—Part XL: The Einstein Hydrogen Atom with "Spin"*, self-published, Amazon Digital Services, 2018, Kindle.

102. N. Schwarzer, *Einstein Had It, but He Did Not See It—Part XLI: Generalized Rotating "Black Holes"*, self-published, Amazon Digital Services, 2018, Kindle.

103. N. Schwarzer, *Einstein Had It, but He Did Not See It—Part XLII: Simplest Way to Quantize the General Theory of Relativity*, self-published, Amazon Digital Services, 2018, Kindle.

104. N. Schwarzer, *Einstein Had It, but He Did Not See It—Part XLIII: A Selection of Einstein-Dirac-Particles*, self-published, Amazon Digital Services, 2018, Kindle.

105. N. Schwarzer, *Einstein Had It, but He Did Not See It—Part XLIV: Fundamental Normalizable Solutions and Their Limits towards Classical Metrics*, self-published, Amazon Digital Services, 2018, Kindle.

106. N. Schwarzer, *Einstein Had It, but He Did Not See It—Part XLV: Inner and Outer Solutions to Schwarzschild and Flat Space*, self-published, Amazon Digital Services, 2018, Kindle.

107. N. Schwarzer, *Einstein Had It, but He Did Not See It—Part XLVI: 3 and More Inner Solutions & Three Generations of Particles?*, self-published, Amazon Digital Services, 2018, Kindle.

108. N. Schwarzer, *Einstein Had It, but He Did Not See It—Part XLVII: Towards A Pauli Exclusion Principle Derived from the Einstein-Field-Equations*, self-published, Amazon Digital Services, 2018, Kindle.

109. N. Schwarzer, *Einstein Had It ... Part XLVIII: Agapornis—Particles One Cannot Separate*, self-published, Amazon Digital Services, 2018, Kindle.

110. N. Schwarzer, *Einstein Had It, but He Did Not See It—Part XLIX: Shear as Origin of the Higgs Field*, self-published, Amazon Digital Services, 2018, Kindle.

111. N. Schwarzer, *Einstein Had It, but He Did Not See It—Part L: Big Bang Inflation Black Hole Pregnant with Universe*, self-published, Amazon Digital Services, 2018, Kindle.

112. N. Schwarzer, *Einstein Had It, but He Did Not See It—Part LI: Why Space-Time Might Be Discrete*, self-published, Amazon Digital Services, 2018, Kindle.

113. N. Schwarzer, *Einstein Had It, but He Did Not See It—Part LII: Why We Might Live inside a Black Hole*, self-published, Amazon Digital Services, 2018, Kindle.

114. N. Schwarzer, *Einstein Had It, but He Did Not See It—Part LIII: They Are Everywhere! Why There Are So Many Sigmoid-Dependencies in This World*, self-published, Amazon Digital Services, 2018, Kindle.

115. N. Schwarzer, *Einstein Had It, but He Did Not See It—Part LIV: Mathematical Philosophy & Quantum Gravity Ethic*, self-published, Amazon Digital Services, 2018, Kindle.

116. N. Schwarzer, *Einstein Had It, but He Did Not See It—Part LV: The Fundamental Connection of Quantum Theory and General Theory of Relativity*, self-published, Amazon Digital Services, 2018, Kindle.
117. N. Schwarzer, *Einstein Had It, but He Did Not See It—Part LVI: What Is Mass? What Is Energy?*, self-published, Amazon Digital Services, 2018, Kindle.
118. N. Schwarzer, *Einstein Already Had It, But He Did Not See It—Part LVII: The Schwarzschild-Hydrogen Atom*, self-published, Amazon Digital Services, 2018, Kindle.
119. N. Schwarzer, *Einstein Already Had It, But He Did Not See It—Part LVIII: The Most Simple Nature of This Funny Phenomenon Called "Spin"*, self-published, Amazon Digital Services, 2018, Kindle.
120. N. Schwarzer, *Einstein Already Had It, But He Did Not See It—Part LIX: A Brief Story about the Matter Antimatter Asymmetry*, self-published, Amazon Digital Services, 2018, Kindle.
121. N. Schwarzer, *Einstein Had It, but He Did Not See It—Part LX: The Hitch Hiker's Guide to the Creation of a Universe*, self-published, Amazon Digital Services, 2018, Kindle.
122. N. Schwarzer, *Einstein Had It, but He Did Not See It—Part LXI: Don't Panic in a Universe with Only One Dimension*, self-published, Amazon Digital Services, 2018, Kindle.
123. N. Schwarzer, *Einstein Had It, but He Did Not See It—Part LXII: The Quantization of Schwarzschild and FLRW-Metrics*, self-published, Amazon Digital Services, 2018, Kindle.
124. N. Schwarzer, *Einstein Already Had It, But He Did Not See It—Part LXIII: Einstein-Field-Equations = Dirac² (+) Klein–Gordon*, self-published, Amazon Digital Services, 2018, Kindle.
125. N. Schwarzer, *Einstein Already Had It, But He Did Not See It—Part LXIV: Origin of the 6 Infinity Stones*, self-published, Amazon Digital Services, 2018, Kindle.
126. N. Schwarzer, *Einstein Already Had It, But He Did Not See It—Part LXV: Swing When You're Singing*, self-published, Amazon Digital Services, 2018, Kindle.
127. N. Schwarzer, *Einstein Already Had It, But He Did Not See It—Part LXVI: The Other Time*, self-published, Amazon Digital Services, 2018, Kindle.
128. N. Schwarzer, *Einstein Already Had It, But He Did Not See It—Part LXVII: Simplest Metric Quantum Operators*, self-published, Amazon Digital Services, 2018, Kindle.
129. N. Schwarzer, *Einstein Had It, but He Did Not See It—Part LXIX: The Hippocratic Oath in Mathematical Form and Why—So Often—It Will Be of No Use*, self-published, Amazon Digital Services, 2018, Kindle.

130. N. Schwarzer, *Einstein Had It, but He Did Not See It—Part LXX: Extended Variation of the Einstein–Hilbert-Action Results in Quantum Theory*, self-published, Amazon Digital Services, 2018, Kindle.

131. N. Schwarzer, *Einstein Had It, but He Did Not See It—Part LXXI: The 3-Generation Problem, Spin, Neutrinos, Electron, Muon, Tauon and the Photon from Fundamental Geometric Principles*, self-published, Amazon Digital Services, 2018, Kindle.

132. N. Schwarzer, *Einstein Had It, but He Did Not See It—Part LXXII: Connection between Schwarzschild Radius and Dirac Rest Mass*, self-published, Amazon Digital Services, 2019, Kindle.

133. N. Schwarzer, *Philosophical Engineering Part 1: The Honest Non-Parasitic Philosopher and the Universal "GOOD" Derived from a Theory of Everything*, self-published, Amazon Digital Services, 2019, Kindle.

134. N. Schwarzer, *Einstein Had It, but He Did Not See It—Part LXXIII: Wave-Particle Duality as a Direct Outcome of the General Theory of Relativity*, self-published, Amazon Digital Services, 2019, Kindle.

135. N. Schwarzer, *Einstein Had It, but He Did Not See It—Part LXXIV: World Formulae-Approaches of Various Orders*, self-published, Amazon Digital Services, 2019, Kindle.

136. N. Schwarzer, *Einstein Had It, but He Did Not See It—Part LXXV: The Metric Creation of Matter*, self-published, Amazon Digital Services, 2019, Kindle.

137. N. Schwarzer, *Einstein Had It, but He Did Not See It—Part LXXVI: Quantum Universes—We Don't Need No ... an Inflation*, self-published, Amazon Digital Services, 2019, Kindle.

138. N. Schwarzer, *Einstein Had It, but He Did Not See It—Part LXXVII: Matter is Nothing and so Nothing Matters*, self-published, Amazon Digital Services, 2019, Kindle.

139. N. Schwarzer, *Einstein Had It, but He Did Not See It—Part LXXVIII: About a Possible Quantum Gravity Alternative to Schwarzschild and Co.*, self-published, Amazon Digital Services, 2019, Kindle.

140. N. Schwarzer, *Einstein Had It, but He Did Not See It—Part LXXIX: Dark Matter Options*, self-published, Amazon Digital Services, 2019, Kindle.

141. N. Schwarzer, *Einstein Had It, but He Did Not See It—Part LXXX: Short Note on the Killing of Dirac*, self-published, Amazon Digital Services, 2019, Kindle.

142. N. Schwarzer, *Einstein Had It, but He Did Not See It—Part LXXXI: More Dirac Killing*, self-published, Amazon Digital Services, 2019, Kindle.

143. N. Schwarzer, *Einstein Had It, but He Did Not See It—Part LXXXII: Half Spin Hydrogen*, self-published, Amazon Digital Services, 2019, Kindle.

144. N. Schwarzer, *Einstein Had It, but He Did Not See It—Part LXXXIII: Quantum Relativity—The Two Sides of the Same Medal*, self-published, Amazon Digital Services, 2019, Kindle.

145. N. Schwarzer, *Einstein Had It, but He Did Not See It—Part LXXXIV: A World formula(?)*, self-published, Amazon Digital Services, 2019, Kindle.

146. N. Schwarzer, *Science Riddles—Riddle No. 1: The Peculiar Quantum Schwarzschild Amplitude—A Coincidence?*, self-published, Amazon Digital Services, 2019, Kindle.

147. N. Schwarzer, *Science Riddles—Riddle No. 2: Quarks—Real or Just THE Greatest Deception the Universe Did to Mankind*?, self-published, Amazon Digital Services, 2019, Kindle.

148. N. Schwarzer, *Science Riddles—Riddle No. 3: Was (IS) There a Big Bang Bomb?*, self-published, Amazon Digital Services, 2019, Kindle.

149. N. Schwarzer, *Science Riddles—Riddle No. 4: What Is the Fundamental Substance of a Physical Potential?*, self-published, Amazon Digital Services, 2019, Kindle.

150. N. Schwarzer, *Science Riddles—Riddle No. 5: What Makes the Quantum Equations So Linear?*, self-published, Amazon Digital Services, 2019, Kindle.

151. N. Schwarzer, *Science Riddles—Riddle No. 6: What Is Time?*, self-published, Amazon Digital Services, 2019, Kindle.

152. N. Schwarzer, *Science Riddles—Riddle No. 7: Is the Theory of Everything THE Most Potent Patent Killing Machine?*, self-published, Amazon Digital Services, 2019, Kindle.

153. N. Schwarzer, *Science Riddles—Riddle No. 9: ...?*, to be published.

154. N. Schwarzer, *Science Riddles—Riddle No. 10: What, Where and Why Is the Higgs Field?*, self-published, Amazon Digital Services, 2019, Kindle.

155. N. Schwarzer, *What Is the Ultimate Good*, www.worldformulaapps.com/wp-content/uploads/2019/12/What-is-the-ultimate-good_final.pdf.

156. N. Schwarzer, *Science Riddles—Riddle No. 11: What Is Mass?*, self-published, Amazon Digital Services, 2019, Kindle.

157. N. Schwarzer, *Science Riddles—Riddle No. 12: Is There a Cosmological Spin?*, self-published, Amazon Digital Services, 2019, Kindle.

158. N. Schwarzer, *Science Riddles—Riddle No. 13: How to Solve the Flatness Problem?*, self-published, Amazon Digital Services, 2019, Kindle.

159. N. Schwarzer, *Science Riddles—Riddle No. 14: And what If Time ...? A Paradigm Shift*, self-published, Amazon Digital Services, 2019, Kindle.

160. N. Schwarzer, *Science Riddles—Riddle No. 15: Is there an Absolute Scale?*, self published, Amazon Digital Services, 2019, Kindle.

161. N. Schwarzer, *Science Riddles—Riddle No. 16: How to Understand the Dirac Equation?*, self-published, Amazon Digital Services, 2019, Kindle.

162. N. Schwarzer, *Science Riddles—Riddle No. 17: How Einstein Becomes First Order and Goes Dirac—Can We Factorize the Einstein-Field-Equations?*, self-published, Amazon Digital Services, 2019, Kindle.

163. N. Schwarzer, *Science Riddles—Riddle No. 18: And What If Space ...? A Paradigm Shift*, self-published, Amazon Digital Services, 2019, Kindle.

164. N. Schwarzer, *Science Riddles—Riddle No. 19: Is There a World formula?*, self-published, Amazon Digital Services, 2019, Kindle.

165. N. Schwarzer, *Quantum Gravity Thermodynamics—And it May Get Hotter*, self-published, Amazon Digital Services, 2019, Kindle.

166. N. Schwarzer, *Epistle to Elementary Particle Physicists—A Chance You Might Not Want to Miss*, self-published, Amazon Digital Services, 2019, Kindle.

167. N. Schwarzer, *2nd Epistle to Elementary Particle Physicists—Brief Study about Gravity Field Solutions for Confined Particles*, self-published, Amazon Digital Services, 2019, Kindle.

168. N. Schwarzer, *3rd Epistle to Elementary Particle Physicists—Beyond the Standard Model—Metric Solutions for Neutrino, Electron, Quark*, self-published, Amazon Digital Services, 2019, Kindle.

169. N. Schwarzer, *Quantum Gravity Thermodynamics II—Derivation of the Second Law of Thermodynamics and the Metric Driving Force of Evolution*, self-published, Amazon Digital Services, 2019, Kindle.

170. N. Schwarzer, *From Quantum Gravity to a Quantum Relative Material Science—A Very First Principle Material Science Concept*, self-published, Amazon Digital Services, 2019, Kindle.

171. N. Schwarzer, *Einstein Had It, but He Did Not See It—Part LXXXV: In Conclusion*, self-published, Amazon Digital Services, 2019, Kindle.

172. N. Schwarzer, *Science Riddles—Riddle No. 20: Second Law of Thermodynamics—Where Is Its Fundamental Origin?*, self-published, Amazon Digital Services, 2019, Kindle.

173. N. Schwarzer, *Science Riddles—Riddle No. 21: Evolution—Where Is Its Fundamental Origin?*, self-published, Amazon Digital Services, 2019, Kindle.

174. N. Schwarzer, *Science Riddles—Riddle No. 22: Anti-Gravity—Is it Possible?*, self-published, Amazon Digital Services, 2019, Kindle.

175. N. Schwarzer, *Science Riddles—Riddle No. 23: Why Does the Dirac Equation Work So Well?*, self-published, Amazon Digital Services, 2019, Kindle.

176. N. Schwarzer, *Science Riddles—Riddle No. 24: What Is This Ominous Dark Energy?*, self-published, Amazon Digital Services, 2019, Kindle.

177. N. Schwarzer, *Science Riddles—Riddle No. 25: How Can We Metrically Understand Spin, Charge, Matter and Antimatter?*, self-published, Amazon Digital Services, 2019, Kindle.
178. N. Schwarzer, *Science Riddles—Riddle No. 26: Is There an Elastic Quantum Gravity?*, self-published, Amazon Digital Services, 2019, Kindle.
179. N. Schwarzer, *Science Riddles—Riddle No. 27: Why Is Gravity so Weak?*, self-published, Amazon Digital Services, 2019, Kindle.
180. P. S. Debnath, B. C. Paul, *Cosmological Models with Variable Gravitational and Cosmological constants in R^2 Gravity*, arxiv.org/pdf/gr-qc/0508031.pdf.
181. P. A. M. Dirac, *The Quantum Theory of the Electron*, Published 1 February 1928. DOI: 10.1098/rspa.1928.0023.
182. J. D. Bekenstein, Information in the holographic universe, *Sci. Am.*, 2003, 289(2), 61.
183. J. D. Bekenstein, Black holes and entropy, *Phys. Rev. D*, 1973, 7, 2333–2346.
184. L. Susskind, *The Black Hole War: My battle with Stephen Hawking to Make the World Safe for Quantum Mechanics*, Little, Brown, 2008.
185. en.wikipedia.org/wiki/Higgs_boson.
186. E. Schrödinger, Quantisierung als Eigenwertproblem (erste Mitteilung), *Ann. Phys.*, 1926, 384(4), 361–376.
187. E. Schrödinger, Quantisierung als Eigenwertproblem (zweite Mitteilung). *Ann. Phys.*, 1926, 384(6), 489–527.
188. É. Cartan, Sur une généralisation de la notion de courbure de Riemann et les espaces à torsion, *C. R. Acad. Sci.* (Paris), 1922, 174, 593–595.
189. É. Cartan, Sur les variétés à connexion affine et la théorie de la relativité généralisée. Part I, *Ann. Éc. Norm.*, 1923, 40, 325–412, and ibid, 1924, 41, 1–25; Part II: ibid, 1925, 42, 17–88.
190. Ch. Heinicke, Exact solutions in Einstein's theory and beyond, Dissertation, Universität Köln, 2005. urn:nbn:de:hbz:38-14637.
191. W. Greiner, J. Reinhardt, *Theoretische Physik, Band 7—Quanten Elektro Dynamik*, 1995, Verlag Harri Deutsch & Thun, Frankfurt am Main, ISBN 3-8171-1426-5.
192. W. Greiner, A. Schäfer, *Theoretische Physik, Band 10—Quanten Chromo Dynamik*, 1998, Verlag Harri Deutsch & Thun, Frankfurt am Main, ISBN 3-87144-710-2.
193. K. Schwarzschild, Über das Gravitationsfeld einer Kugel aus inkompressibler Flüssigkeit nach der Einsteinschen Theorie ["On the gravitational field of a ball of incompressible fluid following Einstein's theory"], 1916, *Sitzungsberichte der Königlich-Preussischen Akademie der Wissenschaften* (In German), Berlin, pp. 424–434.

194. H. Haken, H. Chr. Wolf, *Atom- und Quantenphysik*, (in German), 4th ed., 1990, Springer Heidelberg, ISBN 0-387-52198-4.

195. N. Schwarzer, *Super-Brains via Neuron Entanglement*, self-published, Amazon Digital Services, 2020, Kindle.

196. St. Wolfram, *The Foundations of Mathematics and Mathematica*, online at www.stephenwolfram.com/publications/academic/foundations-mathematics-mathematica.pdf.

197. J. Aditya, P. Shankar Rao, *Quantum Cryptography*, https://cs.stanford.edu/people/adityaj/QuantumCryptography.pdf.

198. F. Grasselli, *Quantum Cryptography—From Key Distribution to Conference Key Agreement*, 1st ed., 2021, Springer, ISBN: 303064359X.

199. D. Stebila, M. Mosca, N. Lütkenhaus, in: A. Sergienko, S. Pascazio, P. Villoresi (eds.), *Quantum Communication and Quantum Networking*, 2010, 36, pp. 283–296, Springer: Berlin Heidelberg, Bibcode:2010qcqn.book..283S, doi:10.1007/978-3-642-11731-2_35, ISBN 978-3-642-11730-5, S2CID 457259.

200. Peter W. Higgs, *Broken symmetries and the masses of gauge bosons. Phys. Rev. Lett.*, 1964, 13, 508.

201. N. Schwarzer, Short note on the effect of pressure induced increase of Young's modulus, *Philos. Mag.*, 2012, 92(13), 1631–1648.

202. A. E. Green, W. Zerna, *Theoretical Elasticity*, 1968, Oxford University Press, London.

203. H. Neuber, *Kerbspannungslehre*, in German, 3rd ed., 1985, Springer-Verlag, Berlin, Heidelberg, New York, Tokyo, ISBN 3-540-13558-8.

204. N. Schwarzer, About holistic optimization—examples from in- and outside the world of coatings, 2015, *Proceedings of the 58th Annual Technical SVC Conference*.

205. N. Schwarzer, Elastic surface deformation due to indenters with arbitrary symmetry of revolution, *J. Phys. D: Appl. Phys.*, 2004, 37, 2761–2772.

205. N. Schwarzer, Some basic equations for the next generation of surface testers solving the problem of pile-up, sink-in and making area-function-calibration obsolete, JMR Special Focus Issue on "Indentation Methods in Advanced Materials Research, *J. Mater. Res.*, 2009, 24(3), 1032–1036.

207. V. I. Fabrikant, *Application of Potential Theory in Mechanics: A Selection of New Results*, 1989, Kluwer Academic Publishers, The Netherlands.

208. V. I. Fabrikant, *Mixed Boundary Value Problems of Potential Theory and Their Application in Engineering*, 1991, Kluwer Academic Publishers, The Netherlands.

209. FilmDoctor, analytical software package for the analysis of complex mechanical problem for inhomogeneous materials, www.siomec.de/filmdoctor.

210. W. Pauli, Über den Zusammenhang des Abschlusses der Elektronengruppen im Atom mit der Komplexstruktur der Spektren, *Zeitschrift für Physik*, 1925, 31, 765–783, Bibcode:1925ZPhy...31..765P. doi:10.1007/BF02980631.

211. https : // demonstrations.wolfram.com/DiracMatricesInHigherDimensions/.

212. N. Schwarzer, *Mathematical Psychology—Tools for the Understanding, Simulation, Mitigation and Potential Prevention of Mass Formation Psychoses*, self-published, Amazon Digital Services, 2022, Kindle.

213. C. A. Sporea, *Notes on f(R) Theories of Gravity*, 2014, arxiv.org/pdf/1403.3852.pdf.

214. N. Schwarzer, *How Systems Evolve—A Red Pill Course on Evolution*, self-published, ASIN: B0B6RGBKQ4, Amazon Digital Services, 2022, Kindle.

215. N. Schwarzer, *How will the Nearby Unification of Physics Change Future Warfare—a Brief Study*, Part 10 of "Medical Socio-Economic Quantum Gravity," self-published, Amazon Digital Services, 2021, Kindle, ASIN: B0971M85JQ.

216. C. Cohen-Tannoudji, B. Diu, F. Laloë, Franck: *Quantenmechanik 1&2*, 2, Auflage, Walter de Gruyter, Berlin–New York 1999, ISBN-10: 3110626004.

217. C. Domb, M. S. Green, *Phase Transitions and Critical Phenomena*, vol. 6, 1976, Academic Press, London, New York, San Francisco, ISBN 0-12-220306-2.

218. N. Schwarzer, *Quantum Gravity Systems Science*, self-published, ASIN: B0BRL8SKSB, Amazon Digital Services, 2023, Kindle.

219. N. Schwarzer, *The Foreseeable Quantum Enigma Disaster*, Part 10b of "Medical Socio-Economic Quantum Gravity," self-published, Amazon Digital Services, June 2021, Kindle.

Index

For Product Safety Concerns and Information please contact our EU representative GPSR@taylorandfrancis.com Taylor & Francis Verlag GmbH, Kaufingerstraße 24, 80331 München, Germany